Vitalism and the Scientific Image in Post-Enlightenment Life Science, 1800–2010

History, Philosophy and Theory of the Life Sciences

Volume 2

For further volumes:
http://www.springer.com/series/8916

Sebastian Normandin • Charles T. Wolfe
Editors

Vitalism and the Scientific Image in Post-Enlightenment Life Science, 1800–2010

Springer

Editors
Sebastian Normandin
Lyman Briggs College
Michigan State University
East Lansing, MI, USA

Charles T. Wolfe
Department of Philosophy and Moral Sciences, Centre for History of Science
Ghent University, Ghent, Belgium

ISSN 2211-1948 ISSN 2211-1956 (electronic)
ISBN 978-94-007-2444-0 ISBN 978-94-007-2445-7 (eBook)
DOI 10.1007/978-94-007-2445-7
Springer Dordrecht Heidelberg New York London

Library of Congress Control Number: 2013940174

Printed on acid-free paper

Springer is part of Springer Science+Business Media (www.springer.com)

Contents

Chapter 1
Vitalism and the Scientific Image: An Introduction

Sebastian Normandin and Charles T. Wolfe

Keywords Biology • Emergentism • Holism • Life science • Organism • Vitalism

To undertake a history of vitalism at this stage in the development of the 'biosciences', theoretical and other, is a stimulating prospect. We have entered the age of 'synthetic' life, and our newfound capacities prompt us to consider new levels of analysis and understanding. At the same time, it is possible to detect a growing level of interest in vitalistic and organismic themes, understood in a broadly naturalistic context and approached, not so much from broader cultural perspectives as in the early twentieth century, as from a scientific perspective – or at least a view lying at the boundaries or liminal spaces of what counts as 'science'.[1] The challenge of understanding and theorizing vitalism in the era of the synthetic is not unlike that prompted by early nineteenth-century successes in chemistry allowing for the

[1] Gilbert and Sarkar (2000) and Laublichler (2000). In the same year, Marc Kirschner and his collaborators published an influential research paper in *Cell* on what they called "molecular vitalism": the suggestion was that, faced with the limitations of genomics, researchers should investigate what the authors "whimsically" termed the "vitalistic" properties of molecular, cellular, and organismal function. They comment in closing that "the organism has fashioned a very stable physiology and embryology. … It is this robustness that suggested 'vital forces', and it is this robustness that we wish ultimately to understand in terms of chemistry. We will have such an opportunity in this new century" (Kirschner et al. 2000, 87).

S. Normandin
Lyman Briggs College, Michigan State University, 919 E. Shaw Lane,
East Lansing, MI 48825, USA
e-mail: vitaliste@hotmail.com

C.T. Wolfe (✉)
Department of Philosophy and Moral Science, Sarton Centre for History of Science,
Ghent University, Blandijnberg 2, B-9000, Ghent, Belgium
e-mail: charles.wolfe@ugent.be

S. Normandin and C.T. Wolfe (eds.), *Vitalism and the Scientific Image in Post-Enlightenment Life Science, 1800–2010*, History, Philosophy and Theory of the Life Sciences 2, DOI 10.1007/978-94-007-2445-7_1,

synthesis of organic compounds (Wöhler), except that now, whether the motivation is molecular-chemical, embryological or physiological,[2] we find ourselves asking fundamental questions anew. What is life? How does it differ from non-living matter? What are the fundamental processes that characterize the living? What philosophical and epistemological considerations are raised by our new understandings?[3] We are driven to consider, for example, what metaphors we use to describe living processes as our knowledge of them changes, not least since some of the opprobrium surrounding the term 'vitalism' is also a matter of language: of which terms one uses to describe a phenomenon (embryo growth, homeostasis, phenotypic plasticity, and so on) such that it might not be a *repoussoir*.[4]

Strangely, as the development of the life sciences moves far beyond observation-based origins into the realms of the applied and technological, vitalism again comes to the fore, as it does whenever the question of boundaries arises. This is not because it is itself a conceptual 'hybrid', like the concept of organism, or a perpetual object of conceptual and experimental appropriation from one 'discursive field' to another, like mechanism. Rather, vitalist approaches, unlike a univocal ontological vision such as materialism, in which there should only be one kind of stuff in the universe (however much it may go through qualitative transformations, and however 'embodied' it may be), "need not claim that every feature of the world is vital . . . ; rather, these categories are 'inclusive', used to name accounts in which at least some vital properties . . . are thought to be required."[5] Vitalism is thus always on the borderland of thought; a term that when invoked reminds us of our ignorance or skepticism. In fact, vitalism nominally implies acceptance of the unknown as a central fact of life. With vitalism, even the known is always in flux. Moreover, its very meaning has changed and evolved over time. Vitalism, then, has its own vitality.[6]

When thinking about vitalism historically, we are often drawn towards the liminal – the spaces in between. This sense of 'between-ness', what was often termed the *juste milieu* (e.g. in mid nineteenth-century France), reminds us that life, history and science are all, in essence, subject to change. That is to say that change is a fundamental element, even the defining element, of all three. Change, dynamism,

---

[2]Respectively, Kirschner et al. (2000), West-Eberhard (2003) (for discussion, Huneman 2010 and Walsh 2010) and Turner (2000).

[3]Morange (2006).

[4]Oyama (2010). Gilbert and Sarkar (2000) are explicit in their intention to discard a 'bad word' ('vitalism') for a good thing (the family of systemic, non-genocentric approaches to development in current life science), and find a more usable one ('organicism'). In his recent discussion of von Uexküll as a thinker of 'biosemiotics', Emmeche makes almost exactly the same terminological distinction, between a (bad) "vitalism" and a (good, naturalistically specifiable) "qualitative organicism" (Emmeche 2001, 653).

[5]Berryman (2003, 346). A 1940 review of Wheeler's history of vitalism actually makes much the same point, although it is phrased in the then-current language of different "temperaments": "The mechanist is the kind of person who feels that everything important is known already, in principle at least, and that only minor details remain to be discovered. The vitalist feels that existing knowledge is only of minor details, and everything of importance is undiscovered" (Ritchie 1940, 7).

[6]Canguilhem (2008) and Greco (2005).

transformation, transmutation, and the constant mutable growth of the living – these are all themes that often lead us to rely on vitalist and emergentist explanations (which are neither identical nor necessarily connected, as discussed in Part II of this volume).

It is with these inconstant constants that we propose a collection of essays on the history and philosophy of vitalism: a moving target, an explanatory and/or metaphysical construct which appears, depending on the context, as a form of overt supernaturalism or as a useful heuristic for biomedical research and theorizing. Of course, there are still landmarks in this shifting sand, and though the lines on the map may move, new cultures and hybridities spring up in this borderland. There is an idea of the new here that occurs in two senses – in the truly novel (vitalism as an avant-garde, including in the farcical sense described by Juan Rigoli (Chap. 4), in which the fascination with new dimensions of life science – notably physiology – can be no more than a "roman de la médecine") and in new interpretations of old tropes and ideas (as in the way the Montpellier vitalists in the eighteenth century reappropriated and reconstructed Hippocratism as a new yet deliberately archaic posture over and against the 'mechanism' of the New Science[7]). Thus we seek to be generative and re-generative. This mutable territory is our scope; a history of attempts to chart the vagaries of life and of souls is conceptually unlimited. But there are shapes and outlines in this broader mindscape, and the works here presented share a harmony in all seeking to puzzle together the patterns of vitalism.

This volume follows in a tradition of other collected close readings of vitalism. One of us recently produced a volume of this nature, but focused on the Enlightenment period.[8] There are numerous other examples of philosophical and literary treatments of vitalism, looking at its presence in science, thought, art and the general culture.[9] More focused book-length work is also present, dealing with various contexts, from the very specific *milieu* of the Montpellier medical school[10] or the vitalist sympathies of the *Encyclopédie*[11] to the broader question of vitalism in the Enlightenment.[12] When we turn to shorter article-length pieces, the list grows dramatically, not least because the term starts to be used in all sorts of ways, extending beyond the context of philosophical, natural-historical and biomedical reflections on the nature of living beings to designate political (or 'biopolitical') positions. To be clear, this is a volume dealing with the former, in the period running from the end of the Enlightenment (starting with Lamarck, in Giglioni's Chap. 2) to the present day (with the essays by Turner, Sonnenschein, Soto et al., Dupré and O'Malley, and Bechtel, in Part III, Chaps. 11, 12, 13 and 14).

---

[7] Rey (1992), Williams (1994) (Chapter 1), Boury (2004, 159–164), and Wolfe and Terada (2008).

[8] Wolfe (2008).

[9] Burwick and Douglass (1992), Duchesneau and Cimino (1997), and Lofthouse (2005).

[10] Williams (2003).

[11] Rey (1987, 2000).

[12] Reill (2005).

What emerges more generally from this historiography of vitalism are a series of themes, the bulk of which will also be explored in our fresh offerings. These include: the non-scientific dimensions of medicine (and conversely, the essential tensions but also interrelations between medicine and 'science', opening on to considerations on the very nature of a 'life science'), both historically and in the contemporary context; the origins of vitalist thought; vitalism in relation to forms of mechanism (mechanistic explanations but also mechanistic ontologies[13]), and the transmutation, in the nineteenth century, of vitalism into fields like evolution, embryology, psychology, and other areas of the life sciences.

Our focus is post-Enlightenment, which also helpfully brackets off past eras for which historians are sometimes too eager to use the label 'vitalist' to describe figures that would have found this confusing at best.[14] Starting in the first years of the nineteenth century, we see a revival of vitalist themes in Romanticism and the rise of *Naturphilosophie* (discussed here in Chap. 3, Steigerwald's essay on 'organic vitality' in nineteenth-century German biology). The nineteenth century also brings with it the very complex of ideas that is biology, and as classic treatments of the period suggest, vitalism is central to the negotiation of this new terrain.[15]

The nineteenth century begins in this context with the relationship between Romanticism and vitalism but goes on to be shaped by vitalist debates in England during the Regency, by new 'alternative' medical ideas (mesmerism, homeopathy, phrenology, etc.), new developments in the laboratory (which give rise to a mid-century materialist-spiritualist debate), new notions of evolution (Darwin) and regenerated ideas about life (spontaneous generation) and finally, a real neo-vitalism post-Claude Bernard, born of philosophy (Bergson), psychology, morphogenesis and embryology (Driesch). That Bernard himself is a complex figure who both criticizes forms of what he calls 'vitalism', and at the same time articulates new models in which the uniqueness of organisms is justified, is another chapter of the story, still to be told (although see some suggestive remarks in Turner's and Bechtel's essays in this volume, Chaps. 10 and 13 and earlier, Roll-Hansen 1976, Normandin 2007). Many of the contributions to this volume touch, at least tangentially, on all these important developments.

---

[13]Wolfe (2012, forthcoming 2013).

[14]In a particularly imprecise way, Walter Pagel was able to describe both Aristotle and William Harvey as vitalists (Pagel 1944, e.g. 147), which is like the historical mirror image of the emptiness of the concept when it is just used to mean the view held by "cranks" throughout the history of biology (as in Francis Crick's rather arrogant pronouncement: "To those of you who may be vitalists, I would make this prophecy: what everyone believed yesterday, and you believe today, only cranks will believe tomorrow"; Crick 1966, 99). In the former case, if we treat Aristotle, Harvey, Montpellier vitalists, Blumenbach, Bernard, Driesch, Bergson and Canguilhem as instances of one view, then 'vitalism' seems to be an *auberge espagnole*, a halfway-house or rumpus room with any possible content; in the latter case, the view from 'mainstream' genetics that 'vitalism' is simply an archaic remainder destined for the rubbish heap, neglects not just historical context but scientific pluralism.

[15]Benton (1974), Coleman (1971/1977), Allen (2005), Gayon (2010/2011), and Wolfe (2011).

Another perspective that has not been dealt with all that extensively is vitalism in the twentieth century, inasmuch as there is an 'x' which traverses all these modifications and heterogeneous contexts. There is work on Bergson and Driesch, true,[16] but the broader texture of vitalism in medicine, philosophy and the life sciences in the twentieth century has been largely ignored, especially in its relation to the topic of emergence in the philosophy of mind, which is the object of detailed analysis here in the essays by Garrett and Malaterre (Chaps. 6 and 7), for emergentism was one of neo-vitalism's central offshoots[17]; a very different vitalist trajectory is sketched out by Dyde in his contribution (Chap. 5) on how unresolved tensions in the scientific attempts to provide a physiological account of the mind in the nineteenth century (from phrenology to the study of reflex action) unwittingly produced vitalistic concepts of mind, appetite and behaviour; as Dyde puts it, "methodologies begot ontologies."

This, along with contemporary debates regarding biological theory and the philosophy of biology, will be one of our central foci, and as such represents a new and exciting direction in scholarship on this subject. We have work within the pages of this volume on theories of emergence, complexity, biological theory (organicism), systems theory, homeostasis, holism, and beyond. Further, much of the prior interpretive discussion of vitalism has been either overly enthusiastic, or overly negative: either it is *the* theory which will make (life, meaning, embodiment, purposiveness, etc.) survive attempts at demystification or deflation in a context of 'physicalism', or it is *the* theory fit for cranks, the scientifically marginal and so on (cf. Normandin's essay on Reich (Chap. 8) for this kind of case). Few interpreters have reflected on either its *meanings* or its *uses* (cf. Oyama 2010 and this volume, *inter alia*, for an attempt to do just this).

We are struck by the idea that vitalism continues to re-emerge in the life sciences in all sorts of fascinating, complex, dynamic, even heretical ways over the period from the Enlightenment to today, and are in accord with the idea that, like its counterpart, mechanism, vitalism is a kind of "meta-theoretical commitment."[18] And yet, at the same time, we are prompted by a methodological heterogeneity (partly the result of the diversity of voices in our chorus) to maintain a level of 'free-play' and anarchism in our theories of knowledge,[19] giving no particular ontological priority to any one epistemological framework: some authors privilege a historicist approach over a naturalistic one (contrast the discussion of Canguilhem in Bianco's essay with that of Goldstein in Ferrario and Corsi's essay, Chaps. 10 and 9 respectively). Perhaps this is an inherently vitalist strategy, but we prefer to think of it as a sage intellectual choice.

---

[16]On Bergson and vitalism, cf. Burwick and Douglass (1992); on Driesch, cf. Freyhofer (1982), and Weber (1999).

[17]McLaughlin (2003).

[18]Hein (1968, 1969, 1972), compare Berryman (2003) and Wolfe (2012) on mechanism and Life.

[19]Feyerabend (1975).

1 Vitalism: Origin, History, and Transformation

Arguably, all understandings of life in antiquity implied a kind of vitalism. Charting the course of vitalism's history brings us from the classical age (where, on the question of souls and *animas*, we might still gesture towards Aristotle,[20] including the way in which his *De anima* was taken up in early modernity) through the core mechanizing forces of modern science (and, in our story, those malcontents on the periphery who criticized this trend[21]) to more contemporary manifestations of 'neo-vitalism' in continental philosophy. While the term 'vitalism' does not come into actual use until the late eighteenth century, many of the ideas and concepts embodied in the word are as old as medical and biological thought. From the *animas* and *pneumas* of Hippocrates, Aristotle and Galen to the ethical inducements towards vitalism found in the French tradition in thinkers like Georges Canguilhem,[22] the idea has a long, multi-faceted history.

Certainly, questions of body (*soma*) and soul (*psyche*) can carry us across broad swaths of space and time.[23] One interesting early history is L. Richmond Wheeler's *Vitalism: Its History and Validity*.[24] Wheeler boldly attempts a panoramic survey of vitalism's rich landscape from the time of Aristotle to the early twentieth century, reminding us that even during ostensibly mechanistic ages, there are vitalist undercurrents. The contrast between Harvey and Descartes' attitudes towards the movement of the heart (in relation to 'life' and to the functioning of the 'body-machine', respectively), has been discussed in various ways, whether to praise Harvey or Descartes – or to call Harvey a vitalist, as Pagel does. The picture looks different if it is not considered from a strictly internalist angle, as for instance here:

> It is interesting to consider … the claim often made in the anthropological and philosophical literature about the 'Cartesian' split between body and mind, dominating Western ethnopsychology and ethnophilosophy as a whole. Dualism is, no doubt, a characteristic feature of traditional 'Western' folk philosophy insofar as Western culture has been, traditionally, a Christian culture. But this traditional dualism has to do with the distinction between body and soul, not between body and mind. …Descartes opposed body, *corps*, to *âme*, and the concept of '*âme*' as used by Descartes was no doubt derived from the folk concept encoded in the French word *âme*, as it was used in the seventeenth-century French. It was certainly different from that encoded in the modern English word *mind*.[25]

This diversion into anthropology is an illustration of the unconventional approach we hope to bring to bear on this subject, pushing beyond the established confines of history

[20] Aristotle (1961/1999).

[21] For ways in which 'marginal' or 'heterodox' figures (who are often viewed as vitalists of a sort) can, should (or should not) be incorporated into a canonical version of the history of the life sciences, cf. Giglioni (2008) (for the case of Francis Glisson), Chang (2004) (for the case of Georg-Ernest Stahl), and Normandin (Chap. 8 this volume, for the case of Wilhelm Reich).

[22] Canguilhem (2008) and Delaporte (1994).

[23] Wright and Potter (2000).

[24] Wheeler (1939).

[25] Wierzbicka (1989, 46).

and philosophy of science. We want to transcend disciplinary boundaries, or better yet, produce new disciplinary hybrids, a vitalist act if there ever was one; we are in search of that flickering oasis in the borderland of ideas, where new notions can coalesce.

Returning to Wheeler's narrative, we follow a pattern of increase in vitalist thought in the second half of the eighteenth century, after the prominence of mechanism in its first half. Indeed, as Reill shows (Reill 2005), there are important elements of vitalism in the late Enlightenment, which blossom even further in the early nineteenth century (see Steigerwald's essay in Chap. 3 this volume). Of course, one must be careful here with the conflation between vitalism, *Naturphilosophie* and Romanticism: there are important distinctions. When speaking of a Romantic science, however, there seem to be clear elements of vitalism supporting it, whether in its actual manifestations (in a focus on sensibility and the passions in medicine, for example) or its cultural and literary importance (in the discussions of the "vital spark" in works like Shelley's *Frankenstein*, or in the public philosophical debate about vitalism between Abernathy and Lawrence in England during the Regency period[26]). Wheeler provides a useful categorization of vitalism in the nineteenth century, suggesting that thinkers of the period can be divided into "naturist" and "chemical" schools, and Ku-ming (Kevin) Chang has quite recently shown the complexity of "alchemical vitalism" in early modern matter theory.[27]

Indeed, in matters of debate as they relate to the question of life, one sees the development of camps increasingly divided by basic epistemological (and even ontological) differences. In this sense the laboratory and the lecture hall come to be more fully divided, and the questions asked by scientists and philosophers are increasingly remote from one another. There is also, connected to this trend, the question of the epistemological variances between medicine and science in the nineteenth century, which only experience a real synthesis in the 'biomedicine' of the later century – itself an episode not without its 'holistic' twists and turns, as described in Sonnenschein, Soto et al.'s contribution to this volume (Chap. 12).[28]

When discussing vitalism in the nineteenth century, the development of experimental physiology and the importance of Claude Bernard cannot be overlooked. Bernard marks the end of 'traditional' vitalisms that insisted on the universal solvent of a "vital force" and the move towards understandings of physiological relationships of the living that accepted complexity and uniqueness as central characteristics. Of course, if we look closer at the situation of medical vitalism in the late eighteenth century – not the topic of the present volume – we can already witness attempts to move away from "metaphysics" towards a more experimental or at least a more heuristically fruitful form of vitalism.[29] In that sense, Bernard and already Bichat,

[26] Jacyna (1983).

[27] Wheeler (1939); compare the distinction between physiological and chemical vitalism in Benton (1974), and Chang (2011) on alchemical vitalism.

[28] Specifically for biomedicine, cf. the essays collected in Lawrence and Weisz (1998).

[29] It is explicit in the later editions of Paul-Joseph Barthez's work *Nouveaux éléments de la science de l'homme* (the 1806 edition being the last one he revised): Rey (1987, 2000) and Wolfe (2011).

who is sometimes his target as an insufficiently experimental vitalist, are part of a process negotiating a shifting terrain of vitalism as a focus on the nature of biological, organismic or embodied life, attempting to do justice to criteria of scientificity which, of course, are themselves in flux, in a process of definitory crystallization in the period. Less historically, and more sharply put, one can also observe that Bernard (like Alexis Carrel in the early twentieth century, as discussed in Sonnenschein, Soto et al.'s essay, Chap. 12) was a "vitalist who practiced methodological reductionism." In Bernard's own terms, however much there may be features unique to the "living machine" (*machine vivante*), nevertheless, "the chemistry of the laboratory and the chemistry of life are subject to the same laws: there is no such thing as two (separate) chemistries."[30]

Post-Bernard, one witnesses a flourishing of new 'vitalisms'; from the biogenesis of Pasteur and the panspermia of Lord Kelvin to the emergentism of Morgan and the *élan vital* of Bergson. In this regard it is interesting to note an overlooked figure like J. H. Fabre (1823–1915), a French autodidact entomologist who developed a notion of instinct and its indefinability. Fabre was another who emphasized the unique character of organic structure.[31] Vitalism, it can be argued, got a boost from new research devoted to understanding the development of that structure.

Moreover, embryological ideas explored by thinkers like Hans Driesch became the basis for new neo-vitalist perspectives. Driesch, in stark contrast to mechanists like Ernst Haeckel (and Driesch's closer contemporary, Jacques Loeb[32]) was influenced by a more nuanced thinker, Emil Du Bois-Reymond, to investigate blastomeres in relation to morphogenesis and embryology.[33] He wrote about philosophy and vitalism, but also about the idea of individuality and even the viability of psychical research.[34] Driesch adopted the term *entelechy*, taken from Aristotle, to describe his belief in a teleological nature in living things that challenged the mechanistic synthesis in biology during this period. He prompted biologists to ask questions about the driving force in development, and helped open the door for research into genetics.

The whole early twentieth-century period could be described as a kind of "vitalist moment." While most of the discussions of vitalism in the era closely connect it to biology, medicine and philosophy, this does not reflect the actual early twentieth-century reality. Bergson's "neo-vitalism" had a wide appeal, extending all the way into the realm of the literary and cultural. There were vitalist themes in modernist art, for example, particularly in the dynamic and motion-based art movements of the

[30]Bernard (1865), e.g. II, 1, § VIII (entitled "Dans les sciences biologiques comme dans les sciences physico-chimiques, le déterminisme est possible, parce que, dans les corps vivants comme dans les corps bruts, la matière ne peut avoir aucune spontanéité"); 136–137; "Le chimisme de laboratoire et le chimisme de la vie sont soumis aux mêmes lois : il n'y a pas deux chimies ; Lavoisier l'a dit" (Bernard 1878, 226).

[31]Fabre (1879–1913).

[32]Loeb (1912/1964); discussion in Allen (2005).

[33]Waisse-Priven (2009) and Normandin (2011).

[34]Driesch (1914a, b, 1933) and Wolffram (2009).

futurists and vorticists. Arguably, the whole context of the period was infused with this sentiment:

> It requires a concerted act of historical imagination to re-create the vitalist moment, a moment which re-enchanted life in the face of mechanist onslaught, sought a reprieve from the more demoralizing implications of evolutionary inquiry and left open a space for spirit or even God in nature. Yet this view, which naturally bled into neighboring fields such as theological speculation and philosophy, attained for George Bernard Shaw's generation a special force that endured until the late inter-war years.[35]

Lofthouse reflects on a notable gap in historiography devoted to vitalism, particularly after 1945. The reasons he gives for this are manifold, not the least of which is the tangential connection of vitalism to fascism. Yet this is an unfortunate and largely unmerited association, one born of specific critiques, particular that of Zeev Sternhell and his discussions of Georges Sorel as a progenitor of fascism.[36]

As we will see in works dealing with the twentieth century, this is a historiographical oversimplification (as are more recent attempts to identify 'holism' in the life science with National Socialism[37]). Perhaps the more convincing reason for vitalism's decline is its inherently complex and nebulous meanings. Our works on contemporary biological theory will explore elements of this theoretic dissonance and dissipation. As to the centrality of malleability and variability in vitalism's variant definitions, there can be little doubt. But this should not dissuade attempts to understand the idea in its broadest terms.

Of course there were other important manifestations of vitalism in the early twentieth century, particularly in the realm of those who studied the *psyche*. Beyond Driesch and Bergson's classic *Creative Evolution* (1907) one is struck by developments in psychology (and, with a figure like Wilhelm Reich, psychiatry, as discussed in Normandin's contribution here). Certainly William McDougall's "hormic theory" was rooted in notions not dissimilar to Bergson's *élan vital* and Jung and Freud's essential fascination with the libido can be seen as the groundwork for vitalism in Reich's "orgone" and "life energy." McDougall was also connected to the philosopher C.D. Broad, and through him and J.B. Rhine, to a larger interest in parapsychology and psychical research, as mentioned above (this was a subject that also drew in Driesch).[38] This link between vitalism and the larger metaphysical questions associated with the nature of the living in the early twentieth century is a fascinating one, suggesting a relationship between vitalism and belief in the idea of a life force that somehow transcends the known material world. We are faced here

---

[35]Lofthouse (2005, 3).

[36]Sternhell et al. (1994, e.g. 24, 32). The identification between Fascism and vitalism is made at greater length in Payne (1995, e.g. 14, 26, 208).

[37]Harrington (1996) (who studies this 'identification' in a series of figures, and then comes to *another* holist of 1920s German life science, Kurt Goldstein, who, she notes, is Jewish ... leaving the aporias and/or fruitfulness of sociocultural contextualist history of science unquestioned or unjustified).

[38]Driesch (1933).

with the realization that maybe not all vitalisms post-Bernard were completely divorced from spiritualist strands on the rise in the early part of the twentieth century.

This idea of vitalism as a kind of spiritual force is, overall, increasingly marginalized in early twentieth-century thought, and the new vitalisms explored herein are multi-faceted examples of this trend – vitalisms of a theoretical or even a material (physical) sort. But we must also come to grips with how vitalism finds occasional expression in the neo-Thomist philosophies associated with Catholicism. Indeed, Catholic philosophy was heavily influenced by Bergson in the early twentieth century, and there is a direct link between Bergson's neo-vitalism and the nascent neo-Thomism of thinkers like Jacques Maritain, which led to various idealist interpretations of biology which labeled themselves 'vitalistic', such as those of Édouard Le Roy (influenced by Teilhard de Chardin).[39]

Such connections between vitalism and Scholasticism hint at a larger link between vitalism and philosophy. Indeed, in the French tradition, it was the historian and philosopher of the life sciences Georges Canguilhem who really made something of vitalism, both as an object of scholarly attention and more curiously, as a viewpoint he rather provocatively claimed for himself, declaring in the Foreword to his book on the development of the notion of reflex action that "it doesn't matter to me if I am considered to be a vitalist" and presenting the book itself as a "defense of vitalist biology."[40] It was due to Canguilhem's influence that thinkers like Foucault[41] and Deleuze also dealt with similar themes. Canguilhem initially applied the historical method to the concept of reflex action and uncovered a wealth of material devoted to understanding the complexities of this question – it is here where he finds the sensible, contractible and irritable, essential elements of vitalist discourse (the roots of which Bianco traces carefully in his contribution, Chap. 10). We will see them revisited herein, notably in Giglioni's study of irritation in Lamarck (Chap. 2).

Canguilhem problematized the categories of the normal and the pathological (inspired by Kurt Goldstein's *Structure of the Organism* [1934], the topic of Ferrario and Corsi's essay, Chap. 9), as well as the causal relationships between agents and disease. Here, then, is the source of portrayals of Canguilhem as a vitalist and individualist.[42] Gayon reminds us of Canguilhem's unique conceptual vision of life. "Life is concept," Canguilhem says, borrowing from Hegel. More forthrightly, Canguilhem suggests that life is not an on/off, normal/pathological, healthy/sick switch mechanism, but an ever transforming, teleological and, one may say, vitalistic reality. We are reminded here of how Canguilhem, under the influence of Bergson, would often return to the complex relationship between "concept" and "life." For Canguilhem there was also always a moral imperative in thinking through vitalism.

[39] Brenner (2011).

[40] Canguilhem (1955/1977), Avant-Propos, 1. For discussion cf. Wolfe, (ms. 2011, forthcoming 2014), and Bianco (Chap. 10, this volume) for a different perspective.

[41] Ransom (1997).

[42] Gayon (1998).

And finally, moving away from history entirely, we are excited about the prospect of a number of papers that give vitalism new vitality, that reintroduces (and reinterprets) some of its central concepts into contemporary biological debate, whether in positive terms (Turner on homeostasis, Bechtel on biological organization, in Chaps. 11 and 14, respectively), in cautiously favorable terms (Dupré and O'Malley's reflections on what it means for an entity to be living, and acknowledgment that there may be such a thing as a "vitalism heuristic" in biology, in Chap. 13); or in cautiously critical terms (Sonnenschein, Soto et al., Chap. 12), just as Garrett's assessment of the concept of emergence and its vitalistic ramifications (Chap. 6) is more philosophically pessimistic than Malaterre's (Chap. 7), and runs counter to the kind of historical productivity described in studies such as Steigerwald's or Dyde's (Chaps. 3 and 5). As we have seen with this introduction, this is the very essence of vitalism – an idea that gets invoked as we search for new understandings, metaphors, and meanings in the life sciences; less the statement of an 'essence' of life and more the realization that Life consists in a series of changing determinations, as Canguilhem might have put it. In both the historical and contemporary sense, then, we hope this collection revitalizes notions of vitalism for the modern academy and, perhaps, even spurs on new debate and discussion.

2 Final Thoughts

You hold in your hands a collected volume on the history and philosophy of vitalism in its relation to the 'scientific image' – the image of what science is but also, *pace* Sellars, the scientific image of the world as opposed to our experiential picture – that moves from historical accounts of the nineteenth century (dealing with, for example, the Lamarckian biology of irritability and its connection to ideas of life and death) and twentieth century (in, for example, reflections on the concept of emergence in the early century), and transitions in later essays towards more contemporary philosophical and theoretical reflections on everything from vitalism and post-modernism to vitalism as "dynamic mechanism."

We are, however, not engaged in unnecessarily convoluted metaphysical considerations. We seek to avoid making programmatic statements about vitalism and its role; this is a practical volume of historical and theoretical texts that take the idea of vitalism as a "meta-theoretical commitment" worthy of consideration, but that also realizes the idea has a rich and sometimes even overwhelming complex of meanings. This volume seeks to clarify rather than obfuscate, but we realize that there are also details and complexities that cannot be ignored. Again, we are aiming for a balanced perspective – something not all vitalists would necessarily agree with.

In the final analysis, we return to the idea of change, and how new images and perspectives on what constitutes 'science' prompt us to reconsider an idea that many too easily dismiss as outdated or merely idle spiritualism and mysticism. Alas, there are elements in this history of vitalism that cannot be divorced from this association. But this misses a key role vitalism has always played in scientific imagining – between the spiritual and the material, the digital and the analog, reductionism and holism, order and chaos, the inert and the animated, the constraining and the liberating, the dead and the living, the closed and the open, the rigid and the dynamic, the structured and the spontaneous, and even, at points, as in the case of our rich collection, the old and new.

Acknowledgments We would like to thank Clare Mason, Claudia Alexandra Manta, Michel Morange, Richarda Normandin, Sorana Corneanu, and Suzanne Wolfe Martin for their assistance and support.

References

Allen, Garland E. 2005. Mechanism, vitalism and organicism in late nineteenth- and twentieth-century biology: The importance of historical context. *Studies in History and Philosphy of Biological and Biomedical Sciences* 36: 261–283.

Aristotle. 1961 [1999]. *De Anima*, ed. David Ross. Oxford: Oxford University Press.

Barthez, Paul-Joseph. 1806. *Nouveaux éléments de la science de l'homme*, vol. 2, 2nd ed. Paris: Goujon & Brunot.

Benton, E. 1974. Vitalism in nineteenth-century thought: A typology and reassessment. *Studies in History and Philosophy of Science* 5: 17–48.

Bernard, Claude. 1865. *Introduction à l'étude de la médecine expérimentale*. Paris: J.-B. Baillière.

Bernard, Claude. 1878. In *Leçons sur les phénomènes de la vie communs aux animaux et aux végétaux*, vol. 1, ed. A. Dastre. Paris: J.-B. Baillière.

Berryman, Silvia. 2003. Ancient automata and mechanical explanation. *Phronesis* 48(4): 344–369.

Boury, Dominique. 2004. *La philosophie médicale de Théophile de Bordeu (1722–1776)*. Paris: Honoré Champion.

Brenner, Anastasios. 2011. Le vitalisme d'Édouard Le Roy entre mathématiques et religion. In *Repenser le vitalisme – Histoire et philosophie du vitalisme*, ed. Pascal Nouvel, 181–190. Paris: PUF.

Burwick, Frederick, and Paul Douglass (eds.). 1992. *The crisis in modernism: Bergson and the vitalist controversy*. Cambridge: Cambridge University Press.

Canguilhem, Georges. 1977. *La formation du concept de réflexe aux XVIIe et XVIIIe siècles*, 2nd revised edition. Paris: Vrin. (First published 1955.)

Canguilhem, Georges. 2008. Aspects of vitalism. In Canguilhem, *Knowledge of life*. Trans. Stefanos Geroulanos and Daniela Ginsburg, 59–74. New York: Fordham University Press. (First published 1952.)

Chang, Ku-Ming (Kevin). 2004. Motus Tonicus: Ernst Stahl's formulation of tonic motion and early modern medical thought. *Bulletin of the History of Medicine* 78: 767–803.

Chang, Ku-Ming (Kevin). 2011. Alchemy as studies of life and matter: Reconsidering the place of vitalism in early modern chymistry. *Isis* 102: 322–329.

Coleman, William. 1977 [1971]. *Biology in the nineteenth century: Problems of form, function, and transformation*. Cambridge: Cambridge University Press.

Crick, Francis. 1966. *Of molecules and men*. Seattle: University of Washington Press.

Delaporte, François (ed.). 1994. *A vital rationalist: Selected writings from Georges Canguilhem*. Trans. Arthur Goldhammer. New York: Zone Books.

Driesch, Hans. 1914a. *The history and theory of vitalism*. London: Macmillan.

Driesch, Hans. 1914b. *The problem of individuality: A course of four lectures delivered before the University of London in October 1913*. London: Macmillan.

Driesch, Hans. 1933. *Psychical research: The science of the super-normal*. Trans. Theodore Besterman. London: G. Bell and Sons.

Duchesneau, François, and Guido Cimino (eds.). 1997. *Vitalisms from Haller to cell theory: Proceedings of the Zaragoza symposium, XIXth international congress of the history of science*. Firenze: L.S. Olschki.

Emmeche, Claus. 2001. Does a robot have an Umwelt? Reflections on the qualitative biosemiotics of Jakob von Uexküll. *Semiotica* 134(1/4): 653–693.

Fabre, Jean-Henri. 1879–1913. *Souvenirs entomologiques*. 11 vols. Paris: Delagrave.

Feyerabend, Paul. 1975. *Against method: Outline of an anarchistic theory of knowledge*. London: New Left Books.

Freyhofer, Horst H. 1982. *The vitalism of Hans Driesch*. Frankfurt: Peter Lang.

Gayon, Jean. 1998. The concept of individuality in Canguilhem's philosophy of biology. *Journal of the History of Biology* 31: 305–325.

Gayon, Jean. 2010. Vitalisme et philosophie de la biologie. *Répha* 2: 7–18. (Reprinted in *Repenser le vitalisme – Histoire et philosophie du vitalisme*, ed. Pascal Nouvel, 15–32. Paris: PUF, 2011.)

Giglioni, Guido. 2008. What ever happened to Francis Glisson? Albrecht Haller and the fate of eighteenth-century irritability. *Science in Context* 21: 465–493.

Gilbert, Scott F., and Sahotra Sarkar. 2000. Embracing complexity: Organicism for the 21st century. *Developmental Dynamics* 219: 1–9.

Greco, Monica. 2005. On the vitality of vitalism. *Theory Culture & Society* 22: 15–27.

Harrington, Anne. 1996. *Reenchanted science – Holism in German culture from Wilhelm II to Hitler*. Princeton: Princeton University Press.

Hein, Hilde. 1968. Mechanism and vitalism as meta-theoretical commitments. *Philosophical Forum* 1: 185–205.

Hein, Hilde. 1969. Molecular biology vs. organicism: The enduring dispute between mechanism and vitalism. *Synthese* 20: 238–253.

Hein, Hilde. 1972. The endurance of the mechanism-vitalism controversy. *Journal of the History of Biology* 5: 159–188.

Huneman, Philippe. 2010. Assessing the prospects for a return of organisms in evolutionary biology. *History of Philosphy of the Life Sciences* 32(2–3): 341–372.

Jacyna, Leon S. 1983. Immanence or transcendence: Theories of life and organization in Britain, 1790–1835. *Isis* 74(3): 311–329.

Kirschner, Marc, John Gerhart, and Tim Mitchison. 2000. Molecular "Vitalism". *Cell* 100: 79–88.

Laubichler, Manfred. 2000. The organism is dead. Long live the organism! *Perspectives on Science* 8(3): 286–315.

Lawrence, Christopher, and George Weisz (eds.). 1998. *Greater than the parts: Holism in biomedicine, 1920–1950*. Oxford: Oxford University Press.

Loeb, Jacques. 1964. The mechanistic conception of life (1912), by Loeb. In *The mechanistic conception of life*, ed. D. Fleming, 5–34. Cambridge, MA: Harvard University Press.

Lofthouse, Richard A. 2005. *Vitalism in modern art, c. 1900–1950: Otto Dix, Stanley Spencer, Max Beckmann, and Jacob Epstein*. Lewiston: The Edwin Mellen Press.

McLaughlin, Brian P. 2003. Vitalism and emergence. In *The Cambridge history of philosophy: 1870–1945*, ed. Thomas Baldwin, 631–639. Cambridge: Cambridge University Press.

Morange, Michel. 2006. Les biologistes moléculaires face au problème de la vie. *Revue des Questions Scientifiques* 177(3–4): 381–394.

Normandin, Sebastian. 2007. Claude Bernard and *an introduction to the study of experimental medicine*: 'Physical Vitalism', dialectic and epistemology. *Journal of the History of Medicine and Allied Sciences* 62: 495–528.

Normandin, Sebastian. 2011. Review of Silvia Waisse-Priven, *d & D: duplo Dilema: du Bois-Reymond e Driesch, ou a vitalidade do Vitalismo. Bulletin of the History of Medicine* 85(2): 307–309.

Oyama, Susan. 2010. Biologists behaving badly: Vitalism and the language of language. *History of Philosphy of the Life Sciences* 32(2–3): 401–423.

Pagel, Walter. 1944. William Harvey: Some neglected aspects of medical history. *Journal of the Warburg and Courtauld Institutes* 7: 144–153.

Payne, Stanley. 1995. *A history of fascism, 1914–1945*. Madison: The University of Wisconsin Press.

Ransom, John S. 1997. Forget vitalism: Foucault and *Lebensphilosophie*. *Philosophy and Social Criticism* 23: 33–47.

Reill, Peter Hanns. 2005. *Vitalizing nature in the enlightenment*. Berkeley: University of California Press.

Rey, Roselyne. 1987. *Naissance et développement du vitalisme en France de la deuxième moitié du dix-huitième siècle à la fin du premier empire*. University of Paris, *thèse d'état*.

Rey, Roselyne. 1992. Anamorphoses d'Hippocrate au XVIIIe siècle. In *Maladie et maladies, histoire et conceptualisation. Mélanges en l'honneur de Mirko Grmek*, ed. Danielle Gourevitch, 257–276. Geneva: Droz.

Rey, Roselyne. 2000. *Naissance et développement du vitalisme en France de la deuxième moitié du dix-huitième siècle à la fin du premier empire*. Oxford: Voltaire Foundation. (Abridged version of Rey 1987).

Ritchie, A.D. 1940. Vitalism: Its history and validity. *Nature* 145: 6–7.

Roll-Hansen, Nils. 1976. Critical teleology: Immanuel Kant and Claude Bernard on the limitations of experimental biology. *Journal of the History of Biology* 9: 59–91.

Sternhell, Zeev, Mario Sznajder, and Maia Asheri. 1994. *The birth of fascist ideology: From cultural rebellion to political revolution*. Princeton: Princeton University Press.

Turner, Scott J. 2000. *The extended organism: The physiology of animal-built structures*. Cambridge, MA: Harvard University Press.

Waisse-Priven, Silvia. 2009. *d & D: duplo Dilema: du Bois-Reymond e Driesch, ou a vitalidade do Vitalismo*. São Paulo: EDUC-Editora.

Walsh, Denis M. 2010. Two neo-darwinisms. *History of the Philosopy of the Life Sciences* 32(2–3): 317–340.

Weber, Marcel. 1999. Hans Drieschs argumente für den Vitalismus. *Philosophia Naturalis* 36: 265–295.

West-Eberhard, Mary Jane. 2003. *Developmental plasticity and evolution*. Oxford: Oxford University Press.

Wheeler, Leonard Richmond. 1939. *Vitalism: Its history and validity*. London: N.F.G. Witherby.

Wierzbicka, Anna. 1989. Soul and mind: Linguistic evidence for ethnopsychology and cultural history. *American Anthropologist* 91: 41–58.

Williams, Elizabeth A. 1994. *The physical and the moral: Anthropology, physiology and philosophical medicine in France 1750–1850*. Cambridge: Cambridge University Press.

Williams, Elizabeth A. 2003. *A cultural history of medical vitalism in enlightenment montpellier*. Aldershot: Ashgate.

Wolfe, Charles T (ed.). 2008. *Vitalism without metaphysics? Medical vitalism in the enlightenment. Science in Context* 21(4).

Wolfe, Charles T. 2011. From substantival to functional vitalism and beyond, or from Stahlian animas to Canguilhemian attitudes. *Eidos* 14: 212–235.

Wolfe, Charles T. 2012. Le mécanique face au vivant. In *L'automate: modèle, machine, merveille*, ed. Bernard Roukhomovsky, Sophie Roux, Aurélia Gaillard, and Jean-Yves Goffi, 115–138. Bordeaux: Presses Universitaires de Bordeaux.

Wolfe, Charles T. ms., 2011. The return of vitalism: Canguilhem and French Biophilosophy in the 1960s.

Wolfe, Charles T. forthcoming 2013. Teleomechanism redux? Functional physiology and hybrid models of Life in early modern natural philosophy. *Gesnerus – Revue Suisse d'Histoire de la Médecine et des Sciences*, special issue : *Entre mécanisme et téléologie : Anatomie, physiologie et philosophie des fonctions (16e–18e siècles)*.

Wolfe, Charles T. forthcoming 2014. Was Canguilhem a biochauvinist? Goldstein, Canguilhem and the status of 'embodiment'. In *Medicine and society, new Continental perspectives*, Philosophy and medicine series, ed. Darian Meacham. Dordrecht: Springer.

Wolfe, Charles T., and Motoichi Terada. 2008. The animal economy as object and program in Montpellier vitalism. *Science in Context* 21(4): 537–579.

Wolffram, Heather. 2009. *The stepchildren of science: Psychical research and parapsychology in Germany, c. 1870–1939*. Amsterdam/New York: Rodopi.

Wright, John P., and Paul Potter (eds.). 2000. *Psyche and Soma: Physicians and metaphysicians on the mind-body problem from antiquity to enlightenment*. Oxford: Oxford University Press.

[illegible]
[illegible]
[illegible]
[illegible]
[illegible]
[illegible]
[illegible]
[illegible]

Part I
Revisiting Vitalist Themes in Nineteenth-Century Science

Chapter 2
Jean-Baptiste Lamarck and the Place of Irritability in the History of Life and Death

Guido Giglioni

M. de Lamarck séparait la vie d'avec la nature. La nature, à ses yeux, c'était la pierre et la cendre, le granit de la tombe, la mort! La vie n'y intervenait que comme un accident étrange et singulièrement industrieux, une lutte prolongée, avec plus ou moins de succès et d'équilibre çà et là, mais toujours finalement vaincue; l'immobilité froide était régnante après comme devant.

Charles-Augustin Sainte-Beuve, *Volupté*, 1834

Et metuunt magni naturam credere mundi
exitiale aliquod tempus clademque manere,
cum videant tantam terrarum incumbere molem!

Lucretius, *De rerum natura*, VI, 565–567

Abstract In the history of philosophy and science, vitalism has a bad reputation, for the very definition of life remains remorselessly murky. And yet life also resists all attempts at reduction carried out by the logic of mechanical reason. While on a superficial level, life smacks of irrational exuberance, on a deeper level, it is in the very uncomfortable company of death. In this chapter, I argue that this ambivalence is particularly evident in Jean-Baptiste Lamarck's natural philosophy. In his chemical, geological, botanical and zoological views, Lamarck advocated a theory of decaying rather than living matter. He characterized orgasm, irritability and sensibility – the forms which life takes on in the physical universe – as momentary interruptions of nature's ordinary course toward death and destruction. This chapter examines Lamarck's notion of irritability, paying special attention to his concept of "intussusception." By intussusception, Lamarck meant a universal mechanism of

G. Giglioni (✉)
Warburg Institute, Woburn Square, WC1H 0AB London, UK
e-mail: guido.giglioni@sas.ac.uk

S. Normandin and C.T. Wolfe (eds.), *Vitalism and the Scientific Image in Post-Enlightenment Life Science, 1800–2010*, History, Philosophy and Theory of the Life Sciences 2, DOI 10.1007/978-94-007-2445-7_2,

organic mutability through which organisms were able to calibrate their reactions to the environment. He argued that through increasingly more complex reactions, living beings could resist the universal tendency to disintegration and breakup by internalizing pressures coming from the environment.

Keywords Death • Evolution • Irritability • Life • Matter • Sensibility

1 The History of Life and Death

In one of the most important natural histories written in the last years of his life, the *Historia vitae et mortis* (1623), Bacon presented death as an essential attribute of all living beings. After all, living things can be said to be alive precisely because they are not dead yet and sooner or later are going to die. Bacon advocated a negative view of life (life as a form of tenacious and secret resistance to death), not a positive one (life as expansion and full actuality). It is for this reason that he saw the natural process of decay in all its forms as part of the very essence of life described as a primordial reaction to a pervasive tendency to restore an original condition of rest and immobility (Bacon [1620] 2004, 412; Bacon [1623] 2007, 346–354). Life, in this picture, manifested the unmistakable traits of uneasiness, discomfort and irritability rather than those of a placidly self-fulfilling activity.[1] Bacon's notion of life as a reaction, diffused at every level in the universe (bodies, souls and minds), against the unstoppable decay of material organizations exercised a deep influence on Francis Glisson (1598–1677), a physician and a philosopher who, especially in his *De natura substantiae energetica* (1672) and *De ventriculo et intestinis* (1677), had Bacon in mind when he defined irritability as a reaction to the force of inertia. The difference was that, while Bacon's account of life and death hinged on blind appetite, Glisson defended the original character of knowledge as prior to any aimless drive in nature.[2] Both Bacon's *Historia vitae et mortis* and Glisson's *De ventriculo et intestinis* demonstrate how closely intertwined the histories of irritability and living matter could be. Glisson, Regius Professor of Physic at the University of Cambridge, who first came up with the term "irritability" (*irritabilitas*), considered the tendency of natural

[1]On Bacon's notions of matter and death, cf. Giglioni (2005, 2011, 74–75). On Bacon's *Historia vitae et mortis* and its later fortuna, cf. Gemelli (2010, 191–205). Francis Glisson, referring to Jan Baptiste van Helmont's treatise *Ignota actio regiminis*, called this tendency to be irritated as 'disquietude' (*inquietudo*). Cf. British Library, MS Sloane 574B, f. 67[r].

[2]Cf. Glisson (1677, 365–368). In defending the primacy of perception over appetite, Glisson sides with Helmont rather than Bacon: "Non me latet, Illustrissimum. D. Dominum *Franc. Baconium*, Vicecomitem *Verulamii*, Tractatu *de Principiis* pag. 211, 212, etc. negare antiquissimum Cupidinem habere causam se priorem. Respondeo, eum proculdubio, sub persona Cupidinis, naturalem quoque perceptionem, absque qua ille subsistere nequit, includi supposuisse. Ipse expressis verbis agnoscit, *dari perceptionem generaliorem ea quae est sensuum*, quae nihil aliud esse potest nisi perceptio naturalis." Cf. Bacon ([1653] 1996), Bacon (1857–1874, I, 610; II, 602).

bodies to respond to external and internal stimuli to be the manifestation of a more general property of matter (i.e. its inherent reactivity). One year after Glisson had published his treatise on irritability, a colleague of his at Cambridge, Ralph Cudworth (1617–1688), Regius Professor of Hebrew, sent his *True Intellectual System of the Universe* to press (1678). Here he coined the term "hylozoism" to expose what he thought was a form of atheistic materialism hidden within the principal arguments of Glisson's philosophical treatise, *De natura substantiae energetica*.[3]

In this chapter, I adopt Cudworth's term "hylozoism" to refer to materialistic views of irritability (such as the one originally set out by Glisson following Bacon's theory of the natural motions of matter). As I argue elsewhere, the term hylozoism is still used in contemporary philosophical discourse to denote any form of thoroughgoing vital materialism and I prefer to employ that term rather than the more problematic "vitalism" (Giglioni 1994).[4] Albrecht von Haller (1708–1777), who in the eighteenth century adopted Glisson's notion of irritability, altered its meaning in quite a dramatic way, cutting its original ties with hylozoism and turning the power of irritable responsiveness into a quasi-mechanical property. Despite the success of Haller's operation, Glisson's suggestion that the irritable reactions of the body evidenced a deeper vital nature in matter was developed in various forms by a number of eighteenth-century natural philosophers and physicians. In different ways, Julien Offray de La Mettrie (1709–1751), Théophile de Bordeu (1722–1776), Pierre-Louis Moreau de Maupertuis (1698–1759), Denis Diderot (1713–1784), Paul-Joseph Barthez (1734–1806), Anthelme Richerand (1779–1840) and Pierre-Jean-Georges Cabanis (1757–1801) all embraced the view that an original source of material vitality displayed itself in nature in various forms and degrees. At the end of the eighteenth century, as the complementary concepts of irritability and sensibility came to shape the field of physiology, Lamarck appropriated the notion of irritability in the revised version that had been popularized by Haller. In this chapter, I focus on those aspects in Lamarck's theory of organic mutability that link the interrelated notions of irritability, evolution and materialism.

During the eighteenth century, Glisson's original attempt to associate hylozoism with irritability developed into broader explanatory frameworks involving questions of evolution, medical environmentalism and geological change. Natural philosophers and physicians focused their attention on innumerable phenomena betraying a pervasive tendency to react: tension, tone, quiver, palpitation, orgasm, erethism, erection, motility, contractility, excitability, incitability and sensibility. In different ways, they all wondered whether irritable responses in matter followed random or teleological patterns, whether they were based on purposive or mechanical organizations of life. What is more, none of them could shun any longer the question concerning the ultimate motor behind vital reactivity in nature. As pointed out by Lester Crocker, "a theory of

---

[3]Cudworth ([1845] 1995, III, 405). Cf. Giglioni (2002).

[4]On the complex issue of the relationship between vitalism and materialism in the early modern period, see the excellent account in Wolfe and Terada (2008). In contemporary discourse, panpsychism is often mistaken for hylozoism. From this point of view, the early modern discussion on life, matter and mind was more accurate than the contemporary one.

transformism could scarcely have been conceived without the belief that matter, and living organisms in particular, possess a self-organizing power or impulse."[5]

A further element contributed to joining doctrines of irritability with theories of evolution, that is, a growing awareness that the environment played a crucial role in turning the universe into one interrelated system. It was certainly no accident that Hippocrates' influential *Airs, Waters and Places* had become an increasingly popular text in the early modern fields of medicine, history and natural philosophy, and the rediscovery of Hippocratism in general reinforced the bond – already well-established – between environmental (geography, earth sciences, and meteorology) and biomedical disciplines. All this led to a synthesis of the earth sciences and the long established tradition of environmental medicine.[6] In stressing the link that connected organisms to their environments, for instance, Barthez had referred to the "endemic modifications" that the *principe vital* has "in each place of the earth" (Barthez 1778, 286–287).

Regarding the relationship between irritability, evolution and earth sciences, it cannot be denied that at the turn of the eighteenth century a distinctive historical understanding of life began to shape geology. As shown in Georges-Louis Leclerc, Comte de Buffon's *Les époques de la nature* (1779), the history of life on earth was closely intertwined with the history of the earth itself.[7] In the same years, advancements in the domain of physiology had led naturalists and physicians to think of the functional complexity of the vital apparatuses in terms of their adaptive histories. It is certainly not accidental that Richerand, whose physiological work Lamarck knew and cited, had compared anatomy and physiology to geography and history (Richerand [1801] 1823, 70). As we will see, one of the critical elements in Lamarck's theory of organic mutability concerned the question of how to reconcile the relentlessly historical nature of life with a mechanical account of vital processes. It was precisely through the historicization of life that early modern beliefs in its continuity and hierarchical gradations were dramatically redefined. Lamarck could thus reject the venerable idea of the chain of being in favour of a vision of organic change in which unbridgeable gaps, interrupted routes, contingent detours and multiple ramifications played a role more important than continuities and similarities. The *circonstances* – a crucial notion in Lamarck's thought – represents the foray of contingent and unforeseen events into any allegedly pre-ordained harmony between structures and functions.

Irritability is central to Lamarck's doctrine of evolution, for the ability to respond actively to the pressures of the environment can be seen as an initial form of adaptation. By espousing the notion of irritability, Lamarck joined a long-established tradition of medical thought. At the same time he provided an original reinterpretation of the phenomenon of irritability. He rejected the hylozoistic implications associated with a materialistic understanding of irritability, while retaining its explanatory potential in regard to organic change. As he did with spontaneous generation, Lamarck both

[5]Crocker (1968, 115) and Omodeo (1997).

[6]Jordanova (1979, 122). Cf. Picavet (1891), Rosen (1946), Temkin (1968), Canguilhem ([1965] 2008, 59–120), Ackerknecht (1967), Desaive (1972), Moravia (1972, 1974), Figlio (1976), Burkhardt (1995, 103), Corsi (1988, 75), and Jordanova (1984, 58–70).

[7]Cf. Corsi (1988, 25, 35–35, 76–84). On Buffon cf. Hoquet (2005).

borrowed and transformed concepts that were part of the medical tradition, purging the notions of generation and responsiveness of their vitalistic assumptions. The difference is that, while the phenomenon of spontaneous generation explained the origin of life in the universe, he used the process of irritability to account for various mechanisms of life adaptation and self-preservation. If spontaneous generation became the ahistorical starting point in the history of life, irritability represented the historical development of life based on the varying circumstances of the environment.

An inquiry into the meaning of irritability in Lamarck's work sheds further light on the role of the environment and of the organizing power (*pouvoir de la vie*) in shaping the characteristics of living organisms. In this chapter, I will argue that Lamarck's system of environmental forces, and not the *pouvoir de la vie*, represents the primary cause of organic development. In a view of a universe where the influences of the environment – the "circumstances" – and the activities of organisms are in direct and constant interrelation, the *pouvoir de la vie* can be seen as the progressive internalization (*intussusception*, to use Lamarck's word) of external stimuli, a process through which the environment is transposed and recreated within the organism itself. Therefore, I use the notion of *intussusception* as a hermeneutical concept to explore Lamarck's view on life and change. It could be said that, in his explanation of vital phenomena, life comes always from the outside; the "outside" here meaning the imponderable and expansive fluids which surround and pervade organisms or any other environmental force. Given these premises, irritability in Lamarck's terms cannot be interpreted as a form of inherent responsiveness characterizing every single aspect of matter – as had been originally proposed by Glisson and in part by Haller. This means that, while Lamarck assigned to the environment the role of activating and nurturing life-forms, reactivity and responsiveness remained problematic in his account of transformism.

In Lamarck's view, the defining characteristic of life is to be intermittent. To use Henri Daudin's pointed phrase, nature seems to progress in "*crises de l'organisation*" that end in relatively stable patterns of vital processes (Daudin [1926] 1983, II, 198). These patterns signal the presence of gaps between the inorganic and the organic, plants and animals, and various levels of activity within the sphere of organic life (orgasm in plants, irritability in lower animals and sensibility in higher ones). Lamarck's explanation of change is premised on the fundamental assumption that inorganic substances do not mutate into organic ones. It is rather the contrary in nature, for organic compounds deteriorate and decompose into inorganic matter. This, in the end, is what our earth is all about: a long history of decay and death.

2 A "Flood of Light": The Notion of *Intussusception* in Lamarck's Account of Organic Change

Lamarck's philosophy of life is based on the existence of specific patterns of reactivity in nature. Two conditions, Lamarck writes in his *Zoological Philosophy* (1809), must be satisfied in order for a body to have life: "a stimulating cause which excites organic movements" and, "the property of responding to the action of the stimulating cause and of producing organic movements" (Lamarck [1809] 1984, 203). Lamarck's evolutionary scheme presupposes in all organisms the ability to receive stimuli from the outside world and to react to them in a wide variety of ways. Irritability is only one of these ways, the one that characterizes the animal kingdom (plants are defined by orgasm and higher animals by inner feeling). The interplay between the organizing tendency inherent in nature and the action of the environment is therefore central in this view. On the one hand, the organization of life forms grows in complexity as a result of the movements of different fluids (Lamarck [1809] 1984, 40, 130); on the other, the environment produces everywhere variations by arousing needs, habits and modifications of organs (Lamarck [1809] 1984, 69). Organic change can thus be traced back to an original dualism between an external factor that acts as a stimulus (the "exciting cause") and an internal force that is capable of responding actively to that stimulus. Here we find the same kind of polarity theorized by Haller in his doctrine of irritability (Giglioni 2008, 2010).

The diverse patterns of organization in nature represent the different ways in which the original relationship between external influences and the organisms' responses to them evolves in nature:

> In animals the exciting cause of organic movements acts powerfully both on the containing parts and on the contained fluids; it maintains an energetic orgasm in the containing parts, puts them in a condition to react on the contained fluids and hence makes them highly irritable; as to the contained fluids, the exciting cause involves them in a kind of rarefaction and expansion, which facilitate their various movements (Lamarck [1809] 1984, 206).

The most elementary living beings display an undifferentiated, nearly osmotic relationship with the external environment, such that fluids penetrate and permeate the organic structures from all sides. The exciting force is "the result of subtle fluids (such as caloric, electricity, and perhaps others) which incessantly penetrate these animals from the environment, set in motion the visible and contained fluids of their bodies, and by exciting the irritability of their containing parts, give rise to the various movements of contraction which they produce" (Lamarck [1809] 1984, 345).[8] The more organized the living being is, the more differentiated from its surroundings its shape and organization become. This observation led Lamarck to a fundamental insight, "a flood of light," which disclosed "the principal cause which maintains movements and the life of organized bodies" (Lamarck [1809] 1984, 5–6). From orgasm in plants to irritability in animals up to the subtlest manifestations of feeling in the higher mammals, a progressive internalization of the originally external animating principle is at work in nature. Lamarck describes in a careful way this shift of energy and organization from the outside to the inside of organisms:

> I remarked that the movements of animals are never directly communicated, but that they are always excited; hence I recognized that nature, although obliged at first to borrow from the environment the *excitatory power* for vital movements and the actions of imperfect animals, was able by a further elaboration of the animal organization to convey that power right into the interior of these beings, and that finally she reached the point of placing that same power at the disposal of the individual (Lamarck [1809] 1984, 6).

[8] In elaborating his notion of *fluides incontenables*, Lamarck drew on Newton's model of experimental physics. Cf. Pichot (1994, 23–24) and Conry (1981).

The forces which stimulate the organic change in lower animals "exist outside of them and yet animate them," they are "subsequently transported and fixed in the animal itself" and, finally, become "the source of sensibility, and last of all of acts of intelligence" (Lamarck [1809] 1984, 11). The process of internalization occurs by borrowing the excitatory power from the outside world. As a result, this power becomes increasingly more available to the organism and allows it to perform its functions in a more suitable way. To explain this process of life borrowed from the environment Lamarck reinterprets in an original way the notion of "intussusception" characteristic of the epigenetic tradition and physiology of nutrition. Contrary to the process of *juxtaposition*, intussusception is "a true development of parts from within outwards" (Lamarck [1809] 1984, 193).[9]

In *Recherches sur l'organisation des corps vivants* (1802), Lamarck expressly rejects Bichat's assumption that the reactive power of the organism is a response against the destructive force of the environment (Lamarck [1802] 1986, 58). Bichat had opened his influential *Recherches physiologiques sur la vie et la mort* (1800) with the famous definition: "Life consists in the sum of the functions, by which death is resisted." The rest of the passage clarifies where the difference lies between Bichat and Lamarck:

> In living bodies such in fact is the mode of existence, that whatever surrounds them, tends to their destruction. They are influenced incessantly by inorganic bodies; they exercise themselves, the one upon the other, as constant an action; under such circumstances they could not long subsist, were they not possessed in themselves of a permanent principle of reaction. This principle is that of life (Bichat [1800] 1827, 10).[10]

Bichat characterizes the interplay of "exterior power" and "interior resistance" as a cycle of assimilation and excretion between the "animal existing within" and the "animal existing without." The result is a "permanent principle of reaction." Georges Cuvier (1769–1832) thought of life in similar terms when he assigned the principle of death to the whole of the external bodies and agents.[11] Contrary to the views carried on by Bichat, Cuvier and other contemporaries, Lamarck regarded the whole system of environmental influences as fostering the vital reactivity of nature. Life is not hindered by, but appropriated from the environment.[12]

---

[9]Cf. Lamarck ([1809] 1984, 185, 246, 343–346). In the *Histoire des animaux*, Buffon compared the process of reproduction to the function of nutrition. In both cases, living beings demonstrate the ability to identify and distinguish 'organic' and 'brute' molecules: "dans la nourriture que ces êtres organisés tirent, il y a des molécules organiques de différent espèces; que, par une force semblable à celle qui produit la pesanteur, ces molécules organiques pénètrent toutes les parties du corps organisé, ce qui produit le développement et fait la nutrition." This "intussusception des molécules" is a process that underlies both nutrition and reproduction (Buffon 1835–1835, III, 394). On the discussion of "intussusception" versus "juxtaposition" in Louis Bourguet (1678–1742), Charles Bonnet (1720–1793) and Carl Linnaeus (1707–1778), cf. Wolfe (2011, 200).

[10]On Bichat and the medical context at the time, cf. Haigh (1984).

[11]Cuvier (1800–1805, I, 3).

[12]As pointed out by André Pichot, Lamarck's "*biologie* est une réponse mécaniste à la *physiologie* vitaliste de Bichat, qui était alors la théorie dominante" (Pichot 1994, 20).

Historians have insisted on the impact that Lamarck's endorsement of spontaneous generation had on his later views about transformism. Spontaneous generation, however, plays the role of a dogmatic starting point rather than that of an actual explanatory scheme (all the more so since spontaneous generation is in Lamarck's view a direct production of well-defined organic structures and not a dissemination of living forms stemming from the universal source of life in matter). By contrast, the process of *intussusception*, which Lamarck himself refers to as a "flood of light," seems to play a more important role than spontaneous generation.[13] *Intussusception* is a mechanism of organic mutability that progressively calibrates the reactions of the organisms and adjusts them to their environment. Through the process of orgasm, the response of plants is still immersed in the energy of the environment; with the emergence of irritable reactions, the external stimulation begins to be gradually internalized; by means of sensibility, finally, the mass of external influences are distinguished, organized and mastered.[14]

3 Irritability in Lamarck's Theory of the Animal Being

Lamarck traces the source of the primordial reactivity underlying all manifestations of life back to a phenomenon of vital tension that is ultimately aroused by heat. "Orgasm" is the first form of reactive energy through which living beings respond to the smallest influences coming from the outside and is defined as "a particular tension in each point of the supple parts" resulting from any given distance among their molecules. Distance is what maintains a state of balance between "erethism" (the highest degree of which is inflammation) and "atony" (Lamarck [1802] 1986, 57, 60–61, 115). In animals, orgasm depends on a subtle and expansive fluid that continuously exudes from the arterial blood. In the case of elementary organisms, this fluid is taken from their environment. However, while no increase of vital energy occurs in plants – and therefore in their case the level of orgasm remains stable – in animals orgasm may raise the tension of the vital system. What distinguishes the animals' vital economy from that of plants is a "gradual acceleration of the fluids' movement" (Lamarck 1815–1822, I, 202–203).[15] Any external body that happens to touch some point in the supple parts of an animal causes a local

---

[13]Burckhardt (1995, 157) hints at this evolutionary mechanism in passing: "[Lamarck's] explanation of the production of the simpler invertebrates demonstrates that in his view the power of life was not opposed to environmental influences but, on the contrary, grew directly out of them. Only after the complexity of animal organization became sufficiently great was 'the productive force of movement' internalized."

[14]As is aptly remarked by Pichot, irritability represents an important phase in the development of the *intussusception* mechanism (Pichot uses the expression *processus auto-catalytique*): "Chez les animaux, la principale conséquence de l'irritabilité des tissus est une intériorisation de la cause excitatrice des mouvements de fluides" (Pichot 1994, 25, 37–39).

[15]Cf. Lamarck (1815–1822, I, 104–105, 115, 175).

dissipation of the subtle fluid, keeping the level of orgasm steady. This discharge forces the molecules of the affected part to contract, resulting in a motion of irritability (Lamarck [1802] 1986, 62–63, 78).

In Lamarck's philosophy of life, irritability represents the defining characteristic of the animal being (Lamarck 1815–1822, I, 111–112).[16] This essential property, however, is in inverse proportion to the complexity of the organization: the more rudimentary the living being is, the more irritable it is (Lamarck [1809] 1984, 47, 51–53, 97, 100, 135–136, 222, 227–229; Lamarck 1801, 20, 40, 358).[17] "Life in the most imperfect animals," says Lamarck, "has no energy of movements; and irritability alone suffices for the execution of vital movements" (Lamarck [1809] 1984, 48). Lamarck also provides an anatomical explanation of why invertebrates are more irritable than vertebrates, based on the observation that animals without a skeleton have a structure that can contract more easily (Lamarck 1801, 7–9). The infusorians, Lamarck argues in *Zoological Philosophy*,

> are only very tiny gelatinous, transparent, contractile and homogeneous bodies, consisting of cellular tissue, with very slight cohesion and yet irritable throughout. These tiny bodies, which look like animated or moving points, feed by absorption and continual imbibition; and they are doubtless animated by the influence of the subtle surrounding fluids, such as caloric and electricity, which stimulate in them the movements constituting life (Lamarck [1809] 1984, 102).

This peculiar relationship between irritability and anatomical structures is particularly evident in the case of polyps, the recently discovered wonder in the organic nature (*les premières ébauches de l'animalisation*). They are highly irritable and capable of contracting. Their movements are "necessary results of impressions received, and are in general carried out without any act of will; they are thus without any possibility of choice, since they cannot have any will" (Lamarck [1809] 1984, 101).[18] More complex organisms react through irritability, but they are also capable of experiencing sensations and "possess a very vague inner feeling of their existence; but they only act by the internal impulse of an inclination which leads them towards some or other object; so that their will is always dependent and controlled" (Lamarck [1809] 1984, 48).

We can therefore say that, in Lamarck's system of nature, irritability develops in inverse proportion to organic complexity.[19] Irritability is diffused and does not

[16]Lamarck (1815–1822, I, 124–125): "Les *animaux* sont des corps vivans doués de parties irritables, contractiles instantanément et itérativement sur elles-mêmes, ce qui leur donne à tous la faculté d'agir, et à la plupart celle de se déplacer."

[17]Lamarck ([1802] 1986, 66): "l'*irritabilité* devient plus grande et plus durable, à mesure que l'organisation animale approche plus sa simplification."

[18]Cf. Lamarck ([1809] 1984, 358): "there is no true will in animals which have a nervous system without any organ for intelligence, and if this is the reason why such animals only act by emotions produced by their sensations, this truth applies still more to animals that have no nerves. It appears therefore that these latter only move by an excited irritability, and as an immediate result of external excitations." Cf. Lamarck (1801, 19, 41–42, 357–359), Lamarck ([1802] 1986, 31–32, 65, 76, 80–84, 124), Lamarck (1815–1822, I, 114–153–154).

[19]Cf. Lamarck (1972, 180): "à mesure que la composition de l'organisation diminue, les facultés animales sont graduellement moins nombreuses mais qu'elles acquièrent proportionellement plus d'étendue."

presuppose any differentiation into organs. In the years during which Lamarck elaborated his view of irritability as confined to the animal kingdom, Richerand was extending the power of irritable reactivity to all living organisms:

> life is more generally diffused, and its phenomena less rigorously and strictly connected, as we descend in the scale of being, from the red and warm-blooded, to the red and cold-blooded animals, from these to the mollusca, the crustacea, worms and insects, to the polypus, who forms the extreme link of the animal chain, and lastly, to plants, of which not a few, like the zoophytes, so similar to them in many respects, are endowed with the remarkable property of reproduction by slips; which implies that each part contains the aggregate of organs necessary to life, and can exist alone (Richerand [1801] 1823, 53).[20]

This view was in line with the thesis, defended by Richerand and other naturalists at the time, that a primordial, all-pervading and decentralized source of physical energy was the distinguishing hallmark of life. It was less consistent with Lamarck's general assumption that life and other vital faculties develop according to their level of structural organization.

With the progress of organization and animalization a new faculty makes its appearance in the scale of nature: sensibility. In Lamarck's view, sensibility is the result of an increased tension in the fibres of particular parts of the body. Instead of keeping steady the level of orgasm that is normally produced by the subtle fluid exhaling from the arterial blood, the increase in tension provokes "revulsion" of the nervous fluid and a contraction in the irritated part. The retrograde motion of the nervous fluid (*révulsion nerveuse*) occurs without any conscious intervention of the will and causes a shock in the organ of feeling, which communicates to the brain the perception of what happened in the irritated part (Lamarck [1802] 1986, 119–120). When sensibility emerges in nature as a result of rudimentary nervous systems, animals cease to be passive machines and become capable of interacting with their environment in a more complex, flexible way (Lamarck [1809] 1984, 346). In keeping with Haller, however, Lamarck separates the sphere of functions pertaining to the exercise of sensibility from those belonging to the domain of irritability.[21] Unlike irritability, sensibility is localized and not diffused throughout the organism (Lamarck 1815–1822, I, 91, 231).[22]

The third part of Lamarck's *Philosophie zoologique* is entirely devoted to a detailed investigation of the physical causes and material conditions of feeling. It is an inquiry into the nature of sensibility that is programmatically carried out from the

[20] In his *On the Relations between the Physical and Moral Aspects of Man*, Cabanis had pointed out the extent to which any increase in organic complexity resulted in a living system's precariousness. Higher sensibility – a result of higher organization – exposed living beings to a greater amount of stimuli, but also to a more complicated network of reactions and dangers (Cabanis [1805] 1981, II, 543–544).

[21] Lamarck (1815–1822, I, 90): "On sait que *Haller* avait déjà distingué ces deux sortes de phénomènes; mais, comme la plupart des zoologistes de notre temps les confondent encore, il est utile que je m'efforce de rétablir cette distinction dont le fondement est de toute évidence." Cf. Ibid., 230, 233.

[22] Cf. Burkhardt (1995, 170, 175), Jordanova (1981), and Gissis (2010).

point of view of a zoologist (Lamarck [1809] 1984, 287). As all faculties depend on organic functions, and therefore on specific organs, feeling, too, must have an organic basis: "feeling is only an effect; that is to say, the result of an organic act and not a faculty inherent in any of the substances, which enter into the composition of a body that can experience it" (Lamarck [1809] 1984, 273). From the organization of separate ganglia, communicating together through nervous threads (in radiarians), to the ganglionic longitudinal cord extended throughout the animal's length (in insects and crustaceans) and then to vital systems organized around a spinal cord, nerves and a brain with two hemispheres, in all these cases a clearly defined organic structure underlies the phenomena of sensibility. From this point of view, the nervous system represents the most sophisticated instance of *intussusception*, in that the stimulating factor from the environment is entirely reconstructed inside the organism.

Lamarck's distinction between irritability and sensibility relies on a clear division between undifferentiated life and forms of sentient activity that depends on the organization of matter. While irritability does not need any special organ and occurs in physiological states in which the boundaries separating the external and internal milieu are thoroughly porous, sensibility requires the presence of reactive mechanisms which are much more elaborate (the nervous system), based on a clear division between external stimuli, peripheral sense organs and a coordinating and unifying centre. By emphasizing the distinction between irritability and sensibility, Lamarck is polemically referring to, among others, Richerand and Cabanis and to their belief in the existence of a form of latent sensitivity diffused everywhere in nature, a view Cabanis connected to a broader and older tradition: "the Stahlians, the semi-animists, the recent solidists of Edinburgh, and the wisest professors of the school of Montpellier" (Cabanis [1805] 1981, I, 84).[23]

In his *Rapports du physique et du moral de l'homme* (1805), Cabanis summed up the question by saying that "to live is to feel" and characterized sensibility as "the general fact of living nature" (Cabanis [1805] 1981, I, 120). He rejected Haller's distinction between irritability and sensibility ("it is hardly more than a question of words") in favour of a complete reduction of all reactive phenomena to one form of sensibility (Cabanis [1805] 1981, I, 83–85; II, 546–550).[24] Likewise, in his *Nouveaux éléments de physiologie* (1801), Richerand equated sensibility and contractility to vital properties extended to every living being, including plants: "the two properties of feeling and motion are indispensable to all the parts of the body. They are properties universally diffused through organized and living matter, but they exist without possessing any peculiar organ or instrument of action." In some cases, as in plant physiology or in basic life operations such as nutrition, sensibility "is always latent or concealed."

Like Glisson in the seventeenth century, Richerand distinguished between two kinds of sensibility: a conscious and "percipient" kind of sensibility and an "obscure and

[23]On Cabanis and the physiological and anthropological doctrines of the *Idéologues*, cf. Moravia (1968, 1970, 1974, 1978), Staum (1980), Williams (1994), and Barsanti (1983).

[24]As Xavier Bichat noted with regard to the distinction between sensibility and irritability, Haller "made them almost insulated properties" (Bichat [1801] 1822, I, 12).

latent" form of perception, "common to all living beings, and without which, it is impossible to conceive life to exist" (Richerand [1801] 1823, 21–30). Xavier Bichat, another contemporary physiologist Lamarck distanced himself from, emphasized the specific status of living matter by referring to a universal form of sensibility: "Chaos was only matter without properties; to create the universe, God endowed it with gravity, elasticity, affinity, etc. and to a part he gave sensibility and contractility" (Bichat [1801] 1822, I, 10). In Bichat's view, sensibility and contractility were the ultimate properties of living matter and could not be reduced to physical properties. More specifically, he listed a series of ascending faculties characterizing living matter: "organic sensibility," "insensible organic contractility"(in plants and organic life in general), "sensible organic contractility" or "irritability" (in zoophytes), "animal sensibility" and "animal contractility" (starting with worms and insects) (Bichat [1801] 1822, I, 14–16).[25] Against Richerand, Cabanis and Bichat, and in full agreement with Haller, Lamarck intended to show that sensibility and irritability were not only "quite distinct faculties," but that "they have not even a common origin and are due to very different causes" (Lamarck [1809] 1984, 208).[26] By distinguishing between orgasm and irritability on the one hand, and irritability and sensibility on the other, Lamarck did not need to grant all forms of living nature perceptive power, for the irritable motions of the vital functions, perfect as they might look, could easily be derived from a sophisticated system of mechanical actions and reactions in the parts involved in the process (Lamarck [1809] 1984, 220).

Lamarck considered orgasm, irritability and sensibility to be forms of organic mutability through which organisms were able to modify and adjust themselves to a physical reality in continuous change without the need to invoke a pre-established harmony of divine origin or to resort to animistic solutions, be those the venerable notion of the "soul of the world" or the view that living beings, even plants, were endowed with desires, tendencies and perceptions (Lamarck 1815–1822, I, 323). The ability to react in all its various forms – orgasm, irritability and sensibility – was seen as unconscious, that is to say, not based on the exercise of the will or on the knowledge of the function performed. Understood as a form of intussusception, it was not even a tendency stemming from within the organism, but a passive property induced and determined by the surrounding milieu.

4 The Interplay of Life and Nature in Lamarck's Work

As a naturalist, Lamarck prided himself on grounding all his arguments on a physical basis:

> It is to the influence of the movements of various fluids in the more or less solid substances of our earth that we must attribute the formation, temporary preservation, and reproduction of all

[25]Richerand accused Bichat of having plagiarized his own theory of sensibility. Cf. Haigh (1984, 10).

[26]Lamarck ([1809] 1984, 218, 228). Cf. Ibid., 322: "the conditions required for the production of feeling are of altogether another nature from those necessary for the presence of irritability. The former demand a special organ which is always distinct, complex, and extended throughout the animal's body, whereas the latter demand no special organ and give rise only to an isolated and local phenomenon."

> living bodies observed on its surface, and of all the transformations incessantly undergone by the remains of these bodies (Lamarck [1809] 1984, 187–188).

In Lamarck's opinion, even thought and the formation of ideas could be traced back to the operations of moving fluids (Lamarck [1809] 1984, 375, Lamarck [1802] 1986, 112–113). According to a recurrent rule in Lamarck's theory of knowledge, nature has to be studied as a purely "physical" phenomenon:

> it is obvious that the phenomena in question are, on the one hand, only direct effects of the mutual relations of different bodies, and only the result of an order and state of things which give rise to these relations among some of them; and, on the other hand, it is obvious that these phenomena result from movements set up in the parts of these bodies by a force whose origin it is possible to ascertain (Lamarck [1809] 1984, 184).[27]

Lamarck's account of living beings in terms of physical phenomena represents a decisive step towards a full secularization of the notion of life. In his philosophy of nature, all bodies are endowed with faculties that are all – "without exception" – physical and dependent on organic structures (Lamarck 1815–1822, I, 4, 13, 121–122, 212–258).[28] Reality is made up of bodies (*la nature ne nous offre d'observable que des corps*) and life is a physical phenomenon (*la vie est un phénomène très-naturel, un fait physique*).[29]

Within this physical account of life, organic changes depend on differences in the states of matter. Lamarck thought that two "co-existing" material conditions were at the origin of life processes in nature: "the one solid, but supple and capable of holding liquids; the other liquid and contained in the first, but quite independent of the invisible fluids which penetrate the body and develop within it" (Lamarck [1802] 1986, 16).[30] Lamarck's insistence on the vital potentialities of the fluids recalls Barthez's doctrine of the sensibility of the humours.[31] Besides the visible fluids, there are other "stimulating subtle fluids," such as caloric, electricity and magnetism, which represent the "special force" distributed throughout the environment, pervading and animating organic structures. While caloric is the fluid which produces and maintains a constant level of orgasm in the supple parts of the living bodies, electricity excites irritability in the animal organisms.[32] The difference with Barthez is that the subtle fluids which act as stimulants of life are of an

[27]Cf. Lamarck (1815–1822, I, 11).

[28]Cf. Russo (1981).

[29]Lamarck (1815–1822, I, 59–60, 260), Lamarck ([1809] 1984, 183, 187, 321–322), Lamarck ([1802] 1986, 56–57): "La vie est un ordre et un état de choses dans les parties de tout corps qui la possède, qui permettent ou rendent possible en lui l'exécution du mouvement organique, et qui, tant qu'il subsistent, s'opposent efficacement à la mort." Cf. Burkhardt (1995, 58–71).

[30]Cf. Lamarck ([1802] 1986, 77, 108) and Lamarck ([1809] 1984, 192, 205).

[31]Barthez (1778, 109): "chaque humeur est formée par une fermentation spécifique vitale; c'est à dire par un mouvement intestin. ... qui anime les mixtes qu'il a produits, et les pénètre toujours plus intimement de l'action du Principe de la vie."

[32]Lamarck ([1809] 1984, 211–218), Lamarck (1815–1822, I, 42, 169–171). On Lamarck's notion of invisible fluids cf. Burkhardt (1995, 65), Jordanova (1984, 49), and Corsi (1988, 152).

unequivocally unsentient nature. In some particular circumstances, these fluids may be able to create primordial rudiments of organization – such as certain kinds of moulds, mushrooms, lichens, infusorians and worms – without involving necessarily a principle of fertilization (unlike viviparous, oviparous and gemmiparous generations). By means of "heat, light, electricity and moisture," says Lamarck, nature "forms direct or spontaneous generations at that extremity of each kingdom of living bodies, where the simplest of these bodies are found" (Lamarck [1809] 1984, 244).[33] What is more, subtle fluids account for the origin of the animal faculties. Heat and light,

> distribute over our earth, at least, the principle of organization and feeling; and since feeling in its turn gives rise to thought as a result of the numerous impressions made on its organ by external and internal objects through the medium of the senses, the origin of every animal faculty may be traced to these foundations (Lamarck [1809] 1984, 245).[34]

The action of the imponderable fluids on the formation of living beings is for Lamarck another way of describing the process of intussusception. Living beings are complex organizations of matter *intussuscepted* by the environment. The environment, in turn, is a reality in constant flux, produced by the continuous activity of living beings, for geological changes and the modification of the earth's surface result from the decomposition of organic materials. In this sense, organic life is, first and foremost, the outcome of geological and geographical processes.[35] Lamarck discusses this view in his *Hydrogéologie* (1802); "Without exception," he points out, "the raw compounds which form most of the earth's external crust and continuously modify it through their changes result from the remains and residues of living organisms" (Lamarck [1802] 1964, 91).[36] Since everything that is inorganic was originally organic, Lamarck is of the opinion that geology should provide the foundations for any investigation into the nature of life. Drawing from ideas common to Buffon, Antoine Baumé (1728–1804), Louis-Jean-Marie Daubenton (1716–1799) and Jean-Guillaume Bruguière (1749/1750–1798), he sees the environment as the result of constant and inappreciably slow transformations of the earth derived from both the action of the waters (which, in the course of centuries, have been carving out mountains and eroding valleys) and the gradual accumulation of debris from dead organisms.[37] It is especially when reading *Hydrogéologie* that one has a clear glimpse of Lamarck's ambitious plan of

[33] Cf. Lamarck ([1802] 1986, 85).

[34] Cf. Ibid., 245: "Cela étant ainsi, il me sera facile de faire voir dans un instant que la *chaleur*, cette mère des générations, cette âme matérielle des corps vivans, parmi lesquels l'homme seul peut être hors de rang et privilégié, que la chaleur, dis-je, a pu être le principal des moyens qu'emploie directement la nature pour opérer sur des matières appropriées, un acte de disposition des parties, d'ébauche d'organisation, et par suite, de vitalisation analogue à celui de la fécondation sexuelle."

[35] Marsh (1864, 68).

[36] Cf. Lamarck (1815–1822, I, 64).

[37] Cf. Burkhardt (1995, 105–114) and Barsanti (1979, 54–64).

investigation. "A sound *Physics of the Earth*," he argues, "should include all the primary considerations of the earth's atmosphere, of the characteristics and continual changes of the earth's external crust, and finally of the origin and development of living organisms." Lamarck's "terrestrial physics" presents itself as a comprehensive system of natural philosophy divided into three essential parts, "a theory of the atmosphere, or *Meteorology*," "a theory of the earth's external crust, or *Hydrogeology*" and "a theory of living organisms, or *Biology*" (Lamarck [1802] 1964, 18). Lamarck's universe is a very complex system resulting from the interactions between the atmosphere, the surface of the earth and the organic units that, through their lives, shape the whole terraqueous system.

It is therefore safe to say that, according to Lamarck's chemical and geological views, no specific force can produce inorganic substances. It is the slow, incessant breakdown of organic bodies that is behind the formation of inorganic elements and minerals. However, if this is the case, the inevitable question becomes: "what is the origin of organic bodies?" In *Recherches sur l'organisation des corps vivants*, Lamarck resorts to the notion of spontaneous generation to explain the presence of life in the universe. Spontaneous generation introduces an element of discontinuity in the history of life, in that nature seems to have produced the first outlines of organized life in a direct way, without any clear transition from inorganic to organic nature.[38] From a metaphysical point of view, however, this is not an explanation. Indeed, it looks like an elegant way of postponing a real explanation. Lamarck is well aware of this:

> The greatest difficulty for us, apparently, is to conceive how nature could establish life in a body which did not have it and was not even prepared for it; and how nature could start even the simplest organization, whether plant or animal, after it produced spontaneous or direct generations (Lamarck 1815–1822, I, 174).

Spontaneous generation had partially regained credibility after John T. Needham (1713–1781) conducted his renowned experiments in the 1750s and 1760s, but is reinterpreted by Lamarck in a very original way: as an immediate, abrupt appearance of organization in the very fabric of matter.[39] There is no attempt on his part to demonstrate that inorganic matter, being somehow already in possession of life, might be able to organize itself into more elaborate structures without being fecundated by an active principle. Far from accounting for the emergence of living forms from a material substratum which is unremittingly throbbing with life, the process of spontaneous generation described by Lamarck results from a rudimentary but self-sufficient system of hydraulic and thermodynamic mechanisms. To interpret Lamarck's spontaneous generation as a way of bridging the gap between inorganic and organic matter – and thus forgetting that Lamarck's nature, structured as it is

[38]On Lamarck's notion of spontaneous generation cf. Daudin ([1926] 1983, II, 176–178, 214), Burkhardt (1995, 151–157), Barsanti (1979, 98–91), Jordanova (1984), 46; Corsi (1988, 67–68, 88), Pichot (1994, 30), and Tirard (2006). On the concept of organization cf. Russo (1981), and Pichot (1994, 21–23).

[39]On Needham and the debate on spontaneous generation cf. Stefani (2002) and Ratcliff (2009).

according to essential gaps, does indeed make leaps – is to ignore one of the basic premises of his system (i.e. the insurmountable divide that separates matter from life, chaos from organization).[40]

When we compare Cabanis' contemporary interpretation of spontaneous generation with Lamarck's, the difference becomes even more apparent. While interpreting spontaneous generation as an agent of organic transformations (rather than the principle of abrupt and intermittent appearances of life), Cabanis argues that,

> there is no known vegetable substance which, placed under the appropriate circumstances. . . does not give birth to individual animalcules, almost always instantaneously. Here we see a manifestation of the nature that is called *dead* being linked by an unbroken chain with living nature; we see the inorganic elements combine to produce different organized bodies: and from the vegetable products derive life and feeling, with their chief attributes (Cabanis [1805] 1981, I, 524).

There is no emphasis here on nature's external interventions. On the contrary, Cabanis sees inanimate matter as being capable, under certain circumstances, "of becoming organized, of living, and of feeling" (Cabanis [1805] 1981, I, 524).

What Lamarck's view has in common with contemporary vitalist accounts of spontaneous generation is the relevant role assigned to matter. The emergence of organic structures out of inorganic materials by way of spontaneous generation, discontinuous a process as it may be, requires a special material substratum. When one examines the history of irritability during the eighteenth century, the idea of a special material substratum is a recurrent theme: Haller's gluten (Haller 1757–1766, I, 7–18), Stahl's *mixtio mucido-pinguis* (Stahl1831–1833, I, 240–241), Bordeu's mucous tissue (Bordeu 1818, 751–752, 802) and Bichat's cellular tissue (Bichat [1800] 1813, 88, 114, Bichat [1801] 1822, I, 199, 227–228). The same is true with Lamarck. He regards the cellular tissue as "the universal matrix of all organization." From infusorians to mammals, from algae to complex dicotyledonous plants, the same tissue is modified in an infinite variety of forms through the action of the subtle fluids. A gelatinous mass (*masse gélatineuse*) represents the most elementary form of organization. It is the smallest structure compatible with the continuous influence of the external environment, capable of maintaining itself while absorbing and exhaling fluids. In this sense, it is the first instance of intussuscepted life to emerge from the milieu.[41] At a later stage, the forces that shape the material universe – from the most elementary ones (caloric and electricity) to the ones prompting life in nature (orgasm and irritability) – transform the gelatinous and mucilaginous masses of matter into the first outlines of life and organization, that is, the "animalised corpuscles" devoid of any kind of organs (Lamarck [1809] 1984, 230, 246, Lamarck

[40] Cf. Barsanti (1979, 82) and Barsanti (1983, 42–43).

[41] Cf. Lamarck ([1802] 1986, 77): "Dans un pareille masse de matières, les fluides subtils et expansifs répandus et toujours en mouvement dans les milieux qui l'environnent, pénétrant sans cesse et s'en dissipant de même, régularisent en traversant cette masse, la disposition intérieure des ses parties, et la rendent propre alors à *absorber* et à *exhaler* continuellement les autres fluides environnans qui peuvent pénétrer dans son intérieur et qui sont susceptibles d'être contenus."

[1802] 1986, 79). In particular climates and in certain times of the year, the earth and the atmosphere swarm with minute organisms (*molécules animées*). In this respect, nature's spontaneous generation causes matter "to become everywhere animalized" (Lamarck [1809] 1984, 216; Lamarck [1802] 1986, 67, 72). Here, however, it is important to keep in mind that Lamarck's "animalization" of matter has nothing to do with a hylozoistic view of nature, in which life is seen as an inherent attribute of matter. There is no doubt that Lamarck's definition of life seems to waver between the idea of a quasi-vitalistic power capable of organizing all beings of nature (the *pouvoir de la vie*) and the notion of a mechanical disposition of parts (*l'ordre et l'état de choses*) (Lamarck 1815–1822, I, 36–37, 134). Sometimes the two aspects are combined in the same definition: "life in a body consists of a series of aroused movements, which renew and maintain themselves to the extent that it is allowed by the order and state of things in the parts of the body, and as long as the arousing cause persists" (Lamarck 1815–1822, I, 55). However, the assumption that nature is responsive to external (and internalized) stimuli questions the very idea that life can be explained as a mechanical succession of physical states.[42]

The picture becomes even more complicated when this notion of life is contrasted with his view of nature.[43] From the *Flore françoise* (1778) to the *Système analytique des connaissances positives de l'homme* (1820), we can observe a marked shift from the notion of nature as a system of relationships pre-ordained by God at the moment of creation to a concept of nature as a process of emergent productions and formations. The *Histoire naturelle des animaux sans vertèbres* (1815–1822) contains the most articulate account of this development. Life is an excitatory state resulting from external (or more or less internalized and intussuscepted) stimuli, and it is of a constitutively transient character. Nature, on the other hand, is a permanent condition of activity and production in the universe. Whereas life "leads to its own destruction," nature, "as well as everything that was created in a direct way, is immutable, unalterable and could not end other than by the supreme will who only made it exist" (Lamarck 1815–1822, I, 321). Life "necessarily dries up," while nature is "inexhaustible" (Lamarck 1815–1822, I, 313). Being the actual productive agent in the universe, nature spreads life to organic bodies through intermittent manifestations of life (i.e. spontaneous generation). By calibrating the forces of attraction and repulsion, it establishes the correct equilibrium between condensation and rarefaction that is necessary to give rise to living bodies (Lamarck 1815–1822, I, 166–169). By sowing the seed of change within matter and promoting the process of organic differentiation throughout the vegetable and animal kingdoms, nature is

[42]This ambiguity is noted by Corsi: "The concept of irritability, a property exclusive to animal fiber, prevented him [*Lamarck*] from explaining in a totally mechanical manner the characteristics of organic movement in animals" (Corsi 1988, 70).

[43]On Lamarck's notion of nature cf. Daudin ([1926] 1983, II, 119), Burkhardt (1995, 131), Barthélemy-Madaule ([1979] 1982, 22–44), Jordanova (1984, 83–88), and Corsi (1988, 175–176), Corsi (2006, 2011).

the ultimate source of activity in the material universe (Lamarck 1815–1822, I, 211–212). There is nothing "metaphysical" in the power of nature, Lamarck insists throughout *Histoire naturelle des animaux sans vertèbres*. It is a force that "does not extend beyond the bodies it incessantly moves, displaces, changes, modifies, varies, destroys and renews." It can only act upon matter and it would not be able "to create, nor annihilate the smallest particle" of matter (Lamarck 1815–1822, I, 219). Whereas the universe is the whole of the existing bodies, a whole that is "immutable, inactive and devoid of a power of its own," nature is

> a true power, subdued in its actions, unalterable in its essence, constantly acting on all the parts of the universe, and it consists of an inexhaustible source of movements, of laws by which the motions are governed, of means that are essential to the possibility of their actions, in a word, it is made up of objects extraneous to the properties of matter (Lamarck 1815–1822, I, 333–334).

Most importantly of all, nature acts throughout the inert universe without following a purpose. In a sense, the power of nature

> is limited and in a way blind, with no intention, goal and will. It is a power that, great as it may be, would not be able to do any more than it does; in a word, it is a power that can only exist as a result of the will of a higher and boundless power which, being the one who established it, is the real *author* of all that comes from it, of all that, in the end, exists (Lamarck 1815–1822, I, 311).

Nature cannot have intentions because it is "not at all a being, an intelligence, but an order of things." What in nature appears as an end is in fact a necessary development (Lamarck 1815–1822, I, 323–324). Only God's power is a real intelligence; nature's power is simply an instrumental cause, driven to act by necessity (Lamarck 1815–1822, I, 216, 304–305, 312–313, 322–325).[44] Madeleine Barthélemy-Madaule described Lamarck's naturalism in terms of a "transfer of God's power to nature." She intimated that "this purely executive power does not include decision making, and therefore does not include setting goals. The execution is 'mechanical'" (Barthélemy-Madaule 1982, 28).[45]

One might say that Lamarck's notion of nature as an unintentional force, with a limited range of productivity, does not seem very different from the view of the universe frequently endorsed by physicians and natural philosophers, starting with Hippocrates. What appears to be original in Lamarck is the way he pieces the parts of the puzzle together, namely, nature, life, matter and God. They represent different levels of reality that in Lamarck's philosophy do not overlap. He never resorts to the idea of a vital force infused into the world by God (panpsychism) and never agrees with the view that matter has an organizing power of its own (hylozoism). He distinguishes the "bodies" from those elements that he calls the "objects of the primordial creation" – matter, movement, the laws governing movement, space and time.

[44]Cf. Lamarck (1972, 57, 77).

[45]On Lamarck's view on God cf. Grasse (1981).

Nature, the "instrument of the supreme power," is "the intermediary between God and the parts of the physical universe and its task is to implement the directions of God's will." In this respect, bodies, with their properties and faculties, can be said to be the "productions" of nature (Lamarck 1815–1822, I, 331). God *created* matter, and created it inert, whereas nature *produced* and still *produces* bodies. The inorganic matter that derives from the accumulation of decayed organic matter is to be considered as already a form of body, otherwise the universe would contain two different kinds of original matter (i.e. the matter that results from the decomposition of living bodies and the matter that is one of the objects of the "primordial creation"). Here it becomes clear how Lamarck uses the concept of spontaneous generation to explain the transition from the kind of matter originally created by God to the type of matter made organic by nature. In this view, God's creative intervention certainly signals the existence of a gap between the transcendent level represented by the "objects of the primordial creation" and the immanent level of "natural productions." Indeed, in Lamarck's universe the gaps are numerous. There is a hiatus between matter (as one of the objects of the original creation) and lack of matter, between created matter and inorganic matter, between life and absence of life, between organic and inorganic bodies, between plants and animals, between irritability and sensibility.

It remains to be understood in what sense nature is opposed to life, and how the *pouvoir de la vie* can also be a principle which tends to its own destruction (i.e. a principle of death). It may sound ironic, but Lamarck advocates a theory of decaying rather than living matter. Orgasm, irritability and sensibility – the forms which life takes in the physical universe – turn out to be momentary interruptions of the ordinary course toward death and destruction.

5 Irritability and Evolution in Lamarck's System of Nature

If there is no real purpose inherent in nature, if the source of activity – nature – is not part of the essence of matter, and if, finally, God's intervention in the universe ended with His production of the "objects of the primordial creation," one might then ask where the ultimate foundation of Lamarck's transformism lies. As I pointed out at the beginning of this chapter, in Glisson's monistic view of life and matter, irritability represented the vital process through which the universal substratum of matter, active by itself, developed into countless material forms, through a source of unremitting activity that preceded any conceptual distinction into active and passive principles, into unconscious vital energy and conscious sensibility. By contrast, Lamarck considers the power of irritability to be only one of the various plans of action followed by nature in prompting organic change, and, in keeping with the principles of Haller's physiology, he thinks that no transformation of orgasm into irritability or of irritability into sensibility can occur in nature. He resolutely rules out the possibility that matter has in itself the property of life and feeling (Lamarck

1984, 3, 294, 321, 326; Lamarck 1815–1822, I, 12–13, 20, 24), that there may be a "gradation" between inorganic and organic bodies (Lamarck 1984, 194), and that irritability and sensibility may derive from the same vital source.[46]

Dismissing the idea of matter as a living and perceptive substratum, Lamarck considers the belief in a gradual "animalization" of matter to be a faulty inference derived from the observation that the earth seems to be teeming with innumerable animalcules and *molécules animées*.[47] He rejects without hesitation any attempt to view the earth as a "living body."[48] His distinction between dead and living matter can be traced back to Buffon's division into inorganic and organized corpuscles. Matter is "inactive in itself and without a power of its own." If there is a natural propensity in matter, this looks like a tendency towards the breaking of the compounds down to their elements. The universe is the totality of passive and physical bodies and, as such, is devoid of specific energy (Lamarck 1815–1822, I, 315–316, 333).

> No kind of matter and no particle of it could have in itself the property of self-motion, nor that of living, feeling, thinking or formulating ideas. And if, in addition to human beings, we observe bodies that are endowed with all or some of the above-mentioned faculties, then we should consider these faculties as physical phenomena which nature was able to produce as a result of the order and the state of things that nature established in each organization and in each particular system of organs, and not by means of a kind that is supposed to be endowed with such and such faculties (Lamarck 1815–1822, I, 121).

While Lamarck had no qualms about stripping matter of all active powers, the role he assigned to nature remains ambiguous. In his opinion, nature seemed to be capable of imparting specific forms of organization upon matter. Lamarck's reluctance to grant life to matter can also be seen in his views on chemistry, which he elaborated in response to Antoine-Laurent Lavoisier (Lamarck 1796, 470–471).[49] Lamarck rejected the idea of elective affinities, for they could not be reconciled with his notion of inert matter. By contrast, it was precisely the vitalistic underpinnings of his thought that had led Cabanis to accept Lavoisier's chemistry, so that it was possible to assume the existence of material properties that were of a different order from purely physical ones. Cabanis argued that attraction was not "a blind force." Rather, it manifested "a sort of will, capable of making choices" (Cabanis [1805] 1981, II, 538). He combined ideas from theories of spontaneous generation, epigenesis and elective affinities to support his view of matter as a principle that could develop around multiple centers of gravity and organizing itself by implementing different levels of conscious activity (Cabanis [1805] 1981, II, 541).

---

[46]Lamarck ([1809] 1984, 208): "sensibility and irritability are not only quite distinct faculties, but. . . they have not even a common origin and are due to very different causes."

[47]Lamarck (1801, 21–22): "Il semble, pour ainsi dire, que la matière alors s'animalise de toutes parts, tant les résultats de cette étonnante fécondité sont rapides."

[48]Lamarck (1815–1822, I, 33): "N'a-t-on pas osé dire que le globe terrestre est un corps vivant; qu'il en est de même des différent corps célestes…n'a-t-on pas osé assimiler la nature même aux êtres doués de la vie!"

[49]On Lamarck's chemistry cf. Conry (1981), Gohau (2006, 10).

Regarding the idea of a gradual shift from inorganic to organic realms of nature, Lamarck's position remained unequivocally negative. He divided the whole system of "natural productions" (*productions naturelles*), traditionally subdivided into three kingdoms (mineral, vegetable and animal), into two main branches (i.e. organic bodies [*corps organisés, vivans*] and inorganic ones [*corps bruts et sans vie*]). Between the two, he assumed that there was "a vast *hiatus*" which prevented the two kinds of bodies from being arranged along a common line of development (Lamarck 1801, 2–5; Lamarck 1815–1822, I, 37–38). Even more significant was the determination with which Lamarck criticized the very idea of the great chain of being (Lamarck 1815–1822, I, 51–52):[50] "Nowhere have I spoken of such a chain," he pointed out in *Histoire naturelle des animaux sans vertèbres*. On the contrary, he recognized the existence everywhere of "an immense distance separating the inorganic from the living bodies," and that,

> plants do not shade into the animals in any point of their series. Indeed, even animals, which are the subject of what I am going to expound, are not at all linked to each other so as to form a series that extends in a simple, gradual and orderly manner. In what I am going to demonstrate, it is therefore pointless to speak of this chain, for there is no chain at all (Lamarck 1815–1822, I, 130).

Cabanis' view again represents a direct alternative – typical of a certain medical tradition – to Lamarck's assumption that inorganic and organic lines of development in matter branch off into divergent directions. According to Cabanis, matter was capable of undergoing "all degrees of organization and animalization" in stages. Zoophites and irritable plants testified to the presence of gradual transitions in material life (Cabanis [1805] 1981, II, 533–536). By contrast, Lamarck warned that to describe the phenomenon of irritability as nothing but a "modification" of sensibility – as Georges Cuvier had done in his entry "Animal" in the *Dictionnaire des sciences naturelles* (1816) – meant that perception was extended to all forms of life. Far from "forming a chain," he insisted, "plants and animals present two distinct branches" with no possibility of mediation. To strengthen the distinction between the two natural orders, Lamarck added experimental evidence of a chemical nature (Lamarck 1815–1822, I, 100–101).[51] In doing so, Lamarck intended to stress his difference from Richerand's, Cabanis' and Bichat's views. In their positions, the medical roots of vitalism were apparent. Like Glisson, they saw in reactive matter an incontrovertible proof of the primordial and universal nature of life. Quite to the contrary, Lamarck used irritability to separate inert matter from the haphazard

[50] Cf. Lamarck (1815–1822, I, 129–130): "On a même supposé que je voulais parler d'une chaîne existante entre tous les corps de la nature, et l'on a dit que cette chaîne graduée n'était qu'une idée reproduite, émise par *Bonnet*, et depuis, par beaucoup d'autres. On aurait pu ajouter que cette idée est de plus anciennes, puisqu'on la retrouve dans les écrits des philosophes grecs. Mais, cette même idée, qui prit probablement sa source dans le sentiment obscur de ce qui a lieu réellement à l'égard des animaux, et qui n'a rien de commun avec le fait que je vais établir, est formellement démentie, par l'observation, à l'égard de plusieurs sortes de corps maintenant bien connus." On the notion of chain of being cf. Lovejoy (1936) and (1968).

[51] Cf. Cuvier (1816–1830, II, 158–159).

appearance of life in the universe. More than once in the *Histoire naturelle des animaux sans vertèbres*, he distinguished between orgasm and irritability by referring to a particular chemical property rather than a special organic structure (Lamarck 1815–1822, I, 84–85, 123, 179), for in his view, irritability was not a characteristic that could be extended to matter in its entirety. He used the distinction between orgasm, irritability and sensibility to shed light on the principal bifurcations in the physical universe, pointing, on one hand, to the two main kinds of organic life (plants and animals), and on the other, to the distinction between inorganic (dead) and organic (living) bodies. Unlike Glisson's notion of irritability, which presents a markedly evolutionary character (irritability being the first degree of vitality in nature, evolving into sensibility and forms of life that become increasingly more aware of their actions and movements), Lamarck's irritability is one of the boundaries necessary to delimit a specific kind of vital activity. This leads to another important difference between the two interpretations of the phenomenon of irritability. In Glisson's case the motor behind organic change is natural and unconscious perception acting everywhere in matter, whereas for Lamarck there is no structural change emerging from the activity of a perceptive power embedded in matter.

It is worth pointing out here that Lamarck's theory of evolution does not assume one continuum of life, but a plurality of developments, each independent of one another. It is true that Lamarck generally maintains linear arrangements for the productions of nature (and never subscribes to forms of reticular organization); however, his model presupposes the existence of fundamental gaps and leaps, of multiple and contingent ramifications (*lignes de démarcation frappante*) (Lamarck 1815–1822, I, 82).[52] As already noticed, Henri Daudin used the appropriate expression *crises de l'organisation* to refer to this phenomenon (Daudin [1926] 1983, II, 198). No seamless transition ever occurs from one to another structural and functional pattern of nature. Organic change takes place only within species types. Nature started from the very beginning with clear outlines of vital developments – plants and animals – divided by a *ligne de démarcation tranchée* (Lamarck 1815–1822, I, 125). Within the animal kingdom, it branched out into irritable, sensitive and intelligent beings. An actual process of transformation is to be found in the way inorganic nature emerged through the sedimentation of decomposed organic matter (which is more a form of degradation than evolution),[53] or in the organic change within the basic self-contained partitions of living nature (i.e., the broad groupings of organisms which Lamarck calls *principal masses*) (Lamarck 1815–1822, I, 126).

One might find Lamarck's opposition to hylozoic forms of vital materialism surprising. Why was he so adamant in resisting the idea that both organic and inorganic

[52] On Lamarck's adoption of the linear arrangement cf. Daudin ([1926] 1983), Burkhardt (1995, 58, 124).

[53] As Burkhardt notes, Lamarck's "supposition that all the different minerals were produced gradually as the elements disengaged themselves from the remains of living things appears to be virtually the inverse of his later idea that all the different organic species were produced gradually as the 'power of life' and modifying circumstances caused simple, spontaneously generated forms to become increasingly complex and diversified" (Burkhardt 1995, 102).

beings might have a common origin? Why did he rule out the possibility of a transition among the main faculties of nature (orgasm, irritability and sensibility)? One possible answer – an easy one – can be found in his desire to avoid charges of pantheism or, even worse, atheism. By devising a metaphysical view based on distinct and hierarchical levels of being, Lamarck avoided being involved in disputes about the most strikingly heterodox consequences of his transformism.[54] As demonstrated by the history of irritability since the publication of Glisson's *De ventriculo et intestinis*, the idea of irritable bodies could end up justifying the most disparate views of nature. While irritability had the advantage of providing an explanation of complex biological phenomena without invoking the distinction between body and soul, as Haller had been able to show with his redefinition of irritability slanted in a quasi-mechanical Newtonian way, the same notion of irritable bodies could likewise be used to vindicate a materialistic view of natural activity. One can certainly interpret La Mettrie's dedication of *L'homme-machine* (1747) to Haller as a sincere acknowledgment of intellectual debt, and not necessarily as a provocative jibe, but it is also a clear proof that the meaning of irritability could span the whole field of natural life in this period (La Mettrie [1748] 1960, 143). In drawing on Haller's ambiguous position, Lamarck defended a view of irritability based on mechanical explanations, in which perception had no privileged status (unlike Glisson, who thought irritable motions to be forms of material perceptions).[55]

Lamarck, however, went beyond Haller's theological precautions. His decision to disassociate irritability and material life rested on clear philosophical reasoning.[56] First of all, Lamarck, the man who contributed decisively to the success of the name and concept of biology, started his intellectual career as a natural philosopher interested in chemistry, physics and geology, and remained, to a certain extent, a 'terrestrial physicist'. It is therefore no wonder that a physical framework (based on the inertia of matter and physical laws) prevailed over a medical view of life (centered on the notion of living matter and the principle of irritability).[57] Secondly, as shown by Daudin, Lamarck's thought is pervaded by a distinctive tension between a conception of organization that is still anthropocentric and a genetic explanation based only on material relations. As a result, Lamarck looked at the lowest levels in the evolutionary series "as if they were born through a reduction of the higher degrees" (Daudin [1926] 1983, II, 135, 199). Clearly, the simplest forms of life in nature cannot have the principle of organization within them and, most of all, do not have the

[54]On this point cf. Barthélemy-Madaule ([1979] 1982, 34–35, 40–41).

[55]Lamarck ([1809] 1984, 220): "Although we are not definitely aware how each vital function is performed, we should not gratuitously attribute to the parts a knowledge and power of choice among the objects which they have to separate out, and retain or evacuate."

[56]In Charles-Augustin Sainte-Beuve's novel *Volupté* (published in 1834), the protagonist remembers with passion Lamarck's lectures he attended in the *Jardin des Plantes*: "At that time he was one of the last representatives of that great school of natural philosophers and general observers who had reigned from Thales and Democritus to Buffon" (Sainte-Beuve [1834] 1995, 105).

[57]On the term "biology," cf. McLaughlin (2002) and Corsi (2006).

ability to ascend to higher levels of life and responsiveness. Thirdly, for precisely the same reasons why Aristotle denied life to states of unorganized matter, Lamarck dismissed the notion of living matter as implausible because all faculties in nature stem from material structures that are defined by a certain level of organization. There cannot be, therefore, forms of orgasm, irritability or sensibility diffused in matter in its inorganic (i.e., undifferentiated) condition (Daudin [1926] 1983, II, 135, 145–146). While nature is prior to structure and is capable of bringing structures back to inorganic matter, life can only appear when structures have been formed by nature.

Richard W. Burkhardt Jr. has noted that "Lamarck's writings have their ambiguities, inconsistencies, and misleading metaphors" (Burkhardt 1995, 144). This, however, should not prevent us from tracing a coherent trajectory in his intellectual development. If Lamarck's system can be regarded as a form of materialism – and already some contemporaries such as Jean-Baptiste Bory de Saint-Vincent (1778–1846) did not exclude this possibility (Corsi 1988, 230) – it is a materialism that rests on mechanical foundations (Lamarck [1809] 1984, 250–251, 255).[58] If this is the case, then the role of irritability as the source of vital change becomes even more problematic.[59] This is particularly evident in the case of animals. Their movements, Lamarck argues in *Histoire naturelle des animaux sans vertèbres*, "are not communicated." They are not the result of "an impulse, a pressure, an attraction or relaxation; in other words, they do not derive from an effect, whether hygrometric or pyrometric, but are movements that have been aroused, and their arousing cause, which acts on parts which are immediately contractile, is not at all proportionate to the effects produced" (Lamarck 1815–1822, I, 116). This means that animals are not machines, and their movements, far from having a mechanical origin, originate from an impulse whose nature is ultimately chemical.

> Although it occurs in various degrees depending on the nature of the supple parts, irritability, being a faculty that is common to all animals, is not the product of a system of particular organs, but of the chemical condition of the substances that enter the composition of these beings (Lamarck 1815–1822, I, 123).

More generally, Lamarck's strong belief in a sense of mechanical necessity pervading the whole system of nature cannot be easily reconciled with the irreducibly historical character of vital change. Growing evidence seemed to demonstrate that life on earth had a long and eventful history. This contributed to secularizing the very notion of life, confirming the image of nature as the result of a series of uneven processes developing through geological time, made up of contingent adaptations

---

[58] See the two "basic principles" in *Histoire naturelle des animaux sans vertèbres*: "1.er *Principe*: Tout fait ou phénomène que l'observation peut faire connaître, est essentiellement physique, et ne doit son existence ou sa production qu'à des corps, ou qu'à des relations entre corps. 2.e *Principe*: Tout mouvement ou changement, toute force agissante, et tout effet quelconque, observés dans un corps, tiennent nécessairement à des causes mécaniques, régies par des lois" (Lamarck 1815–1822, I, 11–12).

[59] Cf. Stahl's objection: "incommodissimum illud schema de irritatione, quod certe de rebus mechanicis praedicari absolute ineptum est" (Stahl [1708] 1831–1833, I, 288).

and external circumstances. The tension between the view of organic development as a serial and mechanical succession of states and the view of evolution as a real history of life (what Daudin has called the "genetic interpretation of the series") runs through Lamarck's whole oeuvre (Daudin [1926] 1983, II, 205).

In *Hydrogéologie*, Lamarck argued that "living organisms form their own constitutive substance through the natural but peculiar action of their organs" (Lamarck [1802] 1964, 91).[60] While organic beings were described in terms of self-replicating structures, inorganic substances were seen as the result of slowly decaying living bodies (Lamarck [1802] 1964, 24–25, 68–70, 81, 122).[61] Contrary to some contemporary mineralogists, Lamarck claimed that there could be no original natural matter. Limestone, for instance, was the result of innumerable superimposed generations of dead coral polyps, millepores, madrepores, astroites and other organisms (Lamarck [1802] 1964, 70).[62] In Lamarck's view, environment derived from a long process of transformation carried out by self-replicating structures. In the "Discours d'ouverture" to his *Système des animaux sans vertèbres*, he pointed out how people could not understand the processes through which living beings

> form their own substance by themselves through the action and the faculties of their organs. And what people do know even less is that these beings, through their remains, give rise to all the bodies compounded of brute matter that we see in nature. These various kinds of matter multiply with time because of the transformations and changes they undergo more or less promptly, according to the circumstances, until they reach their own complete destruction, i.e., the complete separation of their constitutive principles (Lamarck 1801, 3).

Far from looking at matter as a living structure, as Glisson had done while explaining the properties of irritability, Lamarck understood the material universe as an inherently self-decomposing and decaying system in constant need of external stimuli to preserve movement and life (Lamarck [1802] 1986, 128).

All the difficulties examined so far boil down to the most baffling of Lamarck's puzzles – the dualism between a system of external causes, which act as stimulating factors, and a flexible organization of internal responses, which "intussuscepts" life from the environment. The difficulty lies in establishing which of the two types of cause comes before or, even more radically, which of the two can be considered the real cause. As we have seen, Lamarck's characteristic solution is to look at the activating principle as coming from the outside. In this respect, the whole system of environmental influences can be regarded as the activating principle. This is particularly evident when one examines Lamarck's geological theories, for the environment is not seen in terms of static arrangements established from the very beginning of the universe, but is a living, and therefore historical phenomenon. What at first appears to come from within – the so-called *pouvoir de la vie* – is the result of a complex process of internalization of the external stimuli through an elaborate process of organic construction (i.e. Lamarck's *intussusception*) (Lamarck 1801, 366,

[60] Lamarck ([1809] 1984, 51, 250, 255–258). On Lamarck's geological theories cf. Carozzi (1964).

[61] Lamarck ([1809] 1984, 258).

[62] Lamarck (1801, 24–26, 367).

Lamarck 1815–1822, I, 59; Lamarck [1809] 1984, 193, 246, Lamarck [1802] 1986, 53, 76). Paradoxical as this may seem, Lamarck maintains that, from an ontological point of view, the milieu precedes the organism and the habit creates the need. As he points out in the *Système des animaux sans vertèbres*, "it is not at all the form either of the body or of its parts which gives rise to the habits and to the way of living of the animals; on the contrary, it is the habits, the way of life and all the influential circumstances that in time constitute the form of the body and of the parts of the animals" (Lamarck 1801, 15). The environment shapes organic life not as a prolongation of the animal organism, but as a form of "intussuscepted" vital energy. And it cannot be otherwise since the activation of life in the universe comes always from the outside, and not from the very being of the material system.[63]

In the final analysis, Lamarck maintains that for the phenomenon of life to be activated an external stimulus is required. Even the *sentiment intérieur*, which represents one of the most refined forms of organization in higher animals, is triggered by external factors, by "needs which drive them and make them act immediately, without premeditation, and without the contribution of any act of will" (Lamarck 1815–1822, I, 17). It is important to keep this in mind, and not to view plants and animals as endowed with a mysterious will to change, as Julien-Joseph Virey, Georges Cuvier and Charles Darwin, among others, interpreted Lamarck.[64] Admittedly, there are passages in Lamarck's work which seem to assume the primacy of the *pouvoir de la vie* with respect to the role played by external circumstances. While the former is said to be the "first cause" capable of developing the organic structures in a gradual manner, the latter is presented as an "accidental cause," whose function seems to be that of altering the main direction on the process of development.[65] In this case, however, the emphasis is on the organisms' ability to intussuscept life from the environment rather than nature's own way of proceeding. We could define Lamarck's view as characterized by the irreducibly "external" character of life. Life is a property that belongs to organic structures of matter and is induced by the environment. What is intrinsic is the tendency manifested by organic structure to undergo processes of breakup and disintegration – in a word, a

[63]Cf. Daudin [1926] 1983, II, 211: "précisément parce que la différance physique de l'inerte et du vivant est ainsi posée comme inhérente aux modalités d'un fonctionnement, d'une action, elle laisse une place à la notion d'une 'influence' positive des conditions de milieu sur les phénomènes vitaux – et cette notion . . . est sans doute l'idée la plus féconde de toute la 'philosophique zoologique' de Lamarck." Cf. Ibid., 217–218.

[64]Cf. Jordanova (1984, 55, 102) and Corsi (1988, 177).

[65]Cf., for instance, Lamarck (1815–1822, I, 133): "Le plan des opérations de la nature à l'égard de la production des animaux est clairement indiqué par cette cause première et prédominante qui donne à la vie animale le pouvoir de composer progressivement l'organisation, et de compliquer et perfectionner graduellement, non-seulement l'organisation dans son ensemble, mais encore chaque système d'organes particulier, à mesure qu'elle est parvenue à les établir. Or, ce plan, c'est-à-dire, cette composition progressive de l'organisation, a été réellement exécuté, par cette cause première, dans les différens animaux qui existent. Mais une cause étrangère à celle-ci, cause accidentelle et par conséquent variable, a traversé çà et là l'exécution de ce plan, sans néanmoins le détruire, comme je vais le prouver." Cf. Ibid., 160–161.

tendency to death. In this respect, Lamarck's chemical theories, according to which all compounds have a tendency to return to a state of disparate elements, far from having hampered his mature evolutionary theory (Burkhardt 1995, 99), represent an integral part of his theory of organic change.

But if this is the case, another ontological priority – certainly a more disquieting one – looms large in Lamarck's natural philosophy: in the material universe, death appears to be more original than life. Not that this represents a completely new idea. I started this chapter by referring to Bacon's *Historia vitae et mortis*, a work in which life is presented as the domain of lifeless spirits (*spiritus mortuales*). The interplay of life and death is certainly a recurrent theme in medical thought. Georg Ernst Stahl had entrusted the vital principle (the soul) with the task of combating an inexorably disruptive tendency inherent in matter and bodies (Stahl [1708] 1831–1833, I, 240, 443–444, 474). Richerand characterized the power of life as a force meant to counterbalance the effects of material inertia and physical laws (Richerand [1801] 1823, 50). In Bichat's definition, life represented the complex of functions that resist death. The original point in Lamarck is that in his explanation of the contrast between life and death as a primeval opposition between life and nature, death becomes the most natural tendency in the universe (a tendency towards entropic stability), while life is an accidental and temporary deferral of inevitable decay. Life comes from without in the form of an excitatory stimulus and death from within as a return to the original order of things. This means that an unrestrainable tendency to decomposition and death permeates matter and is just postponed by the interplay of excitement and reaction, so much so that it is not unreasonable to define life as an accident which prevents the whole cosmos from collapsing into a state of inertia and immutability (Lamarck [1802] 1986, 15–16, 59–60, 81, 128). In a world without theological anchoring or metaphysical foundation, as Lamarck's openly purports to be, life remains indeed *un accident étrange et singulièrement industrieux*, to use the words of the narrator in Sainte-Beuve's novel *Volupté* (1834):

> Monsieur separated life from nature. Nature, in his eyes, was rocks and ashes, the granite of the tomb, death! Life intervened only as a strange and singularly industrious accident, a prolonged struggle with more or less success and equilibrium here and there, but always vanquished in the end. Cold immobility would reign after as before.[66]

Lamarck did indeed envisage a natural tendency to death in nature. Sainte-Beuve was not referring to the more superficial disposition manifested by living beings to break down into inert matter, but to the deeper tendency always to restore an original condition of stability and permanence. In the history of philosophy and science, vitalism has a bad reputation, for the very definition of life remains remorselessly murky and resists all attempts at reduction according to mechanical reason. While on a more superficial plane, life smacks of irrational exuberance, on a deeper level, as I have argued in this chapter, it shares the very uncomfortable company of death. In the definition of life, depending on whether one stresses the role played by knowledge (life as a certain kind of perception) or by appetite (life as a tendency to

[66] Sainte-Beuve ([1834] 1995, 106).

restore the original condition of rest), death always has a say. This is certainly the case with Lamarck, for whom the history of life is, in fact, a history of death. By characterizing life as a temporary and imponderable *irritation* within the never-changing state of nature and matter, rather than as the domain of the ever-reactive process of *irritability* (as in Glisson, Richerand, Cabanis and Bichat), Lamarck contributed to efface the irreducible power of life in the face of continuous death, by resolving life into a fleeting accident of everlasting matter.

References

Ackerknecht, Erwin Heinz. 1967. *Medicine at the Paris Hospital, 1794–1848*. Baltimore: Johns Hopkins University Press.

Bacon, Francis. 1857–1874. *Works*, ed. James Spedding, Robert L. Ellis, and D.D. Heath. 14 vols. London: Longman. (repr. Stuttgart-Bad Cannstatt: Frommann-Holzboog, 1962; Cambridge: Cambridge University Press, 2012.)

Bacon, Francis. [1653] 1996. *De Principiis atque originibus secundum fabulas Cupidinis et Coeli*. In *Philosophical Studies c. 1611- c. 1619*, ed. Graham Rees, 195–267. Oxford: Clarendon Press.

Bacon, Francis. [1620] 2004. *The* Instauratio Magna *Part II*: Novum Organum *and associated texts*, ed. Graham Rees with Maria Wakely. Oxford: Clarendon.

Bacon, Francis. [1623] 2007. *The* Instauratio Magna *Part III*: Historia Naturalis et Experimentalis: Historia Ventorum *and* Historia Vitae et Mortis, eds. Graham Reese with Maria Wakely. Oxford: Clarendon Press.

Barsanti, Giulio. 1979. *Dalla storia naturale alla storia della natura: Saggio su Lamarck*. Milan: Feltrinelli.

Barsanti, Giulio. 1983. *La mappa della vita: Teorie della natura e teorie dell'uomo in Francia, 1750–1850*. Naples: Guida.

Barthélemy-Madaule, Madeleine. [1979] 1982. *Lamarck the mythical precursor: A study of the relations between science and ideology*. Trans. M.H. Shank. Cambridge, MA/London: Massachusetts Institute of Technology Press.

Barthez, Paul-Joseph. 1778. *Nouveaux éléments de la science de l'homme*. Montpellier: Jean Martel.

Bichat, Xavier. [1800] 1813. *Treatise on the membranes in general, and of different membranes in Particular*. Trans. M. Husson. Boston: Cummings and Hilliard.

Bichat, Xavier. [1801] 1822. *General anatomy, applied to physiology and medicine*. Trans. George Hayward. 3 vols. Boston: Richardson and Lord.

Bichat, Xavier. [1800] 1827. *Physiological researches on life and death*. Trans. F. Gold. Boston: Richardson and Lord.

Buffon, Georges-Louis Leclerc, Comte de. 1835–1836. *Œuvres complètes avec les supplémens, augmentées de la classification de G. Cuvier*. 9 vols. Paris: Duménil.

Burckhardt Jr., Richard W. 1995. *The spirit of system: Lamarck and evolutionary biology*. Cambridge, MA/London: Harvard University Press.

Cabanis, Pierre-Jean-Georges. [1805] 1981. *On the relations between the physical and moral aspects of man*, ed. George Mora. Trans. Margaret Duggan Saidi. Baltimore and London: The Johns Hopkins University Press.

Canguilhem, Georges [1965] 2008. *Knowledge of life*, ed. Paola Marrati and Todd Meyers, Trans. Stefanos Geroulanos and Daniela Ginsburg. New York: Fordham University Press.

Carozzi, Albert V. 1964. Lamarck's theory of the earth: *Hydrogéologie*. *Isis* 55: 293–307.

Conry, Yvette. 1981. Une lecture newtonienne de Lamarck est-elle possible? In *Lamarck et son temps: Lamarck et notre temps*, ed. Madeleine Barthélemy-Madaule, 37–64. Paris: Vrin.

Corsi, Pietro. 1988. *The age of Lamarck: Evolutionary theories in France 1790–1830*. Berkeley/Los Angeles/London: University of California Press.
Corsi, Pietro. 2006. Biologie. In *Lamarck, philosophe de la nature*, ed. Pietro Corsi, Jean Gayon, Gabriel Gohau, and Stéphane Tirard, 37–64. Paris: Presses Universitaires de France.
Corsi, Pietro. 2011. Jean-Baptiste Lamarck: From myth to history. In *Transformations of Lamarckism: From subtle fluids to molecular biology*, ed. Snait B. Gissis and Eva Jablonka, 9–17. Cambridge, MA: The Massachusetts Institute of Technology Press.
Crocker, Lester G. 1968. Diderot and eighteenth-century French transformism. In *Forerunners of Darwin: 1745–1859*, ed. Bentley H. Glass, Owsei Temkin, and William L. Straus Jr., 114–143. Baltimore: The Johns Hopkins University Press.
Cudworth, Ralph. [1845] 1995. *The true intellectual system of the universe*. 3 vols. London: Tegg. (Repr.: Bristol: Thoemmes.)
Cuvier, Georges. 1800–1805. *Leçons d'anatomie comparée*. 5 vols. Paris: Baudouin.
Cuvier, Georges. 1816–1830. Animal. In *Dictionnaire des sciences naturelles*. 60 vols, II, 158–174. Strasbourg and Paris: Levrault and Le Normant.
Daudin, Henri. [1926] 1983. *Cuvier et Lamarck: Les classes zoologiques et l'idée de série animale, 1790–1830*. 2 vols. Paris: Presses Universitaires de France.
de Bordeu, Théophile. 1818. *Œuvres complètes*. Paris: Caille et Ravier.
Desaive, Jean-Paul. 1972. *Médecins, climat, et épidémies à la fin du XVIIIe siècle*. Paris/The Hague: Mouton.
Figlio, Karl M. 1976. The metaphor of organisation: An historiographical perspective on the bio-medical sciences of the early nineteenth century. *History of Science* 14: 17–53.
Gemelli, Benedino. 2010. Introduction. In *Francis Bacon, Scritti scientifici*, ed. Benedino Gemelli, 189–390. Turin: UTET.
Giglioni, Guido. 1994. Panpsychism versus hylozoism: An interpretation of some seventeenth-century doctrines of universal animation. *Acta Comeniana* 11: 25–45.
Giglioni, Guido. 2002. Francis Glisson's notion of *Confoederatio Naturae* in the context of hylozoistic corpuscularianism. *Revue d'Histoire des Sciences* 55: 239–262.
Giglioni, Guido. 2005. The hidden life of matter: Techniques of prolonging life in the writings of Francis Bacon. In *Francis Bacon and the refiguring of early modern thought: Essays to commemorate the advancement of learning (1605–2005)*, ed. Julie Robin Solomon and Catherine Gimelli Martin, 129–144. Aldershot: Ashgate.
Giglioni, Guido. 2008. What ever happened to Francis Glisson? Albrecht Haller and the fate of eighteenth-century irritability. *Science in Context* 21: 465–493.
Giglioni, Guido. 2010. As forças da vida segundo Albrecht von Haller. In *Corpo, poesia e afecto em Albrecht von Haller*, ed. Adelino Cardoso and Palmira Fontes da Costa, 11–26. Lisbon: Edições Colibri.
Giglioni, Guido. 2011. *Francesco Bacone*. Rome: Carocci.
Gissis, Snait B. 2010. Lamarck on feelings: From worms to humans. In *The body as object and instrument of knowledge: Embodied empiricism in early modern science*, ed. Charles T. Wolfe and Ofer Gal, 211–239. Dordrecht: Springer.
Glisson, Francis. 1672. *Tractatus de natura substantiae energetica, seu de vita naturae, ejusque tribus primis facultatibus*. London: Flesher.
Glisson, Francis. 1677. *Tractatus de ventriculo et intestinis*. Amsterdam: Jacob Junior.
Gohau, Gabriel. 2006. Lamarck "Philosophe"? In *Lamarck, philosophe de la nature*, ed. Pietro Corsi, Jean Gayon, Gabriel Gohau, and Stéphane Tirard, 9–35. Paris: Presses Universitaires de France.
Grasse, Pierre-Paul. 1981. Dieu et la nature dans la pensée de Lamarck. In *Lamarck et son temps: Lamarck et notre temps*, ed. Madeleine Barthélemy-Madaule, 203–211. Paris: Vrin.
Haigh, Elizabeth. 1984. *Xavier Bichat and the medical theory of the eighteenth century*. London: Wellcome Institute for the History of Medicine.
Haller, Albrecht. 1757–1766. *Elementa physiologiae corporis humani*. 8 vols. Bern and Lausanne: Bousquet, D'Arnay and Grasset.
Hoquet, Thierry. 2005. *Buffon: Histoire naturelle et philosophie*. Paris: Honoré Champion.

Jordanova, Ludmilla J. 1979. Earth science and environmental medicine: The synthesis of the late enlightenment. In *Images of the earth: Essays in the history of the environmental sciences*, ed. Ludmilla J. Jordanova and Roy S. Porter, 119–146. Aberdeen: The British Society for the History of Science.
Jordanova, Ludmilla J. 1981. La psychologie naturaliste et le "problème des niveaux": La notion de sentiment intérieur chez Lamarck. In *Lamarck et son temps: Lamarck et notre temps*, ed. Madeleine Barthélemy-Madaule, 69–80. Paris: Vrin.
Jordanova, Ludmilla J. 1984. *Lamarck*. Oxford/New York: Oxford University Press.
La Mettrie, Julien Offray de. [1748] 1960. *L'homme machine: A study in the origins of an idea*, ed. Aram Vartanian. Princeton: Princeton University Press.
Lamarck, Jean-Baptiste. 1796. *Réfutation de la théorie pneumatique ou de la nouvelle doctrine des chimistes modernes*. Paris: Lamarck and Agasse.
Lamarck, Jean-Baptiste. 1801. *Système des animaux sans vertèbres*. Paris: Deterville.
Lamarck, Jean-Baptiste. 1815–1822. *Histoire naturelle des animaux sans vertèbres*. 7 vols. Paris: Verdière.
Lamarck, Jean-Baptiste. [1802] 1964. *Hydrogeology*. Trans. Albert V. Carozzi. Urbana: University of Illinois Press.
Lamarck, Jean-Baptiste. 1972. In *Inédits d'après les manuscrits conservés à la bibliothèque centrale du Muséum National d'Histoire Naturelle de Paris*, ed. Max Vachon, Georges Rousseau, and Yves Laissus. Paris: Masson.
Lamarck, Jean-Baptiste. [1809] 1984. *Zoological philosophy*. Trans. Hugh Samuel Roger Elliot. Chicago/London: The University of Chicago Press.
Lamarck, Jean-Baptiste. [1802] 1986. *Recherches sur l'organisation des corps vivants, précédé du Discours d'ouverture du cours de zoologie, donné dans le Muséum d'Histoire Naturelle*. Tours: Fayard.
Lovejoy, Arthur O. 1936. *The great chain of being: A study of the history of an idea*. Cambridge, MA/London: Harvard University Press.
Lovejoy, Arthur O. 1968. The argument for organic evolution before the *origin of species*, 1830–1858. In *Forerunners of Darwin*, ed. Bentley H. Glass, Owsei Temkin, and William L. Straus Jr., 356–414. Baltimore: The Johns Hopkins University Press.
Marsh, George P. 1864. *Man and nature, or physical geography as modified by human action*. New York: Scribner.
McLaughlin, Peter. 2002. Naming biology. *Journal of the History of Biology* 35: 1–4.
Moravia, Sergio. 1968. *Il tramonto dell'illuminismo: Filosofia e politica nella società francese (1770–1810)*. Bari: Laterza.
Moravia, Sergio. 1970. *La scienza dell'uomo nel Settecento*. Bari: Laterza.
Moravia, Sergio. 1972. Philosophie et médecine en France à la fin du XVIIIe siècle. *Studies on Voltaire and the Eighteenth Century* 89: 1089–1151.
Moravia, Sergio. 1974. *Il pensiero degli Idéologues*. Florence: La Nuova Italia.
Moravia, Sergio. 1978. From *homme machine* to *homme sensible*: Changing eighteenth-century models of man's image. *Journal of the History of Ideas* 39: 45–60.
Omodeo, Pietro. 1997. Le transformisme au XVIIIe siècle. In *Jean-Baptiste Lamarck (1744–1829)*, ed. Goulven Laurent, 471–482. Paris: Éditions du CTHS.
Picavet, François. 1891. *Les Idéologues: Essai sur l'histoire des idées et théories scientifiques, philosophiques, religieuses, etc. en France depuis 1798*. Paris: Alcan.
Pichot, André. 1994. Présentation. In *Lamarck, Philosophie zoologique*, ed. André Pichot, 7–49. Paris: Flammarion.
Ratcliff, Marc J. 2009. *The quest for the invisible: Microscopy in the enlightenment*. Farnham/Burlington: Ashgate.
Richerand, Anthelme. [1801] 1823. *Elements of physiology*. Trans. G.J.M. De Lys. Philadelphia: James E. Moore.
Rosen, George. 1946. The philosophy of ideology and the emergence of modern medicine in France. *Bulletin of the History of Medicine* 20: 328–339.

Russo, François. 1981. La notion d'organisation chez Lamarck. In *Lamarck et son temps: Lamarck et notre temps*, ed. Madeleine Barthélemy-Madaule, 119–141. Paris: Vrin.
Sainte-Beuve, Charles-Augustin. [1834] 1995. *Volupté: The sensual man*. Trans. Marilyn Gaddis Rose. New York: The State University of New York Press.
Stahl, Georg Ernst. [1708] 1831–1833. *Theoria medica vera*, ed. Ludwig Choulant. 3 vols. Leipzig: Voss.
Staum, Martin S. 1980. *Cabanis: Enlightenment and medical philosophy in the French revolution*. Princeton: Princeton University Press.
Stefani, Marta. 2002. *Corruzione e generazione: John T. Needham e l'origine del vivente*. Florence: Olschki.
Temkin, Owsei. 1968. The idea of descent in post-romantic German biology: 1848–1858. In *Forerunners of Darwin*, ed. Bentley H. Glass, Owsei Temkin, and William L. Straus Jr., 323–355. Baltimore: The Johns Hopkins University Press.
Tirard, Stéphane. 2006. Générations spontanées. In *Lamarck, philosophe de la nature*, ed. Pietro Corsi, Jean Gayon, Gabriel Gohau, and Stéphane Tirard, 65–104. Paris: Presses Universitaires de France.
Williams, Elizabeth A. 1994. *The physical and the moral: Anthropology, physiology, and philosophical medicine in France, 1750–1850*. Cambridge: Cambridge University Press.
Wolfe, Charles T. 2011. Why was there no controversy over life in the scientific revolution? In *Controversies within the scientific revolution*, ed. Marcelo Dascal and Victor D. Boantza, 187–219. Amsterdam/Philadelphia: John Benjamins.
Wolfe, Charles T., and Motoichi Terada. 2008. The animal economy as object and program in Montpellier vitalism. *Science in Context* 21(4): 537–579.

Chapter 3
Rethinking Organic Vitality in Germany at the Turn of the Nineteenth Century

Joan Steigerwald

Abstract This paper complicates accepted narratives of vitalism in Germany in the years around 1800. The early 1790s were marked by a proliferation of publications arguing for special *Lebenskräfte* to explain the unique properties of organic vitality. These works appeared in reaction to a controversial claim to provide a chemical explanation of the phenomena of life by Girtanner in a 1790 treatise. Despite Kant's critical analysis of the limits of our ability to understand living organisms and his rejection of the possibility of a science of life, several physiologists and naturalists argued for a science of biology based on unique vital principles. But new empirical investigations into the material conditions of excitability and generation from the mid-1790s blurred the boundary between organic and inorganic phenomena. Schelling drew on these new studies to reject a unique vital power or science of life, and instead to conceive living processes as but a stage in the dynamic becoming of nature. Vitalism in Germany thus was not a product of speculative philosophies of nature. Both philosophies of nature and experimental investigations at the turn of the nineteenth century problematized the demarcation of a distinct domain of life, even as they focused attention on organic vitality.

Keywords Biology • Comparative physiology • Excitability • Galvanism • Generation • *Lebenskraft*

J. Steigerwald (✉)
Science and Technology Studies, Humanities, York University,
4700 Keele Street, Toronto, ON M3J 1P3, Canada
e-mail: steiger@yorku.ca

S. Normandin and C.T. Wolfe (eds.), *Vitalism and the Scientific Image in Post-Enlightenment Life Science, 1800–2010*, History, Philosophy and Theory of the Life Sciences 2, DOI 10.1007/978-94-007-2445-7_3,

1 Introduction

A renewed interest in organic vitality at the turn of the nineteenth century was the result of several overlapping concerns. The question "what is life" had, of course, been asked before. But the question acquired a new prominence, if not urgency, in the face of rapid developments in physics and especially chemistry that claimed, at least in part, to answer that question. It also became a question of philosophical import, as philosophers such as Immanuel Kant and F.W.J. Schelling critically examined the limits of our comprehension of organic phenomena and rejected the hypostatization of vital powers to explain the unique characteristics of living organisms. The renewed attention to organic vitality prompted a contestation regarding the location of life; the boundaries between living and nonliving beings, the organic and inorganic; the proper concepts and principles for the explanation of organic vitality; and the appropriate disciplines, methods and instruments for its investigation. It was a debate over the warrant of epistemic claims as well as over who had authority to pass judgment on life.

The new intensity of attention to organic vitality amongst German physicians, physiologists and naturalists was precipitated by the publication of Christoph Girtanner's 1790 "Treatise on Irritability as the Principle of Life in Organized Nature." The existence of a capacity of irritability had been highly contentious since it was introduced by the prominent physician Albrecht von Haller in the mid-eighteenth century. A physician and writer of science and politics based in Göttingen but without an academic position, Girtanner could be more readily dismissed by critics of irritability. But Girtanner's treatise also contained the more provocative argument that the mechanisms for understanding irritability were to be found through new chemical studies. Soon numerous German-language publications appeared providing alternative accounts of the vitality of living organisms, many arguing for the need of a special vital power or *Lebenskraft* to counter chemical forces and preserve the properties of life. The debate in the early 1790s took place in textual terrain, in articles, books and reviews; it consisted of disagreements over concepts and pronouncements on principles. If experiments were enlisted in these disputes, they were largely those of the past; although they were widely contested when first introduced, the experiments of Haller and others from earlier in the eighteenth century were cited as authoritative and their results as decisive. It was only with the controversy over Luigi Galvani's experiments on animal electricity, examined in the German context in the mid-1790s by Alexander von Humboldt amongst others, that experimental reasoning again became the means through which disputes over organic vitality occurred. But galvanic experiments exploring the phenomena of electricity and oxidation in organic bodies raised anew the question of a distinct domain for life. New studies of the generative powers of simple organisms by German naturalists and physiologists such as Carl Friedrich Kielmeyer from the mid-1790s further complicated understandings of living organisms, by demonstrating the material conditions of their remarkable capacities for generation and degeneration. If the debates of the early 1790s made organic vitality a newly important matter of concern, they revolved around abstract principles and fixed concepts. The new experimental investigations of the mid-1790s reanimated the phenomena of life, placing investigators in immediate contact with vital responses to stimulus and presenting them with the complexity of the organic alterations involved in such processes. But if these new investigations gave new life to organic vitality, they also blurred its boundary with inorganic phenomena, by importing techniques and concepts from physics and chemistry.

Kant's *Critique of the power of judgment* also appeared in 1790. From his earliest writings in the 1750s, Kant participated in the debates amongst physiologists and naturalists regarding conceptualizations of living organisms – debates over the differences between organisms and mechanical bodies; the generation and first origin of living beings; and the history of the forms of life, and the variations of species and races arising through that history. But he also made the question of our understanding of organized beings a central concern of his critical project, with the second half of his third *Critique*, the "Critique of the power of teleological judgment," devoted to the interrogation of the form and limitations of our judgments of living organisms. Given Kant's ethos of epistemic modesty, and the strictures he placed on our cognitive claims in general and our knowledge of organisms in particular, and given the widespread influence of his critical philosophy in the German context, the uncritical appeal to principles of life and *Lebenskräfte* in the early 1790s may seem surprising. But Kant's arguments regarding the rightful claims of science and of our judgments of living organisms were only gradually addressed by physiologists and naturalists attempting to develop a science of life. By the end of the 1790s, reflections on the philosophical issues raised by new investigations of organic vitality were extended with the appearance of Schelling's philosophy of nature. Schelling can be regarded as animating the whole of the natural world, conceiving each natural product, inorganic as well as organic, as but a relative stability of opposed active principles. Although Schelling seems to have abandoned Kant's cautions regarding the limits of rational speculation, in his critical analysis of infinite productivity of nature he argued that there is no determinate metaphysical foundations, no simplest or final form, only boundary conditions between different natural processes. Indeed, both Kant and Schelling can be regarded as making a critical posture towards cognitive claims an "epistemic virtue," using organic vitality as a tool for re-conceiving the relationships between natural processes rather than viewing it as a metaphysical power.[1]

This new attention to organic vitality and reflections upon the conditions of a possible science of life resulted in a transformation of both physiology and natural history in German publications in the years around 1800. Physiology in the eighteenth century was conceived within Haller's model of "animated anatomy,"[2] but

---

[1] The expression "epistemic virtue" comes from Daston and Galison (2007). They argue for epistemology wedded to an ethos in the pursuit of scientific knowledge, so that the valuing of scientific objectivity is fused with the valuing of a certain kind of scientific self.

[2] Haller's physiology textbooks were divided according to the main parts of the human body, from the basic animal fibers through to the reproductive parts, providing a description of the anatomical structure and function of each. Haller (1751, 5). Cf. Cunningham (2002, 651–56), Cunningham (2003, 66–67), and Roe (1984).

attention to organic vitality at the turn of the nineteenth century contributed to a shift in attention towards vital functions. Vital functions were also brought to bear on the natural history of living organisms, resulting in a comparative physiology premised on function. Yet, even as biology was named as a distinct science of life by Gottfried Reinhold Treviranus and others, organic processes were also increasingly understood through their material conditions and relationships with inorganic processes, with the boundaries of life and organic vitality complicated rather than fixed through these new inquiries.

2 Vital Principles and a Science of Life

The 1790s began, then, with two works reflecting upon the possible principles and concepts for the study of living organisms, and their relationship to those of the mechanical and physical sciences, which stimulated considerable discussion. Kant's "Critique of the power of teleological judgment" was a purely philosophical work, concerned with the form and limits of our judgments of living beings, but it responded to new investigations of their remarkable capacity for self-organization. Although Girtanner's treatise on irritability was a scientific study of the possible chemical processes underlying organic properties, it also ventured into philosophical terrain in proposing chemistry provided the principle of life. If the response to Girtanner's provocation was immediate, the response to Kant's work amongst German naturalists and physiologists was more gradual. But both works were instrumental in making vital principles and a science of life key topics of debate in the 1790s.

Kant's critical project was concerned with the investigation of the legitimacy of reason's claims and the rights by which it acquires its principles and concepts.[3] The central argument of his transcendental idealism was that the *a priori* concepts of the understanding are the epistemic conditions necessary for human cognition. His claim for the objective validity of these pure concepts rested on his demonstration that through them alone experience is possible; only through concepts can an object of experience be thought.[4] His claim for the objective validity of these concepts also rested on his demonstration of their grounding in the transcendental unity of apperception, the unified self-awareness that provides the basis for the unity of form necessary for all concepts; only through general concepts can an object of experience be thought as a unity of manifold intuitions. But these bold claims for the role of reason in experience were tempered by Kant's tethering of all our cognitive claims to the conditions of our sensible intuitions. The objective validity of the pure concepts of the understanding are only demonstrated when they are also related to what is given under the forms of human sensibility, and shown to be capable of truth

[3] Kant (1902–1983, A xi). In citing the *Critique of Pure Reason*, references are to A and B, the first edition (1781) and second edition (1787), found in volumes III and IV of the *Akademie* edition (Kant 1902–1983), respectively. All other references to Kant's works cite the volume and page numbers of the *Akademie* edition, as is standard in critical editions and translations.

[4] Kant (1902–1983, A93/B126).

or falsity in specific judgments. Kant's critical investigation of the rightful claims of reason thus adopted a posture of relative epistemic modesty. Although he sought to legitimate the necessary *a priori* conditions of experience to counter the skeptical claims of empiricists, he also sought to counter the excesses of rational metaphysics with its pretention to knowledge beyond the conditions of human sensibility. His *Critique of pure reason* exposed the illusions to which pure reason is subject when it severs its ties to sensory intuition. A mature power of judgment, he maintained, should accept how we ought to reason after reflection upon what human cognition can rightfully claim.[5]

Kant argued that the *a priori* concepts of the understanding provide only the conditions for experience in general; the explanation of specific natural phenomena required more specific principles and concepts. Kant, in fact, addressed not only questions regarding a system of knowledge but also those regarding a system of nature, engaging with contemporary issues in the natural sciences throughout his philosophical career. From his earliest writings he wrestled in particular with the problem of the place of organisms in such a system. Kant's model for a scientific explanation of the material world was Newtonian mechanics. In his 1786 *Metaphysical foundations of natural science*, Kant provided a construction of the metaphysical principles of pure natural science corresponding to Newtonian mechanics by analyzing the concept of matter from contemporary empirical science under the guidance of pure concepts of understanding. It was not strictly an *a priori* exercise, as it proceeded via the empirical concept of matter, ensuring that the metaphysical principles applied to corporeal nature could be realized in sensory intuition, as opposed to speculation on the inner nature of matter, in keeping with his critical strictures on the claims of reason. Thus, for example, he demonstrated how the mechanical law that every change of matter has an external cause can be constructed from the empirical concept of matter as moveable in space through the application of the category of causality. The application of such metaphysical principles to particular natural phenomena provided mechanical explanations in the form of particular empirical laws.[6] But Kant contended that living organisms could not be explained through such mechanical laws. His argument was not that the concepts of efficient or mechanical causality determinative of our understanding of nature do not apply to organisms; indeed, we conceive many organic phenomena in terms of causal and mechanical principles. Rather, he argued that these principles are not sufficient to explain their organization and self-organizing capacities. Yet he also rejected the attribution of the immediate formation of individual plants and animals to a supernatural origin as untenable.[7] The critical question posed by living organisms for Kant, given his commitment to a scientific system of nature based on mechanical laws, was how to conceive living organisms as natural products.

[5] Kant (1902–1983, A287/B343, A761/B789).

[6] Cf. Buchdahl (1992, 195–242), Friedman (1992), Carrier (2004) and Breitenbach (2006).

[7] Kant already introduced such strictures in his pre-critical works, his 1755 *General natural history and theory of the heavens* and 1763 *On the only possible argument in support of a demonstration of the existence of God.*

Kant's reflections upon the problems posed by organized and self-organizing beings to scientific explanation led him to suggest theoretical contributions to issues in contemporary natural history and questions regarding generation and degeneration. One of his most substantial contributions was a 1775 essay *Of the different human races*. Drawing on the influential *Natural history* of Georges-Louis Leclerc, Comte de Buffon, Kant argued for a new attention to the history of nature, for a study of "the changes in the earth's form, including those the earth's creatures (plants and animals) have undergone through natural migrations, and their degenerations [*Abartungen*] thereby from the original form [*Urbilde*] of the stem genus [*Stammgattung*]."[8] To account for relatively permanent characteristics of the different races produced in these degenerations, he posited germs [*Keime*] and predispositions [*Anlagen*] in the original organizations of organisms; the predispositions lay dormant until particular environmental conditions induce the unfolding of particular features, which are then passed on in reproduction. Kant cautioned against speculations regarding the first origins or final ends of living beings, however, especially after the appearance of conjectural histories of the world and its living forms such as Johann Gottfried Herder's *Ideas towards a philosophy of the history of humankind* in the mid-1780s. Although he acknowledged the attraction of such "archaeological musings," for Kant they remained outside the reach of human reason and experience.[9] Blumenbach became Kant's exemplar for the correct method for an account of self-organization; starting from an original organization, and attributing a formative drive [*Bildungstrieb*] to the reproductive matter in an organized body, he nevertheless left as much as possible to natural mechanisms.[10] But despite these contributions to the theoretical content of natural history, Kant's primary interest was in establishing the proper principles for our inquiry into the natural world, and with delimiting the form and scope of our judgments of organized beings.

In the *Critique of the power of judgment* Kant critically examined the warrant for our judgments of the apparent purposive organization of living organisms. Our study of these unique natural products demonstrates their remarkable organization and capacity for self-organization; they are forms of natural organization that are contingent in terms of the mechanisms of nature. It seems that we can account for their possibility only by appealing to a concept of reason, namely, the concept of purpose. Yet they are formed by natural processes rather than on the basis of an extrinsic idea. Since organisms are natural products we must conceive of them as self-organizing, as at once cause and effect of themselves, with each part existing not only as the end of all the other parts and the whole, but also reciprocally being the means producing the other parts and the whole. Kant thus named them natural purposes – natural objects judged to be purposive. If enlisting both concepts of

[8] Kant (1902–1983, II, 434n).

[9] Kant (1902–1983, V, 419, VIII, 161–62). On Herder's significance to Kant's history of nature, cf. Zammito (1992, 178–213).

[10] Kant (1902–1983, V, 424). On the relationship of Kant to Blumenbach, cf. Lenoir (1982, 17–35), McLaughlin (1982), Sloan (2002, 246–50), Richards (2002, 329–37) and Bernasconi (2006).

nature and concepts of reason in its formation, the amphibious concept of natural purpose is properly a part of neither theoretical philosophy nor practical philosophy, and thus lacks a domain in his critical philosophy. He acknowledged that the concept of natural purpose does not offer an explanation of the organization and self-organizing capacities of living organisms, accounting for what eludes mechanical explanation; it is not a constitutive concept either of understanding or reason. Rather it is a "regulative concept for the reflecting power of judgment;" it is a "concept of the reflecting power of judgment for its own ends," a means for us to identify and think about these unique natural products.[11] Kant's critical examination of our teleological judgments of organisms thus introduced a distinct mode of judgment, related and yet different from our determinate judgments of objects in general and mechanical explanations, a mode of judgment with only subjective, not objective validity.

Kant proposed that we should pursue mechanical explanations as far as possible in our study of the natural world, and resort to the concept of natural purpose only when necessary to identify and investigate the unique capacities of living organisms. He acknowledged that living organisms thus thwart our attempts to conceive of nature as a unity of empirical laws. Indeed, the latter part of his "Critique of the teleological power of judgment" addressed the antinomy that arises when judgment reflects upon the unity of nature due to the conflicting possible principles that it can give itself. On the one hand, it has the ideal of pursing a single method in the investigation of nature by seeking mechanical explanations for the possibility of all natural products. But on the other hand, certain natural products disrupt this goal by requiring for the judgment of their possibility a concept of purpose that appears at odds with a mechanical concept of nature. Kant emphasized that each maxim serves strictly as a regulative guide for our investigation of nature, "as a merely subjective principle for the purposive use of our cognitive powers."[12] But this critical analysis of the two maxims does not remove the conflict between them. Moreover, if we can reconcile the two maxims to an extent by agreeing to use both, it is not clear how to decide between the two of them in particular instances. It is our attempt to comprehend organisms that gives rise to this antinomy of the power of judgment in its reflections upon the possible unity of nature, the amphibious concept of natural purpose only highlighting the dialectic between these principles.[13]

What Kant did make explicit in his "Critique of the teleological power of judgment" is that his interest in organisms was as an epistemic problem, not a metaphysical one. He was concerned with critically examining how we must judge these unique natural products, not with determining their essential nature or the causes of their original formation. Given the widespread attention to Kant's works in the German context in review journals, and amongst scientists as well as philosophers, it is perhaps surprising that in the years following the publication of his third

---

[11] Kant (1902–1983, V, 375; XX, 236). Cf. Steigerwald (2011), Zuckert (2007), Quarfood (2004) and Ginsborg (2001).

[12] Kant (1902–1983, V, 385–87).

[13] Steigerwald (2013) and Cohen (2004).

Critique so many publications appeared proposing special metaphysical powers to account for organic vitality.[14] But not only did the technical sophistication of Kant's philosophical arguments present considerable challenges to comprehension, his work also contained several ambiguities. Its central concept, that of natural purpose, was comprised of a tension between a concept of nature and a concept of reason. Although he maintained the concept of natural purpose was necessary to grasp the organization and self-organization of living beings, he also argued that mechanical explanations should be pursued as far as possible and provided no rules for when it was required. Moreover, although in his critical analysis of our teleological power of judgment he allowed that the concept of natural purpose was only a regulative concept for our judgments with no objective validity, in several writings during his critical period he appealed to notions of germs, dispositions and formative powers, suggesting constitutive concepts explanatory of the organization of living beings. Perhaps his most influential contribution to subsequent debates was his argument that living organisms are unique natural products that cannot be explicated mechanically. Although Girtanner would argue the new French chemistry could account for organic vitality, most German naturalists and physiologists responding to Girtanner's work in the early 1790s favored Kant's position. But many also transgressed the critical strictures Kant placed on our judgments of organisms, and argued for the need for a distinct vital principle to explain their unique capacities.

Girtanner's "Treatise on Irritability as the Principle of Life in Organized Nature" provoked a much more immediate response amongst naturalists and physiologists than Kant's work. First published in a widely read French journal of physics in 1790 and then in German translation in the *Journal of Physics* in 1791, Girtanner's treatise contained several controversial assertions. He declared that life could be explained through a single principle and that irritability was that principle. Girtanner here drew on the notion of a property of irritability of muscle fibers distinct from the sensibility of the nerves introduced by Haller in the 1750s. Although influential, especially in the German context, Haller's notion of irritability was controversial in the mid-eighteenth century and remained so at the end of the eighteenth century, contested by "nerve patrons [*Nervenpatronen*]" privileging sensibility in physiological function and questioning the possibility of a property of irritability independent of nerves and sensibility.[15] Girtanner, however, now claimed that irritability was the sole principle of life. He affirmed irritability as the original, essential character of a living fiber "generally spread in organic nature," found not only throughout animal bodies, but also in plants.[16] Girtanner also proposed a new and original medical

[14]The general review journal *Allgemeine Literatur-Zeitung* alone carried hundreds of articles on Kant's philosophy. On the post-Kantian philosophical debates, cf. Beiser (2002, 223–259). Numerous physicians, physiologists and naturalists referred to Kant in their works during the 1790s, but with varying degrees of engagement with his work.

[15]Review of Johann Daniel Metzger's *Ueber Irritabilität und Sensibilität, Göttingische gelehrte Anzeigen* 149 (18 Sep. 1794): 1496. The anonymous review was most likely by Girtanner. On Haller's experiments on irritability and sensibility, and the controversies surrounding them, cf. Steinke (2005, 49–174).

[16]Girtanner (1791, 320–22).

system founded on irritability as the principle of life. A Scottish student visiting Göttingen, however, accused Girtanner of plagiarizing the Scottish physician John Brown's medical system, which conceived health and illness in terms of the balance of the excitability or irritability of the human body. Girtanner's treatise thus became entangled not only in a very public controversy over academic dishonesty, but also in the heated debates in German journals over the viability of Brown's medical system.[17] Girtanner further declared that irritability, and thus organic vitality, could be explained chemically, drawing on Antoine Lavoisier's new antiphlogiston chemical theory, making oxygen rather than phlogiston the active element in combustion and respiration. To support French chemistry in the early 1790s when prominent German chemists still wedded to phlogiston theory were dismissing it was already controversial, but to suggest a chemical basis of life was too radical for most German physicians and chemists.[18] His credibility under question, Girtanner's treatise soon became a favored target of attack.

Shortly after the appearance of Girtanner's treatise, several reviews, articles and books appeared providing alternative principles of the vitality of living organisms. The most strident criticisms of Girtanner's treatise came from physicians defending sensibility as the fundamental principle of organic nature, such as the Erfurt medical professor August Friedrich Hecker and the Regensburg physician Johann Ulrich Schäffer. The respected Königsberg professor of medicine, Johann Daniel Metzger, then put forward a compromise position in 1794, positing that sensibility, as a power of the nerves, be complemented by a power of irritability, found in certain fibers. But Metzger was still insistent that special vital principles were needed to explicate the unique capacities of living beings. Naturalists also contributed to the debate, most prominently Kielmeyer at the Karlsschule in Stuttgart, who undertook comparative studies of organic vitality in a variety of organisms. In a widely cited lecture in 1793, Kielmeyer posited each living organism as constituted through a unique interrelation of vital powers, adding generative powers to those of irritability and sensibility, drawing upon Blumenbach's notion of a formative drive.[19] In the 5 years following Girtanner's essay, over a dozen monographs, several physiological and medical textbooks, and numerous articles and reviews appeared in German presenting competing arguments regarding the principles or powers of life. The wave of publications crested in 1795, punctuated by a detailed study by the Weimar physician and professor of medicine at the University of Jena, Christoph Wilhelm

[17]Cf. the letters accusing Girtanner of plagiarism by Mr. Ash, and Girtanner's defense, in the *Allgemeine Literatur-Zeitung* (6 Aug. 1791) Intelligenzblatt 97: 801–2, (3 Sep. 1791) Intelligenzblatt 107: 882, and (10 Sep. 1791) Intelligenzblatt 113: 929–30; and the *Medicinisch-Chirurgische Zeitung* (1791) 4: 369–80. Over the plagiarism controversy, cf. Risse (1971, 148–56) and Tränkle (1986).

[18]Girtanner (1791). The differences between Lavoisier's chemical system and phlogiston theory cannot be reduced to the role of oxygen, although Girtanner focused upon this difference. For very good summaries of the differences between the two, cf. Durner (1994) and Golinski (2003). On the reception of Lavoisier in Germany, cf. Hufbauer (1982).

[19]Hecker (1793), Schäffer (1793), Metzger (1794) and Kielmeyer (1993).

Hufeland. Hufeland, rapidly becoming one of Prussia's most prominent medical professors, was a leading advocate of medical practice based in experience rather than in theoretical systems; yet his contribution to the controversies over organic vitality was a theory of medicine and pathology premised on a *Lebenskraft*. Hufeland accepted the new chemical theories, recognizing their significance for the study of vital processes and disease. Indeed, by 1795 the controversies over the new French chemistry had largely subsided in Germany. But Hufeland criticized the position advocated by Girtanner that oxygen was the principle of organic vitality, and instead sought to establish boundaries for the application of chemistry to physiology and pathology.[20] Like so many of his contemporaries, he insisted that the influence of mechanical and chemical forces must be subordinated to a special vital power in living beings.

Part of the controversy over Girtanner's treatise concerned who had the authority to pronounce on questions of organic vitality. As Girtanner lacked an academic position, his credibility was more readily called into question by established professors and physicians. In the context of a rapidly developing German periodical culture, with new review and specialized journals appearing every year, the rules of engagement for debate were in flux. Certainly the critiques of Girtanner's treatise were often personal or ideological rather than addressed to the specificities of his scientific claims. But this aggravated reaction sparked by Girtanner's treatise is indicative of the significance the issue of organic vitality had acquired. The contestation over organic vitality shows the new politics "life" implied in the 1790s. In part this contestation was scientific and philosophical, in that it concerned founding the study of living organisms on the right principle. But it was also a question of authority – a question not only regarding who had the right theory, but also who had the right to pronounce on the principle of life. In the tribunal of critical opinion, Hufeland could claim the right to both, but Girtanner, apparently, to neither.

Although there was no discipline claiming as its domain the study of living beings in general in the early 1790s, most contributors to the controversies over organic vitality argued for demarcating the study of life from that of the lifeless. The dispute extended beyond the empirical study of organic phenomena and theories of vital processes, as physiologists and naturalists appropriated the terms of philosophy to legitimate their approach to the study of living organisms. Kant had restricted the designation of science to mechanics, contending that there can only be as much science in any doctrine as there is mathematics and determinate principles.[21] Although he agreed that the investigations of living organisms require distinct principles, he also contended that these principles provide only regulative concepts for our judgments and do not warrant determinate explanations of organic phenomena. But in the latter 1790s, several physiologists and naturalists attempted to establish

---

[20] Hufeland (1795). Hufeland founded a *Journal of Practical Medicine* arguing for the import of medical practice based in experience rather than theoretical systems. On Hufeland's career and contributions, cf. Broman (1996, 104–19) and Pfeifer (2000).

[21] Kant (1902–1983, IV, 470).

the study of organic vitality as a science. Johann Christian Reil, the prominent professor of medicine and chief physician in Halle, founded a new journal of physiology in 1795 with this aim. Although he embraced the terms of Kant's critical philosophy, maintaining that powers such as the *Lebenskraft* are purely subjective concepts, the forms through which we think the relationship of appearances to the qualities of matter, he argued that the study of such powers in physiology constitutes a science.[22] In 1797 the Braunschweig physician Theodor August Roose prefaced his *Fundamental traits of the doctrine of the vital power* by representing the work not only as a contribution to science but also as an outline of a biology. Karl Friedrich Burdach also used the term biology in his 1800 guide to medical lectures, using the term for a physiology or doctrine of life of the human being regarding its form, composition and characteristic powers. Then in 1802 the first volume of Treviranus' multi-volume *Biology; or, philosophy of living nature for scientists and physicians* appeared. Encompassing physiology and natural history, his science of biology was premised on a division of nature into the living and the lifeless, and a need to demarcate a distinct domain of inquiry into the question "what is life?"[23] These texts calling for a science of life attempted to establish not only the laws of organic vitality, but also the determinate principles for the investigation of those laws.

The publication of several works attempting to define a science of life in the years around 1800 might seem to suggest the beginning of the formation of biology as a distinct discipline. These developments seem to complement the formation of chemistry and physics as disciplines distinguished by their own communities of scholars with their own journals and textbooks, and the introduction of the natural sciences as distinct areas of instruction in German universities.[24] But such formations remained ambiguous, especially in the areas concerned with the study of living organisms. Professors often held chairs in several areas and regularly taught courses in faculties different from their appointments, and no group of researchers or institutional basis was created for a biological science or a scientific study of living beings in general. Essays on organic vitality continued to be published in journals of chemistry and physics as well as medicine, natural history and physiology, and the application of new concepts and techniques of investigation from both physics and chemistry to the study organic phenomena further blurred the boundaries between organic and inorganic phenomena. Indeed, the introduction of principles and *Lebenskräfte* as ruling powers of life occurred at a time when the laws governing life were becoming increasingly unclear and complicated. The need for a principle of life was felt most urgently, it seems, precisely at a time when the viability of a distinct principle for life was being seriously called into question.

---

[22] Reil (1795). Reil was introduced to Kant's critical philosophy through his contact with the Kantian Marcus Herz in Berlin before his move to Halle. Cf. Broman (1996, 86–88) and Richards (2002, 252–61).

[23] Roose (1797, iii), Burdach (1800, § 195, 162), Treviranus (1802–1822, I, 23, 38). On the first uses of the term biology, cf. Kanz (2006).

[24] Hufbauer (1982), Clark (1997), Ziche (1998) and Bach and Breidbach (2001).

3 Investigating the Material Conditions of Organic Vitality

The many German articles, books and reviews appearing in response to Girtanner's treatise referenced earlier works, concepts and experiments regarding organic vitality. The works of two figures in particular were repeatedly enlisted as authoritative – Haller's studies of irritability and its relationship to sensibility from the 1750s, and Blumenbach's studies of generative processes from the 1780s. Kant also enlisted the authority of Blumenbach to support his philosophical reflections upon the self-organizing capacities of organized beings.[25] Neither Haller's nor Blumenbach's investigations had received general acceptance when they first appeared – their individual experiments, methodologies and interpretations were all widely disputed. Yet in the context of the polemic of the early 1790s, the uncertain results of earlier experiments were repeated with assurance and contested outcomes stated as established matters of fact. Indeed, in these publications the immediacy of the encounter with organic vitality experienced by Haller, Blumenbach and their critics became a distant reference, and the indeterminacies of experimental readings of organic vitality transformed into fixed concepts. One could argue that the moment when the principle of life was most hotly contested was also the moment when its proponents had lost touch with life. In the context of a polemic, principles were reified and positions became entrenched. But fresh empirical investigations in the latter 1790s reinvigorated organic vitality, although they did so by drawing on new techniques from physics and chemistry and thus complicating the boundary between organic and inorganic phenomena. New galvanic experiments and new studies of generative and degenerative processes explored organic vitality in its material conditions, while raising with renewed intensity questions regarding the methods appropriate for a science of life.

Galvanic experiments took their name from the methods devised by Luigi Galvani, Professor of Anatomy in Bologna. Studying contractions in severed frog legs through the application of artificial and atmospheric electricity, investigations that built on Haller's experiments on the irritability of muscles, Galvani was able to produce contractions by the application of a bimetallic arc between a muscle and exposed nerve. He attracted attention to his work by announcing in 1791 that he had discovered a new form of electricity – animal electricity. Galvani's reading of his experiments was soon contested by Alessandro Volta, Professor of Physics at Pavia, who argued that contact between heterogeneous metals provided an external electrical current that stimulated contractions in the frog leg. Interested in the frog leg as

[25] The significance of Haller and Blumenbach to the contestation over organic vitality in the 1790s points to the important influence of the University of Göttingen (cf. Lenoir 1982; Stuber et al. 2005). Haller and Blumenbach, both professors at the prestigious medical school at Göttingen, had extended networks of students and correspondents, edited prominent journals, and authored important textbooks widely distributed in several editions in Latin and German as well as other European languages. Their contributions remained to traditional fields of physiology and natural history, and not a new discipline of biology. But many of the figures developing a new science of life, from Reil and Treviranus to Kielmeyer and Oken, were students in Göttingen.

an instrument for detecting weak electricity in his experiments on metallic chains, Volta resisted claims that the frog leg was generative of electrical phenomena. Reports of Galvani's experiments and his dispute with Volta were closely followed in German periodicals from 1792; their works were translated into German, and German physicists, chemists, physiologists and naturalists published reports on their attempts to repeat their findings.[26]

Alexander von Humboldt undertook one of the most comprehensive studies of galvanic experiments in the German context, beginning his investigations in earnest in 1794 while working as a mining inspector for the Prussian state. He began by investigating the conditions of the chain necessary to effect a stimulus in response to the dispute between Galvani and Volta, but he soon shifted his attention to the conditions of the organic parts receiving the stimulus. He became particularly interested in Galvani's claim in 1794 to be able to produce a contraction in the frog leg by using a chain consisting solely of a nerve and muscle. Then in 1795 he had the opportunity to do galvanic experiments with Volta, who demonstrated how the application of a potash solution instead of water to the nerves increases the responsiveness of a frog leg.[27] The experiment turned Humboldt's attention to the import of the conditions of the organic parts and their receptivity to stimuli for the success of galvanic experiments; he concluded that negative results in experiments demonstrated nothing, as long as the experimenter could not prove his experiments were made with specimens that possess maximum animal excitability. Building on Volta's technique with potash solution, Humboldt found he could increase the excitability of the frog leg chemically by the application of alkaline solutions to the nerves and acidic solutions to the muscles. He further explored the depressing or exciting effects of opiates, alcohols and other drugs as well as different gases. He extended his interest in the quality of specimens to a wide variety of factors that he determined affected their responsiveness to stimuli – the species, age, sex, strength, nutrition and health of the frog as well as the humidity, temperature, light and air quality of the climate in which the experiments were made. The title of his 1797 work detailing the results of these investigations, *Experiments on Stimulated Muscle and Nerve Fibers, next to Conjectures over the Chemical Processes of Life in the Animal and Plant World*, indicated this new direction of his research.[28]

Although in 1793, in the context of debates over organic principles, Humboldt enlisted a *Lebenskraft* to account for the unique capacities of living beings, his attention to techniques of chemically altering the excitability of frog legs in his galvanic experiments convinced him of the significance of the chemistry of vital processes. In making this proposal, Humboldt sought to distance himself not only from his earlier position, but also from Girtanner, who he held had emphasized too

[26]On European debates regarding galvanism, cf. Bresadola (2008), Piccolino (2007), Pincaldi (2003), Bertucci and Pancaldi (2001), Moiso (1994) and Pera (1992). On the reception of the galvanic experiments in Germany, cf. Trumpler (1992).

[27]Cf. Humboldt's letters, 12 December 1792, 29 July 1795, and 26 August 1795, in Humboldt (1973, 219–23, 438–40, 454–56). Also cf. Humboldt (1795, 1797, I, 3, 8, 31–32).

[28]Humboldt (1795, 471–73, 1797, I, 22–27, 34–36, 68–87, 242–48, II, 173–75).

singularly the similarities between combustion and life; both positions he now argued hindered insight into organic phenomena. Instead, he aligned himself with Reil, and argued for a conception of "an important and a *new* branch of natural science," vital chemistry [*vitale Chemie*] – "*the investigation of the chemical alterations of composition* [*chemische Mischungsveränderung*] *that occur in the excitable matter* [*erregbarer Materie*] *during the vital functions*." But if Humboldt was offering an understanding of organic vitality through the importation of techniques and concepts from physics and chemistry, he also still emphasized its distinctive character. He concluded that excitability is the common condition of organic matter, with each organic fibre having a specific excitability dependent on its specific form and composition, and each also able to preserve its specific excitability despite being uninterruptedly stimulated and in a continual exchange of material with the external world. Organic excitability consists of alterations of form and composition that are the common result of a complex entanglement of phenomena – physical, chemical and vital – that cannot be reduced to a single power or principle.[29]

Humboldt's galvanic experiments and his conception of vital chemistry mark a significant shift in understandings of organic vitality during the 1790s. As the results of the experiments on organic excitability of Humboldt and others were published, and as French chemistry became accepted in the German context and its import as a tool for the study of organic phenomena became increasingly clear, new conceptions of organic vitality took shape. Living organisms began to be conceived as not only organized but also as organic. Indeed, only a few years after Hufeland had appealed to a special vital power with such authority, influential works by Reil, Humboldt and others were advocating the investigation of organic processes in relationship to inorganic processes. A new attention to the material conditions of the vitality of living beings also occurred in studies of their generative capacities, even amongst those arguing for a *Lebenskraft* to demarcate living processes from the lifeless, such as Kielmeyer and Treviranus. In these studies, living organisms were not only conceived as organized and self-organizing, but were also shown to have capacities for degeneration and regeneration under changed material circumstances.

Kielmeyer brought a comparative perspective to the study of vital powers, tracing their varied effectiveness in different organisms from simple plants to complex animals under different material conditions. A former student of Blumenbach's at Göttingen, he had positions as a professor of chemistry, zoology and botany as well as responsibilities for natural history collections, first at Karlsschule and then Tübingen during the 1790s. Blumenbach had made comparative physiology as well as comparative anatomy central to natural history. His 1780 *On the formative drive* had provided influential experimental demonstrations of generative processes in a variety of organisms, attributing these generative processes to a formative drive, a *Bildungstrieb*, in organic matter. But although Blumenbach's 1780 work foregrounded experimental demonstrations of generation, in the polemic of the 1790s

[29]Humboldt (1797, II, 41–42). Humboldt's 1793 work appeared in German translation in 1794; compare Humboldt (1797, II, 41–75, 100–47, 430–36).

over vital powers Blumenbach's *Bildungstrieb* was reified into a *Lebenskraft* next to irritability and sensibility. Kielmeyer followed Blumenbach's model of natural history, undertaking empirical research into comparative physiology as well as comparative anatomy. Positing each living organism as comprised of a specific relationship of irritability, sensibility and reproduction, Kielmeyer argued how the predominance of one power in an organism resulted in the decreased influence of the others. He was particularly interested in investigating the relationships of these vital powers in the simplest organisms, where generative powers dominated over sensibility and irritability. But as a professor of chemistry, Kielmeyer also brought the techniques and concepts of chemistry to bear on natural history. His unpublished lectures and manuscripts indicate the extent to which he attempted to demonstrate experimentally the action of organic powers in the material changes occurring in organic forms during reproduction. He also pursued chemical studies of the differences between organic and inorganic compositions of material elements and the relationship between organic compositions [*organische Mischungen*] and organic form.[30] Publishing little, Kielmeyer nevertheless influenced a generation of naturalists though his unpublished lectures and students.

Treviranus' *Biology* had some similar emphases to Kielmeyer's studies, but it also took a different direction. Treviranus based his science of life on a combination of physiology and natural history, like Kielmeyer comparing the organic functions in various living forms, and focusing in particular on simple organisms and their extraordinary reproductive capacities.[31] Like Kielmeyer, he also insisted on a *Lebenskraft* to demarcate the living from the lifeless. But, unlike Kielmeyer, in designating organic nature as a distinct domain, he rejected new chemical studies of the form and composition of matter as able to provide insight into organic vitality, positioning his biology against the model of a science of life proposed by Reil, Humboldt and others.[32] Instead, he posited the existence of an organic viable matter [*lebensfähige Materie*] through which all living beings possess life. In support of this claim, Treviranus enlisted Joseph Needham's infusion experiments from 1750, in which Needham purported to have demonstrated the formation of simple plants or animals from decayed organic matter. Treviranus' former professor at the University of Göttingen, Henri-August Wrisberg, had confirmed Needham's experiments, publishing these results in 1765. In the second volume of *Biology*, Treviranus examined the infusion experiments of Needham and Wrisberg in detail, countered the experiments of Lazzaro Spallanzani and other critics of Needham's results, and presented

[30] Cf., for example, "Ideen zu einer allgemeineren Geschichte und Theorie der Entwicklungserscheinungen der Organisationen," in Kielmeyer (1938, 102–94). Cf. Lenoir (1982, 37–53) and Low (1979, 91–112).

[31] Although Treviranus would later contribute several studies on the anatomy of the reproductive organs of invertebrates as well as fish, amphibians and small rodents (Smit 1976), for his *Biology* he drew on the comparative anatomy and comparative physiology of Blumenbach, Louis D'Aubenton, Georges Cuvier and other prominent naturalists.

[32] Treviranus (1802–1822, I, 52).

his own experimental findings in support of Needham.[33] Treviranus argued that although this organic matter lacks organization, it has the potential to combine into all forms of life, taking determinate and constant forms through the persistent influence of external causes. This viable matter is inseparably bound with the *Lebenskraft*, which ensures the similarities of appearances of living organisms in the face of contingent and changing external influences. Viable matter takes specific determinate forms through its connection with external stimuli whose influences are filtered through the vital power specific to each organic form, which resists certain external influences and allows others.[34] In his *Biology* Treviranus portrayed a continual formation, destruction and new formation of different living forms through the continual interplay between viable matter and physical influences. He was particularly fascinated with simple forms of life, with zoophytes lying at the boundaries between plants and animals (infusorians, polyps, corals, fungi and mosses), which he contended are capable of being continually formed from viable matter and susceptible of radical transformations into new living forms under the changed external conditions.[35] Although such transformations of life are less common in higher plants and animals that propagate through germs, through seeds or eggs and semen, he argued that even these more complex organisms have the potential to take new forms. Treviranus' *Biology* thus offered a theory of the continual generation and degeneration of new forms of life from organic matter under new physical conditions.

Lorenz Oken's 1805 work *Generation* [*Die Zeugung*] framed his contribution to these studies in terms strikingly similar to Treviranus. This work was completed for his habilitation at the University of Göttingen, but the previous year Oken had studied at Würzburg where he took lectures from Schelling and his followers; in 1807 he would become professor of medicine at Jena.[36] In *Generation*, Oken posited simple living beings or infusorians as the primordial matter, the *Urstoff*, of life, from which all living organisms are formed and into which they all decay.[37] Oken turned to the infusion experiments conducted by Needham in the mid-eighteenth century, but also to more recent confirmations by Treviranus and others, to demonstrate that infusorians arise not from eggs or inorganic matter but from putrefying organic matter. All emergence, all growth, all flourishing of all organisms, Oken concluded, occurs through the synthesis of the infusorians spread throughout all of nature, their mass continually rejuvenated through the death and destruction of previous living organisms. He combined this theory of infusorians with more recent accounts of generation, such as Blumenbach's account of the formative drive of

33 Treviranus (1802–1822, II, 267–95, 319–52). On Needham's experiments, cf. Needham (1748), and Roe (1983). Spallanzani's work appeared in German in 1769: Lazzaro Spallanzani, *Physikalische und Mathematische Abhandlungen*. Leipzig: Gleditschens.

34 Treviranus (1802–1822, I, 52, 59–60, 97–103, II, 264–67, 353, 403–4).

35 Treviranus (1802–1822, I, 63–70, 98–99, II, 264–67).

36 Ghiselin (2005).

37 Oken (2007, I, 1–2). Page numbers are from the original edition of *Die Zeugung*, cited in this edition.

seminal matter, by populating the seminal matter with infusorians.[38] Oken represented infusorians as the matter of all forms of life as well as the domain of the simplest forms of organic beings next to plants and animals. He conceived the generation of complex organisms, then, not only as a synthesis of infusorians, but also as a development from infusorians to higher forms, transformations from an infusorians stage through a plant stage to an animal stage. Indeed, he repeatedly emphasized the starting point of all forms of generation in infusorians, and the identity of the mode of generation of even the highest living beings, human beings, with that of the lowest polyps.[39] He was particularly interested in the simplest forms of life, arguing that in simple organisms physiological functions can be more readily understood than in the more complex organizations of higher animals. Oken's emphasis on primary forms of organic vitality enabled him to draw analogies between the generative development of different organisms, and analogies between physiological functions across species, from the lowest to the highest. His emphasis on primary organic functions also offered a conception of how more complex organization developed from and through simpler ones both in individual generation and in the history of life. He then related these organic functions and generative processes to physical and chemical processes. Although Oken offered a theory of generation, a philosophical rather than empirical study of the formation of life, he actually provided more details of the processes of generation than Treviranus through relating his theory to the descriptions of others.

In 1805 Oken published *Outline of the System of Biology*, suggesting an affinity to Treviranus' model for a science of life. But neither in this work nor in *Generation* did Oken enlist a *Lebenskraft* to demarcate the living from the lifeless, as Treviranus did. Oken's *Biology* elaborated the analogies between organic, chemical and physical processes in different kinds of living organisms that he had introduced in *Generation*, thus suggesting far greater continuity between organic and inorganic processes than Treviranus allowed. In his 1809 *Textbook on the Philosophy of Nature* Oken even posited a spontaneous generation of life, albeit as a singular event in the historical formation of the earth. He argued that under specific conditions, with the right admixture of elements, light, and chemical and cohesive processes, a primordial mucous [*Urschleim*] was produced and took the form of infusorians. He maintained that once it was produced, organic life persisted through cycles of generation from and destruction into this primary organic matter.[40] The differences between Treviranus and Oken are indicative of the range of approaches to a science of life at the turn of the nineteenth century.

Although Treviranus' and Oken's biology, and even Humboldt's vital chemistry, showed similar interests in the investigation of the material conditions of life, they offered distinct visions of the relationship of the study of life to the physical sciences. Empirical investigations into the phenomena of excitability and generative

---

[38] Oken (2007, I, 19, 97–107).

[39] Oken (2007, I, 108, 216).

[40] Oken (2007, II, §§833–869, 911–953).

processes shifted emphasis from pronouncement on vital principles to the details of the changes in organic matter produced during the functions of life. But these investigations made the boundary of life more difficult to determine. Indeed, despite calls for a science of living beings, the demarcation between organic and inorganic phenomena had become even less clearly defined in the years around 1800 than around 1790.

4 New Conceptions of Organic Vitality

In naming his textbook a *Philosophy of nature* [*Naturphilosophie*], Oken identified his system of nature with that of Schelling's. Schelling gained prominence while professor of philosophy at the University of Jena from 1798 to 1803, before moving to Würzburg and then Munich. He took as the starting point for his philosophy of nature Kant's transcendental idealism and its concern with "the possibility of nature," with how the connections of phenomena we call nature become actual for us and attain the necessity in our representation with which we are compelled to think of them.[41] But he also argued for pushing beyond the appearances on which Kant's account of cognition rested. If drawing on the most recent studies in the empirical sciences, Schelling argued empirical studies only touched the "surface of nature;" he advocated instead a speculative science penetrating to the "inner spring-work" of nature.[42] But he did not then propose supersensible metaphysical entities or principles as the basis for natural phenomena. Rather his philosophy of nature regarded each natural product as constituting a boundary between the productivity of nature and its material constraint, in an endless process of becoming whose fundamental basis exceeds our cognitive grasp. Schelling also rejected the demarcation between the study of living organisms and the mechanical sciences advocated by Kant as well as proponents of *Lebenskräfte*. Indeed, Schelling blurred the boundary between organic and inorganic processes, positing each natural object as an organized and dynamic whole. The emphasis on productivity in Schelling's philosophy of nature encouraged the rethinking of organic vitality in the early nineteenth century towards a conception of a science of life based on organic functions, towards a comparative physiology rather than a comparative anatomy.

Schelling challenged Kant's model of science based on contemporary mechanics. In his 1786 *Metaphysical foundations of natural science*, Kant excluded chemistry from science because its principles are ultimately merely empirical; although he constructed the pure principles of mechanical science by starting with the empirical concept of matter in contemporary sciences, he deemed chemistry lacking in the necessary determinate principles and mathematical formulations required for

[41] Schelling (1976-, V, 69, 84–85).

[42] Schelling (1976-, VII, 275–80).

science proper.[43] In his 1790 *Critique of the power of judgment* Kant also contended that our concept of living organisms as natural purposes served only as a regulative principle for their investigation, and thus could not constitute a science. Schelling, developing his philosophy a decade later, enlisted new studies of chemical and organic processes to sketch a different model of science. He argued for a continuum between mechanical, chemical and organic phenomena, with the differences between them regarded as differences in degrees of activity and organization rather than differences in kind. Thus, mechanical bodies are open to the conceptualisation of chemistry, which examines how bodies apparently inert can become active under external stimulus. Similarly chemical bodies are open to organic conceptualization, which examines how bodies are capable of self-preservation under continuous external stimulus, and, conversely, organic bodies can become subject to chemical or mechanical conceptualization. It is a question of perspective and method of investigation. Kant had rested his analysis of the principles of mechanics on fundamental attractive and repulsive forces necessary to explain the apparent impenetrability of matter and interaction of material bodies. Schelling argued that even the purported fundamental forces of matter can be subject to further analysis. Rather than resting with determinate metaphysical principles, his philosophy of nature represented each material object as a stage in a becoming that recedes into infinity, beyond our conceptualization.

Schelling posited that the functions of life stand in connection with the general alterations of nature, and that a common principle must be sought for each natural product and nature as a whole. But he had difficulty representing this principle, often invoking figurative language. His 1798 *On the world soul* was framed by this problematic. Schelling was insistent that there is no fundamental transition or reorganization of natural processes in the movement between mechanism and organisms, but also concluded that the world was like an organism in organization, and thus a general organism the condition of mechanism. This work represented all of nature as alive through the play of opposed powers – a positive power or drive of unrestricted protean movement; and a negative power that restricts and guides that movement back into itself and encloses it in a natural product. To represent the organization of each product and nature as a whole, he invoked Kant's conception of organized beings as an "arrested current of causes and effects." He also invoked the antique poetic figure of a "world soul [*Weltseele*]," to indicate the principle for which language had no determinant expression – not to indicate a hyper-physical spirit behind the organization of the natural world, or divine source for the archetypes of creation, but quite the contrary to indicate that the dynamic of opposed powers made each natural product an embodied soul or vitalized matter. He was actually critical of his contemporaries who appealed to a special vital power, a *Lebenskraft*, as the purported cause of the organization and unique capacities of living organisms. He introduced opposed powers not as explanations, but as "boundary concepts [*Gränzebegriffe*]" for empirical natural science, the interplay of powers

[43] Kant (1902–1983, IV, 467–70).

conceived as having infinite possible degrees, evident in a series of natural products of varying degrees of organization.[44] Schelling's 1799 "Introduction" to his *First outlines of a system of the philosophy of nature* similarly represented nature as the mean factor arising out of pure productivity and its constraint. Each natural object was now figured as a whirlpool – not as a fixed product, but as a continual process of production, annihilation, and reproduction. Each object is a finite stage of an infinite becoming, a relative equilibrium of activities, of which none is absolute, an origin or end point, a simplest or final form; each product "never *is*, but only *becomes*."[45]

Living organisms, the varied forms of plants and animals, are particular stages in the infinite productivity of nature, if also the characteristic expression of the organism of nature. To study the unique characteristics of living organisms, Schelling introduced a series of boundary concepts marking the relationships between inorganic and organic phenomena, and between different stages of organic life.[46] He posited that each sphere of organic life has a rudimentary form of excitability that forms its boundary with the inorganic world. The individual organism preserves its own sphere of activity against the activities of the universal organism through excitability as a dynamic relationship between inner and outer activity. Excitability enables the individual organism to remain receptive to the stimulus of external material influences, but also to engage in the activity of resisting them. Here we have the boundary of life represented in terms much like those Treviranus outlined in his *Biology,* except Schelling did not simply posit a vital force as a purported explanation of this boundary, but tried to think what the concept of such a boundary required. Self-preservation requires that the individual organism assimilate or organize all for itself in order not to be assimilated or organized into the rest of nature; this activity then passes over into a continuous process of the production of form.[47] Schelling also conceived these formative activities through boundary concepts that marked the difference of organic matter from inorganic matter as well as their relationships. He appealed to Blumenbach's notion of a formative drive [*Bildungstrieb*] to express the capacity of organic matter to assimilate inorganic matter and to produce itself through nutrition, growth and reproduction. He also appealed to Kant's notion of the predispositions of generative matter, arguing that particular predispositions determining particular directions of the formative drive become fixed under the influence of external causes and passed on through reproduction to generate the distinct kinds of living forms.[48] In higher forms of life, excitability as the dynamic relationship between receptivity and activity becomes distinguished more evidently through organs of sensibility and irritability. In the history of life, both the life of individual organisms and the historical relationships of different organisms, Schelling argued different relative proportions of organic functions – excitability,

[44] Schelling (1976-, VI, 66–77, 61–82, 254, 257).

[45] Schelling (1976, VIII, 311).

[46] Schelling (1976-, VI, 82).

[47] Schelling (1976-, VIII, 117–33, 180–92).

[48] Schelling (1976-, VI, 203–24, 252–54, VIII, 44–48).

generative processes, irritability and sensibility – can be found in a graduated series of stages.[49]

Schelling thus advocated a comparative physiology that analyzed and reciprocally compared living organisms through the boundary concepts and functions correspondent to their stage of development. He rejected the idea that nature had an absolute archetype for all its members before its eyes, so that each organism could be represented as different approximations to this absolute through a comparative anatomy. He insisted no such absolute product exists; there is no absolute origin or end to which individual organisms might be compared.[50] Rather each product is but a relative stage of development or a relative proportion of activities or functions. In developing this conception of organic vitality, Schelling was influenced by disputes over vital principles and powers during the 1790s. But in his reinterpretation of these principles as boundary concepts rather than as metaphysical powers, the influence of critical philosophy on his work can be seen, and in his emphasis on the continuity of organic and inorganic processes, the influence of new investigations of the material conditions of life can also be seen.

Schelling's philosophy of nature in turn encouraged an emphasis on organic functions in German physiology and biology textbooks in the early nineteenth century. This influence can be seen particularly through physicians at the University of Würzburg, where Schelling was a visiting professor of philosophy from 1803 to 1806. Johann Joseph Dömling met Schelling in Jena during his travels after completing his studies in Würzburg in 1797 and before his return to Würzburg in 1799 as a professor of physiology. Although initially unsympathetic to Schelling's ideas, he based his 1802 *Textbook on human physiology* on the principles of Schelling's philosophy of nature, and through his textbook and lectures prepared students for Schelling's arrival the following year. Ignaz Döllinger, appointed professor of anatomy and physiology in Würzburg in 1803, had close contact with Schelling during his stay, and even attended Schelling's lectures. Döllinger produced a textbook to accompany his lectures on physiology in 1805, *Outline of the science of the human organism*, in terms introduced by Schelling, conceiving individual organisms as the product of a pure productivity conditioned by external and material factors. He argued that this representation of the organism resolved the conflict between those who attempted to explicate its existence wholly in terms of general physical laws and those who argued for an indwelling *Lebenskraft* that elevates the organism above physical laws. The body of Döllinger's text was divided into three main sections addressing the three primary forms of organic vitality – generative processes, irritability and sensibility.[51] Textbooks such as those of Dömling and Döllinger indicate a transformation from a physiology premised on anatomy to a physiology premised on function. Physiology in the later-eighteenth century continued to be conceived under the model of animated anatomy advocated by Haller, with

[49] Schelling (1976-, VIII, 180–230).

[50] Schelling (1976-, VIII, 112–13).

[51] Dömling (1802) and Döllinger (1805). Cf. Gerabek (1995, 238–50).

textbooks structured by anatomical parts, each part first described anatomically and then the function of that part elucidated. But attention to organic vitality contributed to a shift in attention to vital functions, which were explicitly represented as involving several organs as well as complex alterations of the form and composition of organic matter. Works such as those of Treviranus and Oken also contributed to this shift, by broadening the investigation of organic vitality from the focus on human physiology, as found in the works of physicians such as Dömling and Döllinger, to conceive a biology including a comparative study of organic vitality in different forms of organic life.

These tentative conceptions for a new approach to the study of living organisms are far from establishing a new discipline of biology based on an accepted set of concepts, practices and institutional allegiances, although they were important for the development of biology during the nineteenth century. The works of Döllinger and Treviranus presented a science of living organisms through the concepts of vital powers that dominated the polemic over a distinct domain for living nature in the early 1790s. But the concept of a biology founded on determinate laws and powers remained contentious in the years around 1800. It was not only a question of whether determinate powers could be found for organic phenomena, which Kant contested, but also whether a definite demarcation could be made between living and non-living processes, which Schelling contested. Indeed, new empirical investigations into the material conditions of vital functions – through galvanic and chemical experiments, and studies of generation and degeneration – increasingly blurred the boundaries between organic and inorganic phenomena. Yet such investigations turned from the debate about abstract principles and dead concepts towards the immediate interaction with vital processes, bringing organic vitality to life even as they complicated the notion of a distinct domain of life.

References

Bach, Thomas, and Olaf Breidbach. 2001. Die Lehre im Bereich der 'Naturwissenschaften' an der Universität Jena zwischen 1788 und 1807. *NTM* 9: 152–176.

Beiser, Frederick C. 2002. *German idealism: The struggle against subjectivism 1781–1801*. Cambridge, MA: Harvard University Press.

Bernasconi, Robert. 2006. Kant and Blumenbach's polyps: A neglected chapter in the history of the concept of race. In *The German invention of race*, ed. Sara Eigen and Mark Larrimore, 73–90. New York: State University of New York Press.

Bertucci, Paola, and Giuliano Pancaldi (eds.). 2001. *Electric bodies: Episodes in the history of medical electricity*. Bologna: CIS, Università di Bologna.

Breitenbach, Angela. 2006. Mechanical explanation of nature and its limits in Kant's *Critique of judgment*. *Studies in History and Philosophy of Biological and Biomedical Sciences* 37: 694–711.

Bresadola, Marco. 2008. Animal electricity at the end of the eighteenth century: The many facets of a great scientific controversy. *Journal of the History of the Neurosciences* 17: 8–32.

Broman, Thomas H. 1996. *The transformation of German academic medicine, 1750–1820*. Cambridge: Cambridge University Press.

Buchdahl, Gerd. 1992. *Kant and the dynamics of reason: Essays on the structure of Kant's philosophy*. Oxford: Blackwell.

Burdach, Karl Friedrich. 1800. *Propädeutik zu Studium der gesammten Heilkunst*. Leipzig: Breitkopf Härtel.

Carrier, Martin. 2004. Kant's mechanical determination of matter in the *Metaphysical foundations of natural science*. In *Kant and the exact sciences*, ed. P. Parrini, 117–135. Dordrecht: Kluwer.

Clark, William. 1997. German physics textbooks in the Goethezeit. *The British journal for the history of science* 35: 219–239, 295–363.

Cohen, Alix A. 2004. Kant's antinomy of reflective judgment: A re-evaluation. *Teorema* 23: 183–197.

Cunningham, Andrew. 2002. The pen and the sword: Recovering the disciplinary identity of physiology and anatomy before 1800. I: Old physiology – The pen. *Studies in History and Philosophy of Biological and Biomedical Sciences* 33: 631–665.

Cunningham, Andrew. 2003. The pen and the sword: Recovering the disciplinary identity of physiology and anatomy before 1800. II: Old anatomy – The sword. *Studies in History and Philosophy of Biological and Biomedical Sciences* 34: 51–76.

Daston, Lorraine, and Peter Galison. 2007. *Objectivity*. New York: Zone Books.

Döllinger, Ignaz. 1805. *Grundriß der Naturlehre des menschlichen Organismus*. Bamberg: Jos Anton Göbhardt.

Dömling, Johann Joseph. 1802. *Lehrbuch der Physiologie des Menschen*. Göttingen: Heinrich Dietrich.

Friedman, Michael. 1992. Causal laws and the foundations of natural science. In *The Cambridge companion to Kant*, ed. Paul Guyer, 161–199. Cambridge: Cambridge University Press.

Gerabek, Werner E. 1995. *Friedrich Wilhelm Joseph Schelling und die Medizin der Romantik*. New York: Peter Lang.

Ghiselin, Michael T. 2005. Lorenz Oken. In *Naturphilosophie nach Schelling*, ed. Thomas Bach and Olaf Breidbach. Stuttgart: Frommann-Holzboog.

Ginsborg, Hannah. 2001. Kant on understanding organisms as natural purposes. In *Kant and the sciences*, ed. Eric Watkins, 231–258. Oxford: Oxford University Press.

Girtanner, Christoph. 1791. Abhandlungen über die Irritabilität. *Journal der Physik* III, 2:317–351, III, 3:507–537.

Golinski, Jan. 2003. Chemistry. In *Eighteenth-century science*, ed. Roy Porter, *Cambridge History of Science*, vol. 4, 375–396. Cambridge: Cambridge University Press.

Hecker, August Friedrich. 1793. Girtanners neues System der Medicin. *Journal der Erfindungen, Theorien und Widersprüche in der Natur- und Arzneiwissenschaft* I, 1: 12–52.

Hufbauer, Karl. 1982. *The formation of the German chemical community*. Berkeley: University of California Press.

Hufeland, Christoph Wilhelm. 1795. *Ideen über Pathogenie und Einfluss der Lebenskraft auf Entstehung und Form der Krankenheiten als Einleitung zu pathologischen Vorlesungen*. Jena: Academischen Buchhandlung.

Humboldt, Alexander von. 1794. *Aphorismen aus der chemischen Physiologie der Pflanzen*. Trans. G.F.V. Waldheim. Leipzig: Voss.

Humboldt, Alexander von. 1795. Ueber die gereizte Muskelfaser, aus einem Briefe an Herrn Hofrath Blumenbach. *Neues Journal der Physik* II, 2:115–116.

Humboldt, Alexander. 1973. In *Die Jugendbriefe Alexander von Humboldts 1787–1799*, ed. Ilse Jahn and Fritz G. Lang. Berlin: Akademie.

Kant, Immanuel. 1902–1983. *Kants gesammelte Schriften*. Berlin: Königlichen Preuschen Akademie der Wissenschaften.

Kanz, Kai Torsten. 2006. ...die Biologie als die Krone oder der höchste Strebepunct aller Wissenschaft. *NTM* 14: 77–92.

Kielmeyer, Carl Friedrich. 1938. Ideen zu einer allgemeineren Geschichte und Theorie der Entwicklungserscheinungen der Organisationen. In *Gesammelte Schriften*, ed. F.-H. Holler. Berlin: W. Keiper.

Kielmeyer, Carl Friedrich. 1993. *Über die Verhältnisse der organischen Kräfte*, ed. Kai Torsten Kanz. Marburg an der Lahn: Basilisken -Presse.

Lenoir, Timothy. 1982. *The strategy of life: Teleology and mechanics in nineteenth-century German biology*. Chicago: University of Chicago Press.

Löw, Reinhard. 1979. *Pflanzenchemie zwischen Lavoisier und Liebig*. Munich: Donau.

McLaughlin, Peter. 1982. Blumenbach und der Bildungstrieb: Zum Verhältnis von epigenetischer Embryologie und typologischem Artbegriff. *Medizinhistorisches Journal* 17: 357–372.

Metzger, Johann Daniel. 1794. *Über Irritabilität und Sensibilität als Lebensprincipien in der organisirten Natur*. Königsberg: Hartung.

Moiso, Francesco. 1994. Magnetismus, Elektrizität, Galvanismus. In *Historisch-Kritische Ausgabe: Ergänzungsband zu Werke Band 5 bis 9. Wissenschaftshistorischer Bericht zu Schellings Naturphilosophischen Schriften. 1797–1800*, ed. Hans Michael Baumgartner, Wilhelm G. Jacobs, and Hermann Krings, 165–372. Stuttgart: Frommann-Holzboog.

Needham, John Tuberville. 1748. A summary of some late observations upon the generation, composition, and decomposition of animal and vegetable substances. *Philosophical Transactions* 45: 615–666.

Oken, Lorenz. 2007. *Gesammelte Werke*, ed. Thomas Bach, Olaf Breidbach, and Dietrich von Engelhardt. Weimar: Hermann Böhlaus.

Pera, Marcello. 1992. *The ambiguous frog: The Galvani-Volta controversy on animal electricity*. Princeton: Princeton University Press.

Pfeifer, Klaus. 2000. *Medizin der Goethe Zeit: Christoph Wilhelm Hufeland und die Heilkunst des 18. Jahrhunderts*. Weimar: Böhlau.

Piccolino, Marco. 2007. The taming of the electric ray: From a wonderful and dreadful 'art' to 'animal electricity' and 'electric battery'. In *Brain, mind and medicine: Essays in eighteenth–century neuroscience*, ed. Harry A. Whitaker, C.U.M. Smith, and Stanley Finger, 125–143. Dordrecht: Springer.

Pincaldi, Giuliano. 2003. *Volta: Science and culture in the age of enlightenment*. Princeton: Princeton University Press.

Quarfood, Marcel. 2004. *Transcendental idealism and the organism: Essays on Kant*. Stockholm: Almqvist & Wiksell International.

Reil, Johann Christian. 1795. Über die Lebenskraft. *Archiv für die Physiologie* 1: 8–162.

Richards, Robert J. 2002. *The romantic conception of life: Science in the age of Goethe*. Chicago: Chicago University Press.

Risse, Guenter B. 1971. *The history of John Brown's medical system in Germany during the years 1790–1806*. Doctoral dissertation, University of Chicago.

Roe, Shirley A. 1983. John Tuberville Needham and the generation of living organisms. *Isis* 74: 159–184.

Roe, Shirley A. 1984. *Anatomia animata*: The Newtonian physiology of Albrecht von Haller. In *Transformation and tradition in the sciences*, ed. E. Mendelsohn. London: Cambridge University Press.

Roose, Theodor August. 1797. *Grundzüge der Lehre von der Lebenskraft*. Braunschweig: Thomas.

Schäffer, Johann Ulrich Gottlob. 1793. *Über Sensibilität als Lebensprincip in organischen Natur*. Frankfurt am Mayn: Gebhard.

Schelling, Friedrich Wilhelm Joseph. 1976. In *Werke: Historisch-kritische Ausgabe*, ed. Hans Michael Baumgartener, Wilhelm G. Jacobs, and Hermann Krings. Stuttgart: Frommann-Holzboog.

Sloan, Phillip R. 2002. Preforming the categories: Eighteenth-century generation theory and the biological roots of Kant's a priori. *Journal of the History of Philosophy* 40: 229–253.

Smit, P. 1976. Treviranus, Gottfried Reinhold. In *Dictionary of scientific biography*, ed. Charles Coulson Gillespie XIII, 260–262. New York: C. Scribner.

Steigerwald, Joan. 2011. Natural purposes and the reflecting power of judgment: The problem of the organism in Kant's critical philosophy. In *Romanticism and modernity*, ed. T. Pfau and R. Mitchell, 29–46. New York: Routledge.

Steigerwald, Joan. 2013. The antinomy of the teleological power of judgment and its significance for the critical project. In *Objectivity after Kant: Its meaning, its limitations, its fateful omissions*, ed. B. Demarest and G. Van de Vijver, 83–97. Hildesheim: Olms.

Steinke, Hubert. 2005. *Irritating experiments: Haller's concept and the European controversy on irritability and sensibility, 1750–90*, The Wellcome series in the history of medicine, clio medica 76. Amsterdam: Rodopi.

Stuber, Martin, Stefan Hachler, and Luc Lienhard (eds.). 2005. *Hallers Netz: Ein europaischer Gelehrtenbriefwechsel zur Zeit der Aufklarung*, Studia Halleriana IX. Bern: Schwabe AG Verlag Basel.

Tränkle, Hans-Peter. 1986. *Der rümlich bekannte philosophe Arzt und politische Schriftsteller Hofrath Christoph Girtanner. Untersuchungen zu seiner Leben und Werk*. Medical dissertation, University of Tübingen.

Treviranus, Gottfried Reinhold. 1802–1822. *Biologie; oder, Philosophie der lebenden Natur für Naturforscher und Aerzte*. Göttingen: Johann Fridrich Röwer.

Trumpler, Maria. 1992. *Questioning nature: Experimental investigations of animal electricity in Germany, 1791–1810*. Doctoral dissertation, Yale University.

von Durner, Manfred. 1994. Theorien der Chemie. In *Historisch-Kritische Ausgabe: Ergänzungsband zu Werke Band 5 bis 9. Wissenschaftshistorischer Bericht zu Schellings Naturphilosophischen Schriften. 1797–1800*, ed. Hans Michael Baumgartner, Wilhelm G. Jacobs, and Hermann Krings, 3–161. Stuttgart: Frommann-Holzboog.

von Haller, Albrecht. 1751. *Primae linae physiologiae in usum praelectionen academicarum auctae et Emendatae*. Göttingen: A. Vandenhoeck.

von Humboldt, Alexander. 1797. *Versuche über die gereizte Muskel- und Nervenfaser, nebst Vermuthungen über den chemischen Prozess des Lebens in der Thier- und Pflanzenwelt*. 2 vols. Berlin: Rottman.

Zammito, John H. 1992. *The genesis of Kant's Critique of judgment*. Chicago: University of Chicago Press.

Ziche, Paul. 1998. Von der Naturgeschichte zur Naturwissenschaft: Die Naturwissenschaftern als eigenes Fachgebiet an der Universität Jena. *Berichte zur Wissenschaftsgeschichte* 21: 251–263.

Zuckert, Rachel. 2007. *Kant on biology and beauty: An interpretation of the critique of judgment*. Cambridge: Cambridge University Press.

Chapter 4
The "Novel of Medicine"

Juan Rigoli

Abstract In early nineteenth-century France, as physiology became a conquering science which laid claim to exclusively describing the entirety of the human experience, it sought to denounce what people had seen in it and what it insisted it no longer was: the "novel of medicine." This expression captures that which was obscure in physiology, its commerce with the invisible and unseen, its compromises with fiction and narrative, everything that the new order imposed by the "observational" paradigm must imperatively transcend. Also stigmatized were the "imaginary entities" and "unintelligible words" of vitalism, to say nothing of the "fantastic anatomy" upon which the physiology of the philosophers relied. If the divide between knowing and "not-knowing" is thus always clearly marked, the schisms that result are not without tensions and conflicts in the field of physiology, which is far from homogeneous. This article reflects on the rejection of the "novel" by physiology and – ranging across a broad range of medical and literary texts – on the links that endured between the two.

Keywords Balzac • Flaubert • Literature and medicine • Nineteenth-Century France • Physiology • Physiologies (literary genre) • Zola

The manifold situation of physiology in nineteenth-century France is not merely an effect of the gradual orientations of a discipline, as would be the case in a teleological history, according to which physiology would abandon obscure doctrinal models in

---

An earlier, shorter version of this chapter appeared in French in Andrea Carlino and Alexandre Wenger, eds., *Littérature et medicine, 1500–1900* (Geneva: Droz, « Recherches et Rencontres », 2007), 199–226.

J. Rigoli (✉)
Département de langue et de littérature françaises modernes, Université de Genève, 5, rue De-Candolle, 1211 Genève 4, Switzerland
e-mail: Juan.Rigoli@unige.ch

S. Normandin and C.T. Wolfe (eds.), *Vitalism and the Scientific Image in Post-Enlightenment Life Science, 1800–2010*, History, Philosophy and Theory of the Life Sciences 2, DOI 10.1007/978-94-007-2445-7_4,

order to become a modern, experimentally founded science. As is well known, this process is actually not so linear, and however much the old order is banished – with its 'hypotheses' and 'reasonings', its 'abstractions' and 'metaphysics', all of which are guilty of having no place in a science of 'facts' – it nevertheless lives on in some forms, with occasional resurgences. The decline and occasional revival of the 'vitalist' current (itself quite heterogeneous) durably sustained the tensions that traversed physiology and rendered its unification more difficult, against the background of the opposition between materialism and spiritualism. At either end of the century, the two great works which serve as monuments to the glory of the 'medical sciences' – Panckoucke's and Dechambre's – and whose differences bear witness to medicine's transformations and reorientations, do concur on this point, despite a long controversy to which they allude:

> It is hardly allowable in this age to not be a vitalist; the progress of the medical sciences has turned us back to this belief, by showing us the emptiness of other opinions, and the power of the forces of life.[1]
>
> Every physician ..., even if he is one of those who assert that at present, or instead as a hope for the future, all vital phenomena will one day be reducible to physicochemical laws, is or becomes a de facto vitalist, once he moves from pure speculation to practice – a vitalist without knowing it, perhaps, or *malgré lui*, and reluctantly, ... but a vitalist nevertheless.[2]

Physiological debates in fact preserve older elements, including philosophical overtones and echoes, and the broader resonance of these debates, as physiology increasingly presents itself as a triumphant discipline, make it a cultural 'fact' well beyond the medical sphere. From moral philosophy to psychology, from the understanding of social facts to literature, "Physiology, this beautiful science which guides man towards self-knowledge,"[3] everywhere extends its scope, in an overarching, unifying ambition which actually generates a number of physiological variations – appropriations, adaptations, accommodations – in each of the areas it invests and which respond to it.[4] Further layers are added to physiology's medical context – extensions it pursued in earnest, or warped reflections that become competitors for its own object: "Life presents itself to the serious mind only through the phenomena and properties particular to the bodies that possess it."[5] But the fact is, this "seriousness" is non-exclusive, and various 'physiologies' multiply in the nineteenth century, defining their object as the study of 'life', whether its principle or its effects, and carried out in varied metaphorical ways, whether medical, philosophical or literary.

---

[1]Mérat "Vitalistes", in Alard, Alibert et al. (1812–1822, LVII, 281).

[2]Brochin (1889, 726–727).

[3]Renauldin (1812, CXXX).

[4]On the literary impact of physiology as the new and henceforth only legitimate "science of oneself" in this period, see Rigoli (2006).

[5]Rullier (1828, 320).

1 The Physiological Obsession

"Physiology, what do you want from me?" Following the famous model of Fontenelle's annoyance invoked by Rousseau or Diderot – "Sonata, what do you want from me?" – the question is repeated like a refrain in the first pages of Balzac's *The Physiology of Marriage*.[6] It is so insistent that, later, when we read "Physiology, for the third and last time, what do you want from me?"[7], the apostrophe occurs for the fourth time. The "author" thus assumes – without counting – the casual and reproachful position of a suspicious "reader" who is uneasy about a fate that could be his, and distrusts, beyond the title, the mode of knowledge claimed in what he is reading, along with the profusion of contradictory opinions that the subject cannot help but suggest. The announced 'physiology' might as well be the response to this confusion of sounds Rousseau detested so much:

> In order to understand what this jumble of Sonatas that assaults us means, we should do like that unrefined painter who was forced to write underneath his images: *this is a tree, this is a man, this is a horse*.[8]

What would Balzac possibly have to "write underneath" his title for his book to have a distinct profile! In any case the "reader" enumerates the clarifications he expects, amplifying with his own questions – always the same one, barely varied – in a long series of disparate and trivial hypotheses, what the *Physiology of Marriage* could possibly teach him that he does not already know. All the disenchanted commonplaces of the union of the sexes, all the antagonistic certainties of its bodily, moral, economic, juridical and political conditions and consequences are reviewed in a kind of inventory of the already-said which the text must confront: "Do you reveal new principles?"[9], asks the reticent reader Balzac invents for himself.

The demand for something new and the risk of repetition – "Here everything is as trite as cobblestones in the street, as ordinary as a crossroads"[10] – are thus immediately inscribed under "physiology." But it must be understood, and Balzac does not leave us any doubt, that this 'physiology' is not the one we believe it to be, even if the title has been able to produce such an illusion: "Legions of doctors," the narrator comments, "produced legions of books on the relation between marriage, medicine and surgery,"[11] and they are not the only ones; but he proudly specifies, in

---

[6]Rousseau (1995, V, 1060). Annoyed by Boucher's painting, Diderot introduces a variant: "Painting, what do you want from me?" (Diderot, "Salon de 1765," in Diderot 1994–1997, IV, 310).

[7]Balzac, *Physiologie du mariage* [1826–1829], in Balzac (1976–1979, XI, 913–916).

[8]Rousseau (1995, V, 1060).

[9]Balzac (1976–1979, 914).

[10]Ibid., 916.

[11]Ibid.

the chapter on "neuroses," that "our Physiology has the most superb contempt for medical classifications."[12] Medicine, seen through Balzac's *Physiology*, is kept at bay, though made present by the power of refusal.

The title of the work could have been indeed different: *study, code, art,* or *monograph* are all available categories that Balzac could have used in this case, all of them largely represented in the bookselling market. René Guise's notes, in his edition of the *Human Comedy* – the *Physiology* is part of the *Analytic Studies* – inform us of the remarks the title provoked after its publication. According to Jules Janin, in *Le Journal des Débats* of February 1830, it could have been called "*The Art of Marriage*," but "given that our century furiously threw itself into medicine and pompously scientific titles, he called his book *Physiology of Marriage*."[13] And again, the following month, in the *Feuilleton des journaux politiques*:

> This word 'physiology' is not as terrible as it sounds. In medicine, it means nothing but the art of knowing the destination of organs, of discovering it as needed, according to the system into which nature has made them. Therefore the physiology of marriage will be the art of finding out the secret of the actions of husbands, and above all those of their wives, according to external appearances, the art of giving a meaning to each of their moves, and of avoiding unfortunate consequences – if ever there is the possibility of protecting oneself from them.[14]

Here we have the methods of medical physiology, at least in the way the writer understands and summarizes them, transposed to marital relations: the "actions of husbands" and "those of their wives" become "organs" with secret functions that one must "find out" for lack of "knowing" them; and if the physiology of doctors extends to therapeutics and hygiene, Balzac's physiology is not less destined to help "protect oneself" from the "unfortunate consequences" of marriage if it is within one's reach. It is a dubious analogy, but one which could be formulated, with a casualness that only slightly diminishes its seriousness.

Balzac's "Physiological Muse"[15] – this is how he names the authority under which he places his book – could be referring to recent models that prepared for the title's acceptance. Alibert's *Physiologie des passions*, published in 1825, was reprinted 2 years later; Brillat-Savarin's *Physiologie du goût* was also published in 1825, 4 years before Balzac's *Physiology* (which Balzac declares had been finished and printed since 1826, but for personal purposes only). Moreover, all three are contemporaneous with some medical physiologies that caused great impact, and also with various standard texts such as Adelon's *Physiologie de l'homme* (1823–1824), reprinted in 1829, Magendie's *Précis élémentaire de physiologie* (1816–1817), reprinted in 1825, and the famous *Nouveaux élémens de physiologie* (1801) by Richerand, which, by 1825, had been reprinted nine times. Whether or not it is justified by a legitimate heritage, the title is insistent, and the corpus diverse. Some physiologists took pride in the discipline's sensation: "Each science has its time of

[12] Ibid., 1166.

[13] Ibid., 1768.

[14] Ibid.

[15] Ibid., 1076.

glory, progress, and maturity; each becomes successively popular. Every century has its favorite science, and ours is physiology."[16] But none of them seemed to realize that this "popular" triumph supposes dispossession and deviation.

The sensational success of Alibert's, Brillat-Savarin's, and Balzac's new "physiologies" was nothing, after all, but the sign of a literary fashion to come between 1841 and 1843, both ephemeral and pervasive, and whose impact can be measured, all "physiologies" combined, in their circulation of more than half a million copies.[17] There were "physiologies" of everything, and especially of everyone, in a venture as homogenous in its form as can be, but very multiple in its subjects, sustained by a typology of the social that threatens – and that is simultaneously its risk and its propeller – to never come to an end. They follow, as we know, in a vertiginous enumeration of roles, places, and objects: the physiologies of the *rentier*, the *concierge*, the employee, the dandy, the *lorette*, the housekeeper, the smoker, the Bois de Boulogne, the theater, the joker, the cuckold, the honest woman, the nameless girl, the doctor, the legislator, the pear, the freemason, the *grisette*, the drinker, the *quartiers* of Paris, the umbrella, etc.; and sometimes they spread out their totalizing ambition in a single title, such as the *Physiologie du bien et du mal, de la vie et de la mort, du passé, du présent et du futur* (1837).

As we should expect, the physiologies above were followed in 1841 by a *Physiologie des physiologies* and a *Physiologie des physiologistes*, conveying, in their reflexive nature, both an enthusiasm and an impasse (the very year the phenomenon reaches its peak): the physiologies hardly stage anything other than their own "type," thus revealing, critically and playfully, the principles of a fabrication whose recipe each of them had already mastered. The anonymous *Physiologie des physiologies* summarizes them all in a very well-known definition:

> Physiology – this word is composed of two Greek words, whose meaning is henceforth the following: an in-18 volume, composed of 124 pages and of an unlimited number of vignettes, tailpieces, foolish remarks and chitchat (*logos*), for the benefit of those who are ignorant of their nature (*physis*)[18].

This is a comical echo of an etymology that medical physiology takes advantage of simply and nobly: "The word *physiology*," Pierre Bérard carefully explains (and he is not the only physiologist to do so), "derives from two Greek words, one meaning *nature*, and the other *to tell*; it is as though we were saying *history of nature*"[19]; and also of calling to mind the vastness of the notion of "nature" – simultaneously *natura naturata* and *natura naturans*, "things created" and "creating power" – to draw the following consequences: "As you see, even if we were to maintain etymological rigor, physiology would still be nothing less than the universal science."[20] Bérard is quick to limit the definition to the study of the "phenomena of living beings"; to the "study" of

[16] Bourdon (1830, VII).

[17] See L'Héritier et al. (1957) and Preiss (1999).

[18] *Physiologie des physiologies*, (1841, 43); cited by Preiss-Basset (1993, 62).

[19] Bérard (1848–1855, I, 1).

[20] Ibid., 2.

their "laws" and "conditions in a state of health."[21] Though limited, the depth and vastness of the task, as well as the deliberately philosophical reach it requires, do not completely silence etymology's voice, which promises physiology a most noble destiny. Jourdan, who translated Müller's great *Handbuch der Physiologie* in 1845, reminds his readers that "nowadays everyone agrees to see it as one of the most important branches of medicine, of general history, and of philosophy."[22]

The various and minute "physiologies" which booksellers profit from could then be understood as the degraded and deviated version of a triumphant knowledge whose success is of such importance, and aims at such a vast reach, that we notice in them a new explanatory order, extended to the whole human – moral and social dimensions included. It is evidently as a parody that such a relationship is established, and if the "physiologies" are in a dialogue with medicine, it is not only by usurping their name and title, but also by playing on a discordance between their smallness and the monumentality of medical works. This is what an exchange of letters stages in place of a prologue to the *Physiologie de la poire* (1832). First the letter of a "big publisher of the *quartier* of the School of Medicine," addressed to the "author of the present treatise," in which the former informs the latter of his disappointment in only receiving from him an insignificant volume instead of the colossal treatise he felt entitled to expect:

> I am a learned publisher of learned works ....
>
> I display at the windows of my spacious store Boyer, Broussais, Thénard, Alibert, Magendie, Richerand, Laënnec, Dupuytren, Récamier, Ampère, Orfila.
>
> I would have displayed your *Physiology of the Pear* with great pleasure if your *Physiology of the Pear* had been more than a meager in-8° volume.
>
> But I should let you know, sir, that I do not like to budge an inch for a single volume; I was born for great things, for gigantic Atlases, for immense Dictionaries, for never-ending Encyclopedias. – I am the Bonaparte of the French bookstores.[23]

The great names of science, and particularly of medicine, who the "learned publisher of learned works" invokes are in fact merely the crushing counterweight that physiologies assume for themselves, as well as the foil of casualness and futility they claim for themselves.

But the literary physiologies maintain with the physiology of doctors a deeper relationship, though always strictly derisive. In fact, physiologies anarchically propose the study or the spectacle of a *social body*, and the project that supports them – if there is one – is marked by the way in which medical physiology conceives, in a series of oppositions with anatomy, the *dynamic* of the living body. This is what Béraud does, not without a didactic effort, in his *Éléments de la physiologie de l'homme*: "the notion of *vital* or *elementary property* corresponds to the notion of *organized substance* or *elements of anatomy*"; "the notion of *attribute* in physiology is linked to that of *system* in anatomy"; "anatomy shows us *organs*," and we "call *use* each particular act of each organ"; "the notion of *function* corresponds to that

[21] Ibid., 3.

[22] "Translator's Preface," in Müller (1851, I, xxv).

[23] "Un gros éditeur du quartier de l'École-de-Médecine, – à l'auteur du présent Traité," in Peytel (1832, IX–XI), in Preiss, ed. (1999).

of *apparatus*"; "the notion of *life* or mode of activity is linked to the notion of *organism*, *economy* and, broadly considered, *body*."[24] This is a lesson that literary physiologies completely assimilated. For in their case, once the transposition from the bodily to the social was acknowledged, the question was to indicate or pinpoint *attributes* and *uses*, recognize *functions*, explain the facts of social "life," capture the vital bustling of crowds in the spectacle of big cities. This is true even if "physiologies" as a genre of "industrial literature" (Sainte-Beuve) develop, with no order and by means of an uncontrolled accumulation, a sort of physiology with no anatomical predecessor or basis, discovering or inventing both the functions and the organs of an expanding social body, whose number of parts no one can anticipate.

If it is thus certainly true that "physiologies" usurp their title, just like numerous "anatomies" were able to do before them, and they are nonetheless much more than what Pierre Larousse defines as the "studies of the customs decorated with the name of *physiologies*,"[25] and very different from what Bouillet believes in when he affirms that "nowadays" physiology has "become a synonym for analytical treatise."[26] For the physiological model works both as the ephemeral genre it designates and as its innumerable occurrences in other texts which welcome, beyond the word itself, at least a share of the mode of thought that physiology relates to. It is true, however, that the authors of physiologies "deliberately" adopt "the classificatory method of zoology *against* the synthetic method of physiology."[27] This confirms the physiological vulgate of the *Larousse* in its medical section:

> Physiology, as an experimental science, is clearly distinguished from zoology and botany, which are natural sciences. Physiology does not try to deduce from its studies any classificatory character; it completely neglects any considerations of class, order, gender, and species, which are the essential object of naturalist, zoological, and phytological studies. For general physiology there are only infinitely diverse vital mechanisms, accomplished with the aid of active common elements. The object of this science is to determine these common elements, with their properties and in their conditions of activity.[28]

Nothing would be further from physiology than the taxonomic data collection of the social that "physiologies" set out to accomplish. But the physiology they agree with is not only this "experimental science" whose principle Pierre Larousse assimilates when referring to the then-recent work of Claude Bernard. The theoretical body of physiology and the modalities of its discourse are indeed much more heterogeneous, and do not categorically turn their back on taxonomic order.

Moreover, under close examination, the classificatory method adopted by the "physiologies" does not lead to a new 'physiology' classified by species: the principles that govern each of them in their presentation of social phenomena are the same, and these phenomena are observed in each "organ" of the social body, according to the principles of a "special or descriptive physiology," still, as Béraud attests,

[24] Béraud (1856–1857, I, 2–3).

[25] Larousse (1874, 918).

[26] Bouillet (1854/1867, 1275).

[27] Preiss (1997, 912).

[28] Larousse (1874, 915).

"that has as a subject the acts whose examination of every part of the organism must be done successively, and whose goal is to find out their connection and mutual dependence."[29] And even when the division of the social body by the "physiologies" conveys the burlesque design of a human zoology, this move is inscribed irreverently in an extension of a physiological specialization: ". . . we conceive that there must be as many special physiologies as particular living species; each one has its own; and since man is a living being, there must be the *physiology of man*."[30] The "physiologies" do nothing but give a supplementary turn of the screw to the physiological machine as well as excessively tune the distinctions amongst "particular living species," including the lorette and the cuckold, the employee and the dandy, and other inhabitants of a social menagerie whose range and variety expand on a daily basis.

From the human body to the social body, from species to sub-species, the transposition may lack rigor, and only sparse lexical traits recall physiology in "physiologies," but that does not mean that their project does not reproduce – or imitate – in many respects the medical model. Moreover, Paul Bourget perpetuated this ambiguity in his *Physiologie de l'amour moderne* – "bearing a grand title due to naïve literary snobbery and the memory of an old, outdated genre" – by affirming that he worked "as a genuine literary physiologist with more or less justified pretensions to the title of scientific physiologist."[31]

2 The Life of the Social Body

There is another version of this cultural expansion of physiology by means of homology, which is non-parodic and definitely more serious: that of Saint-Simon, for whom "physiology is . . . the science not only of the individual life, but also of the social life, of which the lives of individuals are nothing but the wheels."[32] Naturally, the call for medical physiology to expand followed:

> Enriched by all the facts that have been discovered by precious works carried out in these different directions, general physiology can dedicate itself to considerations of a higher order; it hovers over individuals who are, for it, nothing more than organs of the social body whose organic functions it must study, the way special physiology studies those of individuals.[33]

A continuity of terms, as well as a dislocation, clearly indicates a power relation: the "general physiology" of doctors should acknowledge another above itself, and thus the entire discipline of medicine is lowered to the level of a "special physiology."

[29] Béraud (1856–1857, I, 4).

[30] Adelon, "Physiologie," in Adelon et al. (1821–1828, XVI (1826), 486).

[31] Bourget (1891, III and 132).

[32] Saint-Simon, *De la physiologie appliquée à l'amélioration des institutions sociales*, in Saint-Simon (1865–1878, XXXIX, 180).

[33] Ibid., 177.

This surpassing is latent in the relation that medical physiology itself establishes with hygiene; in a tension between the individual and the collective; it is also fundamentally inscribed in the "education" and the "law of perfectibility" of "sensations" and "ideas," whose echoes are heard weakly in the physiology of Magendie, but very significantly in others, such as Béraud, who himself refers to Cabanis: "the causes which enable the intellectual and moral faculties to develop are indissolubly linked to those that produce, conserve and set forth organization, and it is in the organization of the human race itself that the principle of its perfectibility can be found."[34] The formerly Saint-Simonian Buchez – Morel's friend who co-authored the *Précis élémentaire d'hygiène* with Ulysse Trélat in 1825 – develops this idea of a "social physiology" in an important series of articles on physiology in *Le Producteur* (1826):

> Physiology, as a science of man, is called to provide a positive base to individual morality, and, as a science of the conditions of existence, to organize and supervise public hygiene; this double relation will make it useful for all times.[35]

All this provided that, he adds, a "science of the species" is able to erect itself on suitable foundations, for "society is not only the expression of individual tendencies" and "the species is subjected to particular laws that are different from those of physiology."[36] Buchez's entire work is thus a search for "the terms of the transition from individual physiology to social physiology," against a background of certainty with regard to the obsoleteness of the division between psychology and physiology: "Cabanis and Destutt de Tracy trace manifest phenomena to hidden ones; thus physiology takes over psychology, and the science of man becomes one."[37] This more fundamental level supports the medical approach to moral facts, but the goals it incorporates evidently go beyond medicine, even if this remains its starting-point; furthermore, physiology partly carries out the philosophical ambition of medicine (but running contrary to it).[38]

Medical physiology thus finds itself dispossessed of its object and encompassed by philosophy, in the name of a mode of knowledge that it in fact contests, when, without targeting Saint-Simonian physiology (which was never considered an adversary), it disavows the *abstraction* of philosophical physiology, which is willingly vitalistic. And thus Béraud affirms again, but for that matter without denying the name "physiology" to this branch of science, whose "manner" distances itself from medicine:

> We have sometimes, but erroneously, given the name *general physiology* to *abstract* physiology, that is, the one which, without having any specific living species as a point of departure, touches upon the phenomena of life philosophically, independently of any application.[39]

---

[34] Béraud (1856–1857, II, 828).

[35] Buchez (1826, 132).

[36] Ibid., 132–133.

[37] Ibid., 274.

[38] Cf. Jacyna (1987).

[39] Béraud (1856–1857, I, 5).

The relation between medical physiology and social physiology becomes evident. And if the literary physiologies seem to distance themselves from social physiology to the point of becoming strangers to it; if their dispersal turns them into a "shattered mirror" of society under Louis-Philippe, their common recipe does not fail to emphasize, if not a cohesive project, then at least a clearly oriented movement, parallel to Saint-Simonian physiology and founded, not in spite of, but *because of* their parodic casualness, on the exaltation of social norms. Indeed, these literary physiologies particularly highlight the behavioral missteps of the species they carve out from human beings. Literary physiologies, as both close and turbulent cousins of hygiene, signal the multiple gaps, the imbalances of social health, and turn them into objects of mockery and enjoyment.

3 The Body of Thought

From the early nineteenth century onwards, 'physiology' benefits from an extraordinary semantic plasticity that allows it to function as a banner of decidedly heterogeneous works without ever completely losing sight of the medical flourishing that supports it. This begins with Alibert, whose *Physiologie des passions ou nouvelle doctrine des sentiments moraux* is the very example of a generic diversity in its dullest and most diluted version. It is the work of a doctor, notes the *Dictionnaire encyclopédique des sciences médicales*, who, "though dedicating himself to scientific research truly responsible for his glory, ... was unable to break free from literary tendencies to which he was drawn in his youth;"[40] hence his *Physiologie des passions* also "gave him, maybe more in the world than among doctors, a great reputation as a writer."[41] Alibert presents his project with his usual emphasis, and also in continuity with the subordination of the moral to medicine:

> To study moral feelings is to study man in the most precious and the noblest attributes of his being. What other science is more worthy of human spirit! Aren't doctors the ones who should particularly dedicate to it? Little can we imagine how much a profound knowledge of our physical infirmities could open up routes leading to the true theory of the passions. Descartes meditated on the organization of the living body, but imperfectly. He barely had the physiological data that had been acquired in his time. It thus undoubtedly follows that the majority of his explanation is generally considered defective and insufficient . . . .
>
> In spite of the infinite appeal of such studies, few people pay attention to them. Man avoids observing himself. Could it be that he is afraid of knowing himself? And yet we are saddened to pass away without piercing the darkness of ignorance, without penetrating the wonders of spirit, without diving into the depth of soul, without going back to the primitive source of sensations and ideas, without explaining the secret of our own emotions, without

[40] Émile Beaugrand, "Alibert (Jean-Louis)," in Raige-Delorme et al. (1864–1889, III (1869), 8).
[41] Ibid.

> having applied our faculties to this immense study of the nature of the intellect, which encompass the highest meditations of speculative philosophy; without having lifted some of the veils that still cover the great enigma of existence. Socrates was right in envisaging this science as the most worthy of occupying our reason with . . .[42]

There is no higher "science" than the one that leads to self-knowledge; what had been left to "speculative philosophy" must now return to "doctors," whose knowledge has been enriched at last by an understanding of certain "physiological data." There is nothing in this project that had not been already stated many times before Alibert decided to make himself its public defender. It should be added that the status of "physiology" in his discourse is purely metaphorical; his book was hardly nourished by medicine, and conveyed a considerable openness to common wisdom and inspiring stories: the study of the "moral" is here reduced to the expression of a received morality. The first sentence of the treatise could not be more opposed to its title: "In order to know man," declares Alibert with dualist enthusiasm, "we must seek him in his soul, and not in the material organs of his bodily envelope." It is medical spiritualism that decidedly turns its back on medicine.

A dermatologist converted into a moralist (for his inability to "break free from literary tendencies to which he was drawn in his youth"), Alibert's is certainly "one of the most original physiognomies of contemporary medicine"[43]; but we would be wrong to understand the extreme latitude with which he considers "physiology" as the sign of a radical exception. The physiological metaphor will durably be medicine's lot, and also notably that of psychiatry, in its distress before a slippery "moral" etiology: various projects of different statures and natures present themselves under the same heading, from Frédéric-Joseph Bérard's *Doctrine des rapports du physique et du moral, pour servir de fondement à la physiologie dite intellectuelle et à la métaphysique* (1823), to Scipion Pinel's *Physiologie de l'homme, appliquée à l'analyse de l'homme social* (1833), to Lélut's *Physiologie de la pensée. Recherche critique des rapports du corps à l'esprit* (1862), or even to Maudsley's *Physiology and Pathology of Mind* (1867); all of them tend to dissertate on the extension of "physiology," consolidate a profession of faith, and build up the utopia of an impending dawn during which the noble discipline – without adopting, for that matter, Cabanis's "certainty," also pointed out by Maudsley, "that the brain digests in a certain way our impressions; that it organically produces the secretion of thought"[44] – would finally reach a complete physiological understanding of the phenomena of thought.

And even when it does not provide them with a title, "physiology" is nonetheless present in the heart of the discourses on the relations between the "physical" and the "moral," but with different degrees of metaphor. This does not mean that there are not rhetorical contortions or even the upheaval of announced "physiological"

[42] Alibert (1827, I, 1–2).

[43] Beaugrand, in Raige-Delorme et al. (1864–1889, 7).

[44] Cabanis (1802/1805, I, 154) and Maudsley (1870/1891, 71).

projects, as in Lélut's reorientation of his study of the "organic conditions of the exercise of thought" and modification of the initial plan of his book:

> In this kind of suggested correlation and in a kind of equation of thought and its details with the body and its organs, instead of putting them in the foreground, of making them the hinges of these studies, I gave this first place to the subject itself, as it was required, I should say, I gave this first place to the mind and the various groups of its faculties. I did this substitution . . . .[45]

I should add that the philosophical and anatomico-physiological parts of his treatise never establish a dialogue with one another. The "physiology of mind" is thus not a physiology, unless we consider that it could practically result from the simple juxtaposition of a physiology of the body and a philosophy of the mind.

Serious or casual, encased in medicine or decidedly medical, Alibert's, Bérard's, Lélut's and Scipion Pinel's attempts sufficiently show, in any case, how much the force of physiology – the *wish* that leads it to the understanding of everything: "Physiology, what do you want from me?" – can be urgently felt in the most comprehensive domain of knowledge and culture: from medicine to philosophy and literature, in as many back-and-forth movements as transversal paths, it was at their crossroads that the "physiologies" of Brillat-Savarin and Balzac certainly found their place.

4 The Style of Physiology

The literary inclinations of certain doctors, physiologists by metaphor, promised to open the doors of success to them. Drawing out a sinuous path between "true" medical "glory" and "a reputation as a writer," Alibert engages in authorship with a pleasure inversely proportional to the boredom that his book provokes in us nowadays. But he also pays the price, for he is guilty of an abuse that medicine forbids which even the *Dictionnaire encyclopédique* ferociously denounces:

> First and foremost *l'homme d'esprit*, – but superficially – he seems, in general, to be more worried about the shine of form than the solidity of content. His polished and correct style generally covers up the pretentious and emphatic aspects so much in vogue at the beginning of this century. Thus these pathological descriptions, sometimes strikingly true, sometimes imbued with an exaggeration that borders on ridicule.[46]

Armed with both literary and medical criteria, the condemnation is indisputable, inscribed in the continuity of a norm with regards to "taste" and "style," to which "medical literature" submits; Philippe Pinel had already stated its necessity in the same "beginning of this century": medical texts must be the fruit born by the supervised relationship between the singular and the general, and distinctly bear the seal

[45] Lélut (1862, xx).

[46] Beaugrand, in Raige-Delorme et al. (1864–1889, 8).

of a discipline that demands all at once the precision of the gaze and the sobriety of discourse.[47] Alibert's book, therefore, doubly appears as a book of a different time: the offspring of an outdated "fashion" which the norms in vogue had already condemned, although visibly without success.

A similar break, regarding both language and time, is highlighted in the same *Dictionary*, but much more kindly, concerning the great Richerand, who already acquired impressive medico-literary stature in 1801, a year during which:

> ...this twenty-one-year-old author published his *New Elements of Physiology*, a book destined to one of those unprecedented successes that mark an epoch in the annals of the bookselling market, a book that had eleven printings almost one year after another, a book translated seventeen times, a book which greatly influenced the youths in schools, and which made the study of a science regarded as mostly severe and arid promising and easy; at last, it was a very seductive read, sprinkled with flowers along the most arid paths, letting out a perfume of literature in all of its pages, and which one gladly carried under one's arm in order to open, leaf through, and read during a stroll, smelling spring's morning scents.[48]

An extraordinary harmony here between "nature," the physiologist, and "literature," in which the origin and the ancient use of the word *physiology* resounds very well; bygone days giving way to an era when undoubtedly no one expects the budding of "flowers" in between the pages of a physiology book. The reminiscence of this past time, tinged with nostalgia, rests on the (at once happy and miserable) certainty that this ancient happiness will not happen again, since physiology now "lets out" a completely different "perfume."

The remoteness of this "seductive reading," both obsolete and acclaimed, evidently has little to do with the individual stylistic qualities of Richerand and his successors, and more with the profound mutations of physiology, as well as the long purification of its language, which, without being linear, is not in the least irreversible. It suffices to confront Richerand and Magendie to assess the measure of a revolution of physiological style and thought. Richerand still avowedly belongs to a culture from which his heirs will have the self-imposed duty of escaping. The preface to the first edition of his treatise gives it special attention:

> With regards to the spirit in which these *New Elements* are written, I have constantly sacrificed elegance for the sake of clarity, well convinced that the latter is the first merit of an elementary book. Moreover, I think I have observed at all places the same order in the succession of objects and applied to the science of the living man the principle of the natural connection of ideas so well developed by Condillac, in his *Treatise on the Art of Writing*, to which this philosopher made evident that all the rules of this art refer. Despite the rigor I have imposed on myself, I believe I have employed metaphorical expressions as needed, following the example of the ancients, of Bordeu, and of many other doctors and physiologists not less preeminent among the moderns, because concision, as a woman of our times who renders the highest honors to her sex has said, is not the art of reducing the number of words, and it is even less that of the removal of images. The concision one must wish for is

[47] Cf. Pinel, "Goût," in Alard, Alibert et al. (1812–1822), XIX (1817), 60–66, and Pinel and Bricheteau, "Littérature médicale," in Ibid., XXVIII (1818), 474–488.

[48] Achille Chéreau, "Richerand (Anthelme-Balthazar)," in Raige-Delorme et al. (1864–1889, V, 21).

that of Tacitus, which is all at once eloquent and vigorous; and images, far from damaging this brevity of style rightly admired, are the figural expressions that retell the most thoughts with the fewest terms. [49,50]

For Richerand, situating the discourse of science and determining its impact amount to a weighing of "rigor" and "metaphor," of the medical and the literary: this is what justified the presence of Tacitus and Madame de Staël in a medical work.

But this does not stop Richerand from calling for a strict control of the words and objects he allows himself to use in his treatise; for him, this means a division between science and the "novel" – and here the word means a mixture of fiction, culpable narrative and abuse of style – which will be long present in the spirit of physiologists, even when they reach the certainty that their language is no longer compromised by the literary. Richerand himself seems to be worried about it when he announces the disappointment his book will necessarily cause in "those who still insist on seeing nothing but the novel in physiology, instead of the history of animal economy"[51]; and Béraud, in the mid-century, will be just as careful with regard to the appetite of a "bookselling market" curious about physiology and "always avid":

By filling our two volumes with anatomical and physiological particularities that are curious, albeit useless, with facts borrowed from anatomy with more or less ingenious deductions and various notions about the natural history of humankind, we could have written one of those books that captivate audiences and for which the bookselling market is always avid. But it is a false opinion that physiology is composed primarily of curious data and seductive hypothetical deductions, and that it requires, to be done and understood, more imagination than labor.[52]

Jacques Lordat, an avowed vitalist and adept at metaphor, has the same reproach, in the same terms, as he contemptuously recalls the time of iatromechanical and iatrochemical doctrines:

This method of philosophizing, accredited mainly by Descartes, took over Physiology: its hypotheses were taken from physics and chemistry and were made believable by means of a fantastical anatomy.

But there always were some austere practitioners who professed little regard for such a futile science incessantly in opposition with thousands of facts ignored by the speculators who created it. This is how it was then tarnished with the name *The Novel of medicine.*[53]

As for Magendie, he makes sure to keep at bay what he assures his reader he will not find in any page of his book:

Natural sciences had, like history, their mythical time. Astronomy started as astrology; chemistry was formerly nothing but alchemy; physics was for a long time a vain collection of absurd systems; physiology, nothing but a long and fastidious novel ….

[49] *De la littérature considérée dans ses rapports avec les Institutions sociales*, by Madame de Staël-Holstein, v. II. [Richerand's footnote].

[50] Richerand (1825, I, xi–xii).

[51] Ibid., xiii.

[52] Béraud (1856–1857, I, iv).

[53] Lordat (1813, 39–40).

Such was the state of natural sciences until the seventeenth century.[54] He nevertheless challenges this comforting distance by evoking the present of medicine, in which "mythical time" and "the novel" seem to persist within the remnants of vitalism:

> [next] to the phenomena of CIRCULATION, BREATHING, MUSCULAR CONTRACTIBILITY, we also see, placed on the same line and in the same degree of importance, simple metaphors, such as ORGANIC SENSIBILITY, certain imaginary beings, such as NERVOUS FLUID, certain unintelligible words, such as FORCE or VITAL PRINCIPLE.[55]

But Magendie, locating himself between two eras, immediately affirms the imminence of a definitive epistemological and linguistic reorientation:

> Now the systems of organic functions are no longer favorably welcomed; and to bring to light a work of ROMANTIC PHYSIOLOGY, we are forced to carry out, or to say we carry out, some experiments.[56]

Isidore Bourdon does not hesitate to identify a physiology conducted by "imagination" rather than experimentation, both guilty and "agreeable," as the herald of a "genre" to come, under the patronage of Walter Scott:

> Treated with this condemnable lightness, the most useful of all sciences becomes the most harmful to man; unfortunately, this is how many of our distinguished doctors have written about it; yes, they were aware of the mixed genre of the historical novel before a Scottish writer made it fashionable in Europe. Instead of only demonstrating to you what it is, they deemed it convenient to tell you without any hesitation what they suppose it is: you asked them to retrace to you the history of life, and they offered you an agreeable fiction.[57]

From the novelistic (*romanesque*) to the "romantic," a similar attitude is denounced, whose origin is obscurely perceived as literary; but the insistence to proclaim this exclusion shows how hard a time these two cultures in the process of separating, have abandoning each other. These stylistic recommendations are, most fundamentally, putting vitalism on trial – seeking to purify the language of medicine in order to better assert the rigor of its thinking. But this leaves aside the fact that what is "properly physiological" in physiology calls for a language of approximation, which no positive reference can replace:

> If anatomical descriptions are merely a representation of objects that immediately meet our senses, it isn't the same for the objects that constitute physiology. The foundation for these is material since they depend on the formation of organs and the totality of our constitution (*appareils*); but the part that is properly physiological comes almost entirely from reasoning. From there comes the great capacity for being misled that made the best minds confuse truth with error and others even deny that physiology is anything other than a continuous illusion.
>
> However, allowing this branch of medicine its true value, we see what degree of confidence it deserves. These expressions, 'vital force', 'vital motions', 'sensibility',

---

[54] Magendie (1825, I, v).

[55] Ibid., viii.

[56] Ibid., x.

[57] Bourdon (1828, I, 4).

> displease certain doctors who seek to give their thoughts an algebraic quality. But what are these expressions, if not the statement of facts of unknown nature, which we only portray by attributes; because the inner nature of attraction is unknown, must we deny it and reject the explanations for which it serves as a foundation? [58]

Besides, the days of physiological metaphor are not over, not only because it survives its banishment, but also because it happens that the least vitalist physiology validates it by discovering in it a truth founded on "experience." As Claude Bernard, a disciple of Magendie, writes: "The ancient fiction of life compared to a flame that shines and fades ceased to be a simple metaphor and became a scientific reality. It is the same chemical conditions that feed both fire and life."[59] The suggestion is ambiguous: it decomposes or dissolves the metaphor by sending it back to its proper meaning, but also rekindles its prestige by welcoming it back honorably into medical discourse.

There is also a prevailing of the "novel" (again, *roman*) in that which is assumed; a narrative, beyond the imaginary apprehension associated with it. Physiologists vainly tried to protect themselves from it – narrative is inherent to their discipline – and physiology could barely dissociate itself, for itself and in the eyes of others, from the image of an *animated anatomy* set into motion, and therefore into narration. Warnings, however, are never lacking, attached or not to Haller's evocation: "once anatomy has been learnt, physiology as a whole is left to learn. Nothing as inexact as to call physiology *anatome animata;* it must be developed, in its turn, on new and completely special experimental bases."[60] But the idea persists – that of a physiology in charge of not only explaining "life," but also of representing it, at last of unfolding, one day, the continuous narrative of that which anatomy only provides a fragmented description. It is a narrative vocation, waiting for this future narrative which physiologists know to be very distant, manifesting itself in the urgency for physiology to tell its own story, constantly measuring its present to its past accomplishments, writing and rewriting the narrative of the "experiments" of which it is the fruit, like so many prestigious anecdotes that are part of the statement of acquired knowledge.

However, this profusion of narratives that escort physiology and comfort it in its status as a developing discipline are undoubtedly due to the mode of reading it generates and the circulation it had. It is, moreover, the narrative reach of physiology that animates the movement of vulgarization of which it is the object, sometimes even on the part of physiologists. Isidore Bourdon is one of them; he is the author of both the *Principes de physiologie médicale* (1828) and a work of very open vulgarization in epistolary form, the *Letters on Physiology to Camille*, published in 1829 and reprinted in 1847. Also part of this group are Jean Macé and his *Histoire d'une bouchée de pain* (1861), reprinted in 1865 and 1869, and the distinguished Louis Figuier, physiologist turned vulgarizer, who published in 1879 the *Notions de*

---

[58] Nacquart, "Description", in Alard, Alibert et al. (1812–1822, v. VIII (1814), 508–509).

[59] Bernard (1872a, 5).

[60] Béraud (1856–1857, I, 6).

physiologie à l'usage de la jeunesse et des gens du monde, whose subtitle, *Know yourself*, clearly indicates the moral ambition of the project. All of them accumulate narratives – physiological functions or the findings of physiology – and reinstitute them in a plot in which the "romanesque" reclaims, or rather maintains, its rights. Another sign of narrative's grip, under the title of an updated vulgarization proposed almost a century later by André Senet: *The Novel of Physiology* (Senet 1956)[61]; a tradition prevails, which makes of the latter the mirror that reflects the human in all its truth, by effect of a representation in which anyone can recognize oneself in one's own physiological becoming.

5 Romances of Physiology

If physiology does not cease to distance itself from the "novel" – and to find it again – more than one novel welcomes physiology with open arms. In fact, numerous novelists will look into the mirror physiology offers them throughout the century. Some will see their own reflection; others will not; some will be dazzled, and others will desperately try to shatter them. Zola will recognize himself in it so well that he will try to turn himself into a mirror so that others can recognize themselves in their turn: the poetics of *Le roman expérimental* (1880) is entirely shaped by this reflection that it seeks to amplify. What is at stake, for Zola, is to extend the physiological project and to complete it by enlarging its application to social and psychological fields (something it always sought to do):

> [t]his is what makes the experimental novel: to understand the mechanism of phenomena found in the human, to show the machinery of intellectual and sensory manifestations the way physiology will explain them to us, under the influences of heredity and the circumstances of our surroundings, and then to show the living man in the social milieu that he himself has produced and modifies every day, and in which he goes through a constant transformation. Therefore we are leaning on physiology, we are taking the isolated man from the hands of the physiologist, in order to continue the solution to the problem and scientifically solve the question of knowing how men behave from the moment they are in society onwards.[62]

And Zola so much intends to be *reflective* that not only will he transcribe Claude Bernard, rejoicing in their similarity ("Again, put the word "novel" here in place of the word *medicine*, and the passage is still true"[63]), but will also absorb and assimilate him in a writing which is *physiological* in itself: "I have often written the same words, given the same pieces of advice, and I will repeat them here,"[64] and that is exactly what Zola does by appropriating Claude Bernard's texts and adapting them

---

[61](or *Romance of Physiology*, Ed.). Senet wrote *Homme à la recherche de ses ancêtres. Roman de la paléontologie* 2 years before.

[62]Zola, *Le Roman expérimental*, in Zola (1966–1969, X, 1184–1185).

[63]Ibid., 1196.

[64]Ibid.

to the needs of his own theoretical body. And if physiology sometimes resists him and refuses assimilation, Zola makes a great effort to reject its residues:

> I have remarked that many learned men, and the greatest of them, protective of the scientific certainty they possess, want to confine literature to the ideal. … I do not accept the following words of Claude Bernard: "As for the arts and letters, personality dominates everything. It is the question of a spontaneous creation of the spirit, and that has no longer anything in common with the assessment of natural phenomena, in which our spirit must not create anything." Here I surprise one of those most learned men caught up in the need of denying others access to the scientific domain. I don't know which Letters he wants to talk about when he defines a literary work: "A spontaneous creation of the spirit that has nothing in common with assessing natural phenomena." He is undoubtedly thinking of lyric poetry, for he would not have written this phrase if he had thought of the experimental novel, or the works of Balzac and Stendhal. I can only repeat what I have said: if we set form and style apart, the experimental novelist is nothing but a special learned man who employs the tools of other learned men: observation and analysis. Our domain could be the same as the physiologist's, except that it is broader.[65]

Claude Bernard's objection is reduced to nothing – even worse, his *protectiveness* is punished by means of the subordination of physiology to the novel.

Another, or in fact two other readers will not experience the same happiness in front of the mirror of physiology. It is between 1840 and 1850 that, perceiving a lack (preceded and followed by many others), they obtain at a "secondhand bookseller's" "Richard's and Adelon's treatises, very famous at the time;"[66] these two are among the rare books they acquire. But it is between 1875 and 1877 that Flaubert makes an effort to conceive the relation of two "fellows" with medicine, as he was himself "lost in the combinations" of a "chapter" for which he takes and retakes "notes on physiology and therapeutics in a comical point of view, which is not an easy task at all."[67]

The work is indeed exorbitant, and Flaubert is bound to it in a subtle rewriting of physiology, of which he "crams many volumes"[68] – and not only those which the tale tells us fell in Bouvard's and Pécuchet's hands: from Richerand to Claude Bernard, Flaubert's journey leaves few texts aside and mobilizes the sharpest critical effort. Physiology is thus disrobed, exposed to everyone's eyes, its philosophical ambition made visible, and far beyond the real powers it has. It is in fact the seductions and promises of the physiological narrative, the certainties that the discipline bestows upon its future by repeating the history of its achievements – it is, all in all, the rhetorical armor of physiology that produces paradoxical effects as one reads *Bouvard and Pécuchet*. All the famous experiments that the treatises refer to (those of Spallanzani on digestion, of Sanctorius on perspiration, and some others), the two aspiring physiologists scrupulously dedicate themselves to reproduce, thus following their nature as copyists, but with a demand for knowledge which, albeit clumsy and ridiculous, is nonetheless founded on the principle of experimental repetition.

[65] Ibid., 1200–1201.

[66] Flaubert (1880/1979, 122).

[67] "À Madame Roger des Genettes" [April 2 1877], in Flaubert (1930, 25–26).

[68] "À Léonie Brainne" [October 5, 1872], in Flaubert (1973-, IV, 582).

Nothing in their attempt, however, is according to the original. Bouvard and Pécuchet are not more successful at physiology than at the other sciences they study: all their experiments fail, in spite of their attention to follow to the letter the treatises they put to the test. Thus, with regards to "cutaneous absorption":

> One could see them run along the great road, dressed in wet clothes, under the burning sun. That was to verify if thirst is quenched when water is applied on the epidermis. They went back home, huffing and puffing; and both had a cold.[69]

The same happens shortly thereafter concerning "animal heat," put to the test by Bouvard, while Pécuchet, sitting on a scale, tries to attest to the decrease of his "weight" provoked by the release of a "subtle steam":

> Learned men argue that animal heat develops by means of muscular contractions, and that it is possible, by moving one's thorax and one's pelvic members, to increase the temperature of a lukewarm bath.
>
> Bouvard went to get their bathtub – and, when all was ready, he immersed himself in it, holding a thermometer.
>
> ...since everything was going well, they chatted with serenity.
>
> Bouvard, however, was feeling a little cold.
>
> "Move your limbs!" said Pécuchet.
>
> He moved them, but the thermometer did not change; "It is decidedly cold."
>
> "I am not hot either," replied Pécuchet, himself overcome by a shiver; "but move your pelvic limbs! Move them!"
>
> Bouvard pulled his thighs apart, twisted his flanks, shook his belly, and breathed like a sperm whale; then he looked at the thermometer, which was still going down. "I don't understand it! But I am moving!"
>
> "Not enough!"
>
> And he went on with his gymnastics.
>
> It lasted for three hours, and one more time he grabbed the tube.
>
> "What? Twelve degrees! Oh, good night! I am leaving!"[70]

Failure is their faithful companion. Though the copyists keep trying to copy, the experiments cannot be reproduced. And yet Michel Lévy's *Traité d'hygiène*, reprinted several times, never ceases to promise its reader, and therefore Bouvard and Pécuchet, who read it attentively, a success accessible to all: "contraction without movement" is enough "to increase" body "heat," "one can warm up a bath several degrees by the movement of the pelvic muscles."[71] The "thermometer" Bouvard is "holding," his "pelvic limbs" – and we cannot accuse him of doing a lesser job if compared to anyone else – all seemed ready for a dazzling verification, which, however, refuses to present itself. As for the "cold" they get when they want to verify "if thirst is quenched by the application of water on the epidermis," it is certainly Adelon's *Physiologie de l'homme* that is behind it – and their reading cannot, under any circumstances, be deemed wrong: "according to the accounts of trustworthy travelers," explains Adelon, "thirst is appeased by baths, and by the application of humid clothes on the skin."[72] The setback is very cruel, and the confidence in

[69] Flaubert (1880/1979, 123).

[70] Ibid., 124–125.

[71] Lévy (1869, II, 209–210).

[72] Adelon (1829, III, 10).

physiology would rightly feel betrayed, if Adelon had not remarked "there is debate among physiologists regarding this absorption. Some say it is both frequent and easy, and invoke various facts. .... However, other physiologists deny this cutaneous absorption, or at least believe it is less frequent or easy than it has been said."[73]

The experiments on animals [*bêtes*] will follow, whose only end will be to convey Bouvard and Pécuchet's "stupidity" [*bêtise*], at least in their servant's view, herself a witness of the martyrdom of the poor dog on which the two physiologists vainly tried to prove "the magnetization of steel by contact with the spinal cord" – before their guinea pig escapes, "all bloody, with strings around its paws."[74] Virey's *Physiology* seems nonetheless to admit its possibility ("Prévôt, from Geneva, says he was able to magnetize soft-iron needles by putting them both very near and in the electric current from an animal's spinal cord"[75]), even though it seemed to exclude it completely before: "it could not be verified that steel needles, implanted in the nerves of a living animal, become magnetized and attract iron filings, as Vavasseur, Béraud, Jules Croquet etc. say."[76] The floating of statements – mockingly denounced by Flaubert: "Contradiction? Quid?"[77] – extends far beyond: it is the entirety of Virey's discourse that flickers over the course of a sinuous argumentation that signals his disturbance:

> We certainly accept neither *animal magnetism*, nor its prestige, according to our article on the subject in the *Great Dictionary of Medical Sciences*; but one must recognize the facts of nervous communication among individuals. If the experiments carried out with galvanometers do not provide any concluding evidence of the electricity of nerves, they did not indisputably show that they can never happen. ... The descending (*nervimotor*) and the ascending (*nervisensitive*) currents, despite their analogy with the electric fluid, cannot be assimilated to it; however, their transmission to neighboring living bodies is not without example, especially in the most intimate sexual communications.[78]

Bouvard and Pécuchet, both naked some instants before, one on a scale, the other in a bathtub, did not push their inquiry to that limit. And not having been able to experiment on that "fellow"[79] dog, they will not be able to help Virey to settle the issue either. But that does not put a final stop to their works in the "laboratory," nor to the cruel series of their failures:

> The other experiments failed. Contrary to the authors, the pigeons whose stomachs they bled (whether full or empty), died in the same period of time. Little kittens pushed in water died at the end of five minutes – and a goose they had filled with madder conveyed periostea of a complete whiteness.[80]

---

[73] Ibid., 11–12.

[74] Flaubert (1880/1979, 126).

[75] Letter to the Academy of Sciences, January 2, 1838, cit. in Virey (1844, 321n).

[76] Ibid., 290n.

[77] Cf. Flaubert's preparatory notes, published and annotated by Norioki Sugaya (Sugaya 1999, II, f° 104).

[78] Virey (1844, 289–290n); *sic*.

[79] "You, my fellow, will be used in our experiments!" (Flaubert 1880/1979, 126).

[80] Flaubert (1880/1979, 127).

The divorce of "authors" and "pigeons" is clear, and so is the one of the other species, none of them consenting to die in front of Bouvard and Pécuchet the way the books declared. And it is Adelon who disappoints them again: "Duhamel, having fed animals with food mixed with madder tincture, noticed that, during this time, the bones of these animals became pink; when he stopped feeding madder to these animals, he saw their bones go back to their initial color."[81] That is good enough evidence, for Adelon, of the reality of an "interstitial absorption"; he nonetheless remarks "we cannot *yet* discard it as doubtful,"[82] thus not discarding the possibility of its being refuted one day. That could have given Bouvard and Pécuchet a chance, had they not messed it up perhaps by not feeding their goose long enough. As for the other experiments, Flaubert gathered them, as his preparatory notes show, from Küss's *Cours de physiologie*, which this time leaves no room for doubt: "it is enough to bleed a rabbit in the ordinary state by 30 g of blood to cause death by hemorrhage; at the end of three days of starvation, 7 g is enough to obtain the same result."[83] And elsewhere: "We know that newborn puppies can be immersed for half an hour in lukewarm water and be removed from it still alive."[84] These are statements which, in view of Flaubert's text, denounce numerous experimental variants introduced by Bouvard and Pécuchet, which we can conceive as part of their experienced failures: "pigeons" replace a "rabbit," and worse for the performance of their experiment, Bouvard and Pécuchet confuse "empty … stomach" and "starvation," just like they *bleed* their animals without any precautions or accuracy, abandoning the strict dosage of a quantity in favor of a rough estimation of duration: the "pigeons" stop living and being useful for physiology "in the same period of time." The same for the "kittens" that take the place of "newborn puppies": the substitution does not guarantee them the same age, nor does it assuredly give them the same chance, and perhaps even less of a chance if they are "pushed in water" instead of "immersed in lukewarm water." The protocols of the experiments suffer a distortion of which the books are not the cause. But how can we find them completely not guilty when the physiological literature is as profuse as it is contradictory?

Germaine, the servant, is probably right in her reproach of her masters, even though it is "in their spirit," more than it is in hers, that the "pitiful faculty of seeing stupidity and no longer tolerating it"[85] develops. Bouvard and Pécuchet are faulty readers, but their "stupidity" is here extraordinary because it contaminates everything it touches, or rather because it always reveals another one as it occurs. But not always. The boundaries of common sense are not certain, and physiology, this "Science of Ourselves"[86] which the century has such a massive confidence in, is not

---

[81] Adelon (1829, III, 6).

[82] Ibid, my italics.

[83] Küss (1876, 136).

[84] Ibid., 405n.

[85] Flaubert (1880/1979, 319).

[86] Bourdon (1828, I, 19).

more protected than others from Bouvard's and Pécuchet's attacks. But it is less physiological science that their reading puts into question – they are certainly not fit to prepare its trial – than the way it presents itself, both in its ineluctable course towards a definitive knowledge and in its ambition to subject all of the human sphere to the reason of the body, including when, by anticipation and by principle, the material understanding of phenomena escapes it completely.

Furthermore, the gaps opened in the physiological edifice by Bouvard's and Pécuchet's "stupidity" are nothing but the "comical," and thus critical, counterpart of the fissures that certain physiologists sometimes took pleasure in acknowledging, in a sublime enthusiasm that led them to celebrate their discipline by means of the glorious confession of its weaknesses:

> Oh, difficult science of life, *in which the order of phenomena is such that one can distinguish neither beginning nor end*; in which the influences are so complex that we are always taking the risk of confusing causes and effects! For there is not a single effect, in the acts of life, that does not influence its own cause; not a single cause that does not incalculably become entangled in itself and in its own effects; not a single phenomenon, however unimportant it may seem, which, suffering a perceptible change, does not venture through many inextricable paths and modify the life of all its acts.[87]

The canvas of this obscurity, on which the elementary laws of logic could never have a hold, is in any case in accordance, if not in intensity, then at least in range, with the lesson that Bouvard and Pécuchet learn from their physiological adventure: "we know nothing about it!"[88] Hence it is the end of the line for their studies, and that leads them, in their ignorance of life and of their own body, to the final assessment that science accumulates contradictions and uncertainties:

> We do not even know what the force of the heart really is. Borelli suggests the one necessary to lift up a weight of one hundred eighty thousand pounds, and Keill assesses it at about eight ounces. Hence they concluded that Physiology is (according to an old expression) the novel of medicine. Having not been able to understand it, they did not believe in it.[89]

Bouvard's and Pécuchet's very deceptive *conclusion* is borrowed, as the preparatory notes indicate, from Claude Bernard's lectures on experimental pathology (*Leçons de pathologie expérimentale*), published in 1872, much closer to Flaubert than to Bouvard and Pécuchet:

> Physiology is, among all experimental sciences, the last one to appear, and it was the last one to develop, for sciences develop according to their level of simplicity. . . . But for a long time physiology was considered as an ideal, and even romanesque science, for we used to call it the 'novel of medicine'; it had not yet won its place among the experimental sciences.[90]

As for Claude Bernard, perhaps he owes this witticism to Jourdan, the translator of Müller's *Handbuch der Physiologie*: "the time when we had to recommend the

[87] Ibid., 7–8.

[88] Flaubert (1880/1979, 127).

[89] Ibid.

[90] Bernard (1872b, 471).

study of physiology is over. Back then we were probably not entirely wrong in regarding it as the novel of the natural sciences, and in granting it very little interest . . ."[91] Bernard thus rejoices – just like Jourdan in 1851, or Magendie in 1825, or Lordat in 1813, or even Richerand in 1801 – that physiology is no longer, and not since long ago, the "novel of medicine."

But the "old expression" (belonging to Bouvard, Pécuchet, or even Flaubert), signaled in parentheses in the text in such a way that we are unable to know who is speaking, is undoubtedly much older than Bernard supposes it to be when he transcribes it, in a new variation of the formula by which his predecessors expressed their extreme fear of the "romanesque." For, perhaps without his knowledge, as Flaubert points it out, it is Molière and his *Imaginary Invalid* who Bernard rewrites in rewriting the physiologists. The word comes out of Béralde's mouth when he vainly tries to persuade his brother Argan of the completely rhetorical, and thus artificial, powers of medicine.

> When a doctor speaks of helping, rescuing, relieving nature, of removing from it what harms it and giving it what it lacks, of restoring it and repairing it to the full fluency of its functions; when he speaks of rectifying blood, of tempering the guts and the brain, of reducing the size of the spleen, of mending up the chest, of restoring the liver, of fortifying the heart, of restoring and conserving natural heat, and of knowing the secrets of making life many years longer, he is precisely repeating to you the novel[92] of medicine [*le roman de la médecine*]. But when you face truth and experience, you will not find any of that, and it is like those beautiful dreams which, after you wake up, leave nothing behind but the displeasure of your having believed in them.[93]

Subjected to a novelistic treatment by Flaubert, the "novel of medicine" becomes a complicated statement. It certainly echoes Claude Bernard's voice, as well as that of his precursors, recalling a time which physiology continues to see as bygone; but since it is Bouvard and Pécuchet who get to formulate this *conclusion*, we must ourselves conclude that their "stupidity" does not make them speak less like physiologists: deep down, not "believing" without "understanding" is nothing but sound physiology. And behind one of these voices, that of Béralde discreetly speaks of an old literary mistrust with regard to medicine, which will become, by means of a curious inversion, the sign of a rejection of the literary by doctors.

At the conclusion of their fruitless study of physiology, the hapless Bouvard and Pécuchet arrive almost accidentally, despite themselves, at the practice of therapeutics.[94] They are taken aback by patients they visit ("the symptoms noted by the authors were not those they had just seen"), just as they are disconcerted by nosography: "As regards the names of diseases, Latin, Greek, French, a mishmash of all languages. They number in the thousands, and the Linnean classification is quite handy, with its genera and species; but how do we decide on the species?" This

[91] Translator's preface, in Müller (1851, I, xxv).

[92] *Translators' note*: even though the word "romance" would undoubtedly be more accurate here, our choice of "novel" makes more sense in the context of this article.

[93] Molière, *Le Malade imaginaire*, in Molière (1971, II, 1154).

[94] Flaubert (1880/1979, 127–129), for all the following references.

leads them to lose themselves in "the philosophy of medicine": "They daydreamed on Van Helmont's archaeus, vitalism, Brownism, organicism ..." Their chaotic reverie leads them both to produce unhoped-for cures, and to put their patients' lives in dangers; but the genuine doctor, with whom they compete, does no better than them. To be sure, he reproaches them for their outdated vitalism ("What tommyrot is that – talk of a vital principle! How is it? Who has seen it?"), but he – the observer who speaks out against "systems," is sometimes a "bad observer" and cannot cure as well as his ignorant rivals can.[95]

The 'novel' of medicine once again points medicine back to the 'novel' it no longer wishes to be. The 'medical philosophy' in which the two clinician-partners get bogged down, is hardly decisive on the issues they struggle with. Bouvard and Pécuchet are in fact just the heirs – and muddled ones at that – of medicine's own awkwardness, from which it strives to extricate itself.

Translated from the French by Bruno Penteado and Sonja Stojanovic, revised

References

Adelon, Nicolas-Philibert. 1823–1824 [1829]. *Physiologie de l'homme*, 2nd ed. Paris: Compère Jeune.

Adelon, Nicolas-Philibert, et al. (eds.). 1821–1828. *Dictionnaire de médecine*. Paris: Béchet Jeune.

Alard, Alibert, Bayle, Barbier, Boyer and Biett, et al. 1812–1822. *Dictionnaire des sciences médicales, par une société de médecins et de chirurgiens*. Paris: C.L.F. Panckoucke/Crapart/Le Normant, then C.L.F. Panckoucke.

Alibert, Jean-Louis. 1827. *Physiologie des passions, ou nouvelle doctrine des sentiments moraux*, 2nd ed. Paris: Béchet jeune.

Anon. 1841. *Physiologie des physiologies*. Paris: Desloges.

Balzac, Honoré de. 1976–1979 [1826–1829]. *Physiologie du mariage*. In *La Comédie humaine*, ed. Pierre-Georges Castex. Paris: Gallimard, "Bibliothèque de la Pléiade."

Bérard, Pierre-Honoré. 1848–1855. *Cours de physiologie fait à la Faculté de médecine de Paris*. Paris: Labé.

Béraud, Bruno-Jacques. 1856–1857. *Éléments de physiologie de l'homme et des principaux vertébrés, répondant à toutes les questions physiologiques des programmes de fin d'année*, 2nd edn, Revised by Charles Robin. Paris: Germer Baillière.

Bernard, Claude. 1872a. *De la physiologie générale*. Paris: Librairie Hachette.

Bernard, Claude. 1872b. *Leçons de pathologie expérimentale*. Paris: J.B. Bailllière et Fils.

Bouillet, Marie-Nicolas. 1867 [1854]. Physiologie. In Bouillet, *Dictionnaire universel des sciences des lettres et des arts,* 8th ed. Paris: Hachette.

Bourdon, Isidore. 1828. *Principes de physiologie médicale*. Paris: J.-B. Baillière/Gabon.

Bourdon, Isidore. 1830. *Principes de physiologie comparée, ou histoire des phénomènes de la vie dans tous les êtres qui en sont doués, depuis les plantes jusqu'aux animaux les plus complexes*. Paris: Gabon/J.-B. Baillière.

[95] This is Flaubert's comic extension of Balzac's own effort to critically assimilate medicine, a half-century earlier, in the final pages of *La peau de chagrin* (1831), where he staged a bedside "consultation" to a dying patient, in which the then-fashionable medical systems, notably 'vitalism' and 'organicism', challenged each other and ultimately cancel each other out (see Neefs 1979).

Bourget, Paul. 1891. *Physiologie de l'amour moderne. Fragments posthumes d'un ouvrage de Claude Larcher recueillis et publiés par Paul Bourget, son exécuteur testamentaire*. Paris: Alphonse Lemerre.

Brochin, Robert-Hippolyte. 1889. Vitalisme. In *Dictionnaire encyclopédique des sciences médicales*, Fifth series, vol. III, ed. Amédée Dechambre, 719–728. Paris: Masson/Asselin.

Buchez, Philippe-Joseph-Benjamin. 1826. De la physiologie. *Le Producteur. Journal philosophique de l'industrie, des sciences et des beaux-arts*, Paris: Sautelet et Cie., 1825–1826, v. III.

Cabanis, Pierre-Jean-Georges. 1805 [1802]. *Rapports du physique et du moral de l'homme*, 2nd ed. Paris: Crapart, Caille et Ravier.

Diderot, Denis. 1994–1997. Salon de 1765. In *Œuvres*, ed. Laurent Versini. Paris: Robert Laffont.

Flaubert, Gustave. 1930. *Correspondance*, new expanded edition, eighth series (1877–1880). Paris: Louis Conard.

Flaubert, Gustave. 1973. *Correspondance*, ed. Jean Bruneau. Paris: Gallimard, Bibliothèque de la Pléiade.

Flaubert, Gustave. 1979 [1880]. *Bouvard et Pécuchet*. Introduction and notes by Claudine Gothot-Mersch. Paris: Gallimard-Folio.

Jacyna, Leon Stephen. 1987. Medical science and moral science: The cultural relations of physiology in restoration France. *History of Science* 25(67): 111–146.

Küss, Émile. 1876 [1872]. *Cours de physiologie d'après l'enseignement du Professeur Küss*, published by Mathias Duval [1872], 3rd ed. Paris: J.-B. Baillière et fils.

L'Héritier, Andrée, Jean Prinet, Claude Pichois, Antoinette Huon and Dimitri Stremooukhoff. 1957. Les Physiologies. *Études de presse*. Paris: Université de Paris - Institut français de Presse, IX:17.

Larousse, Pierre. 1874. Physiologie. In *Grand Dictionnaire universel du XIXe sièclel*, vol. XII, 915–918. Paris: Librairie Classique Larousse et Boyer.

Lélut, Louis-Francisque. 1862. *Physiologie de la pensée. Recherche critique des rapports du corps à l'esprit*, 2nd ed. Paris: Didier et Cie.

Lévy, Michel. 1869 [1844]. *Traité d'hygiène publique et privée*, 5th ed. Paris: J.-B. Baillière et fils.

Lordat, Jacques. 1813. *Conseils sur la manière d'étudier la physiologie de l'homme, adressés à Messieurs les Élèves de la Faculté de Médecine de Montpellier*. Montpellier: Delmas.

Magendie, François. 1825. *Précis élémentaire de physiologie*, 2nd ed. Paris: Méquignon-Marvis.

Maudsley, Henry. 1891 [1870/1874]. *Le Crime et la folie*, 6th ed. Paris: Félix Alcan.

Molière. 1971. *Œuvres complètes*, ed. Georges Couton. Paris: Gallimard, "Bibliothèque de la Pléiade."

Müller, Johannes. 1851. *Manuel de physiologie*, 2nd ed. Trans. A.-J.-L. Jourdan. Paris: J.N. Baillière.

Nacquart, Jean-Baptiste. 1814. Description. In *Dictionnaire des sciences médicales*, *op. cit.*, vol. VIII, 506–512.

Neefs, Jacques. 1979. La localisation des sciences. In *Balzac et La peau de chagrin*, ed. Claude Duchet, 129–142. Paris: CDU & SEDES.

Preiss, Nathalie. 1997. Physiologies en France. In *Dictionnaire du XIXe siècle européen*, ed. Madeleine Ambrière. Paris: P.U.F.

Preiss, Nathalie. 1999. *Les Physiologies en France au XIXe siècle. Étude historique, littéraire et stylistique*. Mont-de-Marsan: Éditions Interuniversitaires.

Preiss, Nathalie (ed.). 1999. *De la poire au parapluie. Physiologies politiques*. Paris: Honoré Champion.

Preiss-Basset, Nathalie. 1993. Les physiologies, un miroir en miettes. In *Les Français peints par eux-mêmes*. Paris: Réunion des Musées Nationaux, "Les Dossiers du Musée d'Orsay.".

Raige-Delorme, J., A. Dechambre, and L. Lereboullet, eds. 1864–1889. *Dictionnaire encyclopédique des sciences médicales*, ed. Amédée Dechambre. Paris: Masson/Asselin.

Renauldin, Léopold-Joseph. 1812. Introduction. In *Dictionnaire des sciences médicales, par une société de médecins et de chirurgiens*, vol. I, XIX–CLXIX. Paris: C.L.F. Panckoucke/Crapart/Le Normant, then C.L.F. Panckoucke.

Richerand, Balthazar Anthelme (baron). 1825 [1801]. *Nouveaux élémens de physiologie*, 2nd ed. Paris: Béchet Jeune.

Rigoli, Juan. 2006. Digérer sans le savoir. In *Être et se connaître au XIXe siècle. Littérature et sciences humaines*, ed. John E. Jackson, Juan Rigoli and Daniel Sangsue, Pref. Alain Corbin, 35–83. Geneva: Métropolis.

Rousseau, Jean-Jacques. 1995. *Dictionnaire de musique*. In *Œuvres complètes*, vol. V, ed. Bernard Gagnebin and Marcel Raymond. Paris: Gallimard, "Bibliothèque de la Pléiade".

Rullier, Pierre Joseph. 1828. Vie. In *Dictionnaire de médecine*, vol. XXI, 320–323. Paris: Béchet Jeune.

Saint-Simon, Claude-Henri de Rouvroy, comte de. 1865–1878. *Œuvres de Saint-Simon et d'Enfantin précédées de deux notices historiques et publiées par les membres du Conseil institué par Enfantin pour l'exécution de ses dernières volontés*. Paris: Dentu/Leroux.

Senet, André. 1956. *L'Homme à la découverte de son corps. Roman de la physiologie*. Paris: Plon.

Sugaya, Norioki. 1999. *Les Sciences médicales dans* Bouvard et Pécuchet *de Gustave Flaubert*. PhD dissertation, Docteur de l'Université Paris VIII en Littérature française.

Virey, Julien-Joseph. 1844. *De la physiologie dans ses rapports avec la philosophie*. Paris: J.-B. Baillière.

Zola, Émile. 1966–1969. *Œuvres complètes*, ed. Henri Mitterrand. Paris: Tschou/Cercle du livre précieux.

Chapter 5
Life and the Mind in Nineteenth-Century Britain

Sean Dyde

Abstract It might be said that vitalism has two histories. As a metaphysical admonishment – to treat "life" as a special class of phenomena – its place in the historical record is already assured. But as a way to explain bodily processes, its history seems more complicated and uncertain. Perhaps this stems from classificatory problems: which theories should be described as vitalistic, and in what ways these use vitalistic notions is unclear. In this chapter, I will argue for a history of vitalism as a series of conceptual tools that were used as researchers in early- to mid-nineteenth century Britain attempted to explain the mind in physiological terms. Phrenology (George Combe), reflex action (Marshall Hall) and cerebral reflex function (Thomas Laycock) all provided a model for how the mind operated, yet all three in some way failed to persuade their scientific colleagues. From their efforts, however, a satisfactory account emerged that explained the mind as a series of abstract and teleological processes: a vitalistic account of the mind.

Keywords Phrenology • Reflex action • Cerebral reflex function • Psychological Physiology • Combe, George • Hamilton, William • Hall, Marshall • Grainger, Richard • Carpenter, William • Laycock, Thomas • 'Vindex'

S. Dyde (✉)
Department for the History and Philosophy of Science, University of Cambridge, Free School Lane, Cambridge CB2 3QA, UK
e-mail: skd33@cam.ac.uk

S. Normandin and C.T. Wolfe (eds.), *Vitalism and the Scientific Image in Post-Enlightenment Life Science, 1800–2010*, History, Philosophy and Theory of the Life Sciences 2, DOI 10.1007/978-94-007-2445-7_5,

1 Introduction

Traditionally understood, vitalism is a dirty word. If the task of science is reduce the complexities of the world into the simplest possible laws then vitalistic theories, which posit that a special class of phenomena – "life" – cannot be analysed in terms of physics and chemistry alone, are seen as an unwarranted flight into scientific obscurantism. In medicine and biology, vitalism has been seen as a philosophically-charged term, a pseudoscientific gloss that corrupted scientific practice and allowed concepts such as "spirit" and "soul" to creep into genuine understandings of the body. Describing Claude Bernard's physiological views, Georges Canguilhem has written that "The theories that Bernard condemned were systems such as animism and vitalism, that is, doctrines that answer questions by incorporating them."[1] Scientifically dubious and conceptually suspect, vitalistic theories were seen as an embarrassing relative of experimental physiology.

However, in the twentieth century this image has become a little more complicated. In his 1914 work *The History and Theory of Vitalism*, Hans Driesch developed a new philosophical position – "neo-vitalism" – which not only borrowed Aristotelian metaphysics (in particular the concept of entelechy), but which also outlined many characteristics that vitalistic theories share: irreducibility, temporality, autonomy, purposiveness, wholeness.[2] While entelechy may be outside the purview of contemporary biology, in recent decades this shift in emphasis has been adopted by many historians and philosophers of biology, with interesting results. On the one hand, studies by Owsei Temkin, Timothy Lenoir and Randall Albury (to name but three) have shown that vitalistic notions were much more prevalent in biology and medicine than previously acknowledged. Indeed, in many cases vitalistic theories have been found to have much more explanatory power than purely mechanistic or animistic theories.[3] On the other hand, E. Benton has shown that a single spectrum of philosophical classifications – with "mechanism" on one side and "animism" on the other – is untenable. There are many different typologies of vitalism, or rather, a great number of objects to which the term can be applied. These may be chemical or physiological (to name but two), and refer to some vital agency ("nomological") or not ("anti-nomological"). As Benton has said, these theories "do not admit of any placing on a continuum of greater or lesser extremism. Was Bichat more extreme than Müller? If we decide that he was, have we made a distinction of any importance?"[4] During the course of the twentieth century, therefore, the history of vitalism has found itself in a predicament: just when its significance was being rehabilitated, it lost its value as a coherent notion. As Charles Wolfe has written, simply labelling scientific theories as "vitalistic" would yield "various theories which have little in common with each other, with entirely different empirical bases and/or metaphysical commitments."[5] Clearly then, identifying

[1] Canguilhem (1994, 108).

[2] Driesch (1914, 185–239).

[3] Temkin (1946), Lenoir (1982), and Albury (1977).

[4] Benton (1974, 18).

[5] Wolfe (2011).

vitalistic elements in scientific theories can only be part of a meaningful historical analysis.

Yet if vitalism can only be conceived as a conceptual instrument, rather than as an object of historical study in its own right, then this raises further questions: which aspects of scientific discourse can historians describe as vitalistic? Are we using the right tools for the job? There is some guidance on this issue. Within the philosophy of biology, vitalism has been extended from a supposedly barren explanation of biological processes to an epistemic principle. This transformation stems from Canguilhem's explanation of vitalism as "an imperative, rather than a method and more of an ethical system, perhaps, than a theory."[6] Put simply, describing the body in terms of teleology and holism reflects a caution with which vitalists approach their subject; life is not necessarily mystical, but it is certainly complex. This, according to Canguilhem, is the impetus for vitalistic terminology in biology and physiology. Faced with the task of isolating and studying a complex world, Monica Greco has posited that scientists may look to vitalistic theories as "providing a form of resistance or antithesis to the recurrent possibility of reduction, and to the temptation of premature satisfaction."[7] Regarded in these terms, if vitalism is more than just an explanatory measure – if it can also be regarded as an epistemic warning – then this has significant implications for how its history should be written. The history of vitalism is the history of a particular type of response to certain logical and social complications. Historical attention, therefore, can be given to the scientists who deployed vitalistic notions, as well as (or even separate from) medical philosophers that invoked vitalism as a metaphysical principle. Such a history must, for instance, examine the ways in which physiologists responded to empirical evidence or scientific debate, or how biologists developed their scientific ideas in light of other failed attempts to explain the natural world. Histories of vitalism, in other words, also study the fortunes of scientific theories.

This chapter will examine the works of three nineteenth-century scientist-physicians and their accounts of the mind and brain: George Combe, the father of British phrenology; Marshall Hall, the advocate of reflex action; and Thomas Laycock, the architect of the cerebral reflex function. At first glance, these three scientists seem to have little in common; although they aimed for a physiological account of the mind and nerves, their work forms three distinct approaches to the issue rather than one overarching project. However, their attempts were significant for two reasons. First, even though they were certainly not the first to provide a biological account of the mind, their work was given much closer scrutiny than that of their predecessors. Prior to the nineteenth century, the mind held a distinct place within natural philosophy: still analysed within Locke's framework of the "association of ideas" and concerned with precisely how simple ideas were transformed into more complex.[8] But as a result of possessing a complete and indivisible mind – a

[6]Canguilhem (1994, 288).

[7]Greco (2005,18). This is already a well-established theme in the history of medicine: cf. Lawrence & Weisz (1998).

[8]This is not to say that the behavioural effects of the mind were left unexamined. Cf. Hatfield (1995).

self-evident point in introspection – the nervous system was left undivided as well; a vehicle for the "sensorium commune."[9] However, by the beginning of the nineteenth century this position had to be abandoned; indeed, mental philosophy was changing. Advances in physics, chemistry and more importantly, physiology, gave those who were interested in studying the mind a new standard by which they judged their efforts; in the words of Thomas Dixon, "the 'science of mind' methodology and rhetoric was designed to draw on the success and status of the physical sciences."[10] A new field of study was thereby legitimised, though it was up to physiologists and physicians to provide it. How the mind – once thought an immaterial substance – interacted with matter became the question that all three theories attempted to answer. Secondly, these researchers were active during a period of social unrest. As Adrian Desmond has shown, calls for reform in politics were echoed by calls for reform in medicine. Within this charged atmosphere, such biological accounts of the mind easily led to charges of materialism, atheism or political heterodoxy.[11] Physiologists and physicians interested in studying the mind had to convince their peers that their theories were not only correct, but respectable.

Combe, Hall and Laycock, therefore, had much more in common then perhaps their theories would suggest: all had to address these concerns if their scientific theory was to be successful. Moreover, within these social constraints these scientists had to fulfil the empirical criteria that they themselves had set: the validity of their theory, and the relevance of any particular experiment. Taken together, these standards were quite daunting and it is perhaps unsurprising that these theories met with different levels of success. Indeed, every theory studied in this paper met with detractors. In order to examine these debates and the role that vitalism played in them, this chapter will provide a detailed analysis of one particular dispute for each theory, and the techniques used to advance it: the arguments deployed, the scientist's rhetorical flair and the medium in which these ideas were discussed. As this chapter will show, a vitalistic theory of mind was not the explicit goal of each of these researchers, but through a process of conjecture and argument, this was the direction they took. In other words, nineteenth-century accounts of the mind fumbled towards vitalism.

2 Phrenology: George Combe Versus William Hamilton

Phrenology is already well-known within the history and philosophy of science. It has been co-opted into debates around science and pseudoscience, and has served as a testing ground for the sociology of scientific knowledge. Yet while much has been written on phrenology, this has tended to focus on the social dimensions of the discipline; it is analysed in terms of political reform, social mobility or personal aggrandisement.[12] These are no doubt important aspects in the history of phrenology,

[9]Cf. Figlio (1975).

[10]Dixon (2003, 118). Cf. also Young (1973).

[11]Cf. Desmond (1989).

[12]Cf. Cantor (1975a, b) Shapin (1975), Gieryn (1983), Cooter (1984), and Van Wyhe (2004).

but this tends to overlook the nature of the ideas that were debated. Its central idea was disarmingly simple: character traits were associated with protuberances on the skull, which could be "read" by a skilled practitioner. The mind operated within a straightforward, isomorphic structure; ideal for medical or physiological study.[13] But if phrenology seemed to be an uncomplicated doctrine, proving or disproving it was not; both phrenologists and anti-phrenologists had an armoury of rhetorical techniques designed to refute their opposition. This section will investigate just one of these debates, that between George Combe and Sir William Hamilton.[14]

Hamilton, like Combe, was Scottish, a Whig and a lawyer whose main pursuits lay outside the legal profession. His interests focussed on academia; philosophy and history in particular. He had been denied the Professorial Chair of Moral Philosophy at the University of Edinburgh for political reasons, but he accepted the less prestigious Professorship of Universal History in 1821.[15] In the late 1820s, Hamilton gave a number of lectures warning of the fallacies of phrenology. Combe was urged to respond to these claims. This began a series of letters, first to the *Caledonian Mercury* and later in private (all these letters were later republished in the *Phrenological Journal and Miscellany*) and thus began a correspondence within the pages of the *Caledonian Mercury* that tried, spectacularly but unsuccessfully, to prove or disprove the value of phrenology.

According to Hamilton, phrenology was littered with conceptual and methodological problems. Although phrenologists claimed that they had discovered an incontrovertible fact of nature, Hamilton pointed to the arsenal of techniques they used to avoid contradictory evidence. One of these was the distinction between "power" (the capabilities of any mental organ) and "activity" (how a mental organ was actually used). This distinction was baseless, Hamilton argued, but it allowed phrenologists to explain any discrepancy between their cranial readings and the character traits of the person they were studying. Moreover, if the anatomical discoveries of phrenology were true, they contradicted the findings of other, more prominent physiologists: Christoph Hufeland, Alexander Monro, François Magendie, Karl Rudolphi and Pierre Flourens. In short, Hamilton stated that phrenological doctrines "were diametrically opposed to all that nature manifested, and other physiologists had observed."[16] To Hamilton, phrenology was nothing more than an elaborate confidence trick.

Even though he believed the entire phrenological system was faulty, Hamilton knew his refutation had to be airtight; were there any room for doubt or vacillation, the phrenologists would exploit it. He constructed a list of 14 propositions that he believed would prove incontrovertible. In reality, the phrenologists were harder to pin down. Johann Spurzheim, the German physician who had brought phrenology

[13] Introspection, according to Combe, did not belong in "The True Philosophy of Medicine." Cf. Combe (1826).

[14] As Shapin (1979) points out, there were many areas of technical debate, including the convolutions of the cerebral cortex, the cerebral fibres and the cerebellum.

[15] An account of Hamilton's life, politics and philosophy can be found in Veitch (1869).

[16] Republished in the *Phrenological Journal and Miscellany*. Hamilton (1828–1829a, 3).

to Britain, denied that he ever held three of the proposals (a claim that Hamilton repudiated, to little effect).[17] Very quickly, other propositions were repudiated or simply ignored and the debate focussed on a single topic: the existence of frontal sinuses. If these cavities were present amongst most of the population and differed in size, then phrenologists could no longer claim that they were reading the brain (and therefore the mind) through bumps on the head.

Or rather, the frontal sinuses *should* have become the crux of the issue. Instead the correspondence between Combe and Hamilton became increasingly devoted to experimental protocols. For example, Hamilton had wanted to use the 50 skulls that Spurzheim had collected from the Catacombs of Paris. Yet on the 22nd of November 1827 when the two decided to meet, these skulls were rejected by the group of judges that Hamilton had assembled. Combe labelled this as proof that Hamilton's criticism was baseless. Hamilton pointed out that it was Combe who had wanted these skulls (collected by a fellow phrenologist) rejected.[18] As a result, the 'arbitors' decided to procure their own samples. But as Combe was quick to point out, "it may require years before they find a sufficient number to enable them to arrive at any general result." This left the umpires in a difficult situation. Determining whether a number of skulls contained sinus cavities could have been a relatively simple affair, but with two angry rivals contesting each piece of evidence and every measurement, the chance for a speedy resolution were slim. This was recognised in their report published on the 8th of February 1828:

> after a desultory conversation on the best method of procuring accurate facts for deciding the points at issue between Hamilton and Combe, arbiters proposed to attend pathological dissections at the Infirmary and Fever Hospital. In a few months, a sufficient set of correct observations was hoped to be procured.[19]

Not that Combe was concerned about the outcome either way. He dismissed the whole process, saying "even after they shall have come to a conclusion, their decision will still be that of only three individuals."[20] Indeed, since no further meetings were held, the arbitors might have come to the same conclusion.

From this point on, the disagreements between Combe and Hamilton continued as a series of private correspondence and quickly flared into a volley of personal attacks. Hamilton feigned astonishment that Spurzheim would not join any professional demonstration "by which the *assertions of Phrenology* would be held up in collation with the *facts of Nature*."[21] Spurzheim responded, accusing Hamilton "of being 'without exception, the most erroneous of observers recorded in the whole history of science'."[22] Hamilton was criticised for not attending Spurzheim's public lectures, a prospect Hamilton described as "an irrelevant lecture to an idle mob."[23]

---

[17] This rather fruitless disagreement continued throughout the correspondence published.

[18] Combe (1828–1829a, 13) and Hamilton (1828–1829b, 19).

[19] Scott and Christison (1828–1829, 34).

[20] Combe (1828–1829a, 12).

[21] Hamilton (1828–1829c, 38).

[22] Spurzheim (1828–1829a, 41).

[23] Hamilton (1828–1829d, 42).

After a number of unsuccessful attempts to meet publically, Spurzheim wrote to Hamilton that he "declares all correspondence to any other purpose in vain. Dr S. takes no notice of the puerile boasting, and so little *professor-like* expressions, in which Sir W. indulges in his letters."[24] Spurzheim's departure from Edinburgh was viewed as an escape by Hamilton, yet as further proof to Spurzheim and Combe that Hamilton never had any evidence to contradict their claims. To Combe's accusation that he lectured against phrenology simply for financial gain, Hamilton replied "This is either the veriest twaddle, or there is an insinuation (which I shall not stoop to characterize) intended for those who know nothing of the circumstances, that my lecture was for any profit of my own."[25] Very slowly, the scientific question about the frontal sinuses began to fade from consideration.

The result (if it can be called that) of this acrimonious dispute was two books. Hamilton intended to produce a volume entitled *Fictions of Phrenology and Facts of Nature*, but it was never published. Such a work would be inconsequential, Hamilton concluded, since phrenology was "a doctrine, which, from its own futility, and the mode in which it is defended, I now regard as all too contemptible for controversy."[26] In 1828 Combe published his *The Constitution of Man Considered in relation to External Objects*, a copy of which Hamilton had procured (through dubious means) before publication.[27] This manuscript only seemed to confirm to Hamilton that phrenology was a lot worse than merely unscientific. It was atheistic, reminiscent of the radicalism that imbued Robert Grant's morphology in the 1820s.[28] For Hamilton, it had no place within civilised society. Phrenology remained popular in certain circles long into the nineteenth and twentieth centuries, but for Combe and Hamilton the debate was certainly over. They left these discussions assured of their own veracity, yet completely unable to sway their opponents.

But rather than noting the disintegration of these talks, it is just as noteworthy that this debate continued for so long in the first place. The correspondence between Combe and Hamilton certainly degenerated into abuse, yet snide remarks had been a recurrent feature of their letters from the beginning. Relatively early in the debate, Combe had told the editor of the *Caledonian Mercury* that "Your readers will have discovered long before this time, that philosophy or facts in nature attract, in a very subordinate degree, the notice of Sir William Hamilton."[29] A little later, Combe took offense on behalf of Spurzheim and Franz Joseph Gall (the Austrian founder of

[24] Spurzheim (1828–1829b, 47).

[25] Hamilton (1828–1829e, 54).

[26] Hamilton (1828–1829e, 55–6).

[27] In turn, Combe published these letters in the *Phrenological Journal* in its entirety. On the relationship between published works and private correspondence, cf. Desmond (1989) and Winter (1997).

[28] Combe fervently denied that phrenology implied atheism: Combe (1828–1829c). Previously, Hamilton had excluded such theological objections from his more scientific argument. Hamilton (1828–1829a), 2. For an analysis of *The Constitution of Man* and its aftermath, cf. Van Wyhe (2004). On Robert Grant, radical morphology and its possible connections with phrenology, cf. Desmond (1989).

[29] Combe (1828–1829b, 26).

phrenology), objecting to Hamilton's portrayal of them as "the most worthless of observers."[30] A little over a month after the arbitors had dismissed them, Hamilton wrote to Combe that "the irrelevant personalities with which you have of late attempted to screen the weakness of your cause, if not to disgust your opponent with his controversy."[31] Replying the next day, Combe labelled Hamilton's language "exceedingly discourteous and unhandsome," further stating (apparently without irony): "In my own letters, I have watchfully endeavoured to avoid being drawn into the imitation of such conduct."[32] What is striking in these letters is their resemblance to an early nineteenth-century courtroom, where insulting the opposing counsel was a common legal technique. David Bentley has documented incidents in which witnesses were harassed, judges insulted, courts recorders reduced to tears, and QCs came to blows.[33] Given that Combe and Hamilton were trained as lawyers, this is perhaps unsurprising. But unfortunately for both of them, this was not a court case. They could not control their witnesses: the arbitors had urged patience until new evidence could be gathered. More importantly, a verdict was never handed down. Readers of the *Caledonian Mercury* and the *Phrenological Journal* were no doubt entertained by these angry missives, but they were never obliged to reach a consensus. To their detriment, Combe and Hamilton did not recognise the differences between their legalistic fantasy and social reality.

As public mistrust turned into personal hatred, both Combe and Hamilton lost sight of the original reason that initiated this debate. Indeed, both believed that their physiological arguments stemmed from sound scientific practice. Part of phrenology's appeal, Combe noted, was that it accorded with the "well-established principle in physiology, that *different* functions are never performed by the *same organ*."[34] But this had created a problem for phrenology, one Hamilton was quick to spot. In order to create their new physiology, phrenologists had elided character traits with the mind, cranial bumps with the brain. This ensured a very direct relationship between mind and brain. Many phrenologists found this relationship useful, as it provided firm anatomical foundations on which their science could be built. But as Hamilton suggested, this isomorphic structure left its experimental basis relatively bare. Phrenology was a straightforward doctrine, and it could be brought down by means just as simple.

The phrenology debates of the early nineteenth-century proved inconclusive for a variety of reasons. Many found the phrenologists too fervent and too controversial to be worth engaging with. At the 1834 meeting of the British Association for the Advancement of Science, phrenology was excluded from the official proceedings; a clear sign an emerging scientific establishment dismissed the practice.[35] Conceptually

---

[30] Combe (1828–1829b, 30).

[31] Hamilton (1828–1829f, 58).

[32] Combe (1828–1829d, 63–4).

[33] Bentley (1998, 100–01).

[34] Combe (1826, 35).

[35] Morrell and Thackray (1981, 276–81).

speaking, many found phrenology's simple correlation between anatomy and character unpalatable: too materialistic to be theologically secure and too mechanistic to withstand scrutiny. Although phrenologists had reduced the mind to matter in motion, they did not speculate on what motion this view entailed. For all their talk of providing a physiological account of the mind, they had produced only a dubious anatomical map. The question of whether the mind could be explained in physiological terms, and exactly what terms such an explanation would take, remained unanswered. But this wait was short-lived; a few years later another theory was developed that, it was hoped, could account for the mind in a scientific way. The theory was "reflex action," and its champion was Marshall Hall.

3 Reflex Action: Marshall Hall Versus the World

Marshall Hall holds a contested place within the history of physiology. Since at least the 1890s, his concept of reflex action – functions of the nervous system that did not require conscious participation – has been seen as a significant advance in neurophysiology.[36] More recent works, however, have re-examined the import of Hall's research. One historian, Ruth Leys, has concluded that Hall did not "discover" reflex actions in the traditional sense of the word, but that his work helped to reconceptualise mental physiology away from quasi-mystical notions of the soul.[37] In their 1987 publication *Nineteenth-century Origins of Neuroscientific Concepts*, Edwin Clarke and Stephen Jacyna acknowledge Hall as the discoverer of reflex action, but also note he was "a man of small stature but immense conceit. He was aggressive, quarrelsome, and of a rebellious nature, and he demonstrated excessive possessiveness concerning his discoveries."[38] Moreover, "his publications are voluminous, repetitive, and tediously vituperative."[39] Hall's talent for alienating people, they assert, blinded his fellows to the brilliance of his work even to the point where they charged him with plagiarism. Yet according to Clarke and Jacyna, these factors should be disregarded if we are to provide a fair assessment of Hall's scientific work.[40]

Whether as a scientific discoverer or merely as a theoretical reorganiser, these historians all acknowledge Hall as an important contributor to the history of physiology. But these pronouncements (gained with the benefit of over 150 years hindsight) still deeply contrast with the near-total dismissal of Hall's work during his lifetime. Leys, Clarke and Jacyna suggest that since Hall's doctrines were correct,

[36]Hodge (1890) and Fearing (1964, 122–145).

[37]Leys compares Hall's notion of reflex action with W.P. Alison's "spinal soul" (Leys 1990).

[38]Clarke and Jacyna (1987, 114–15).

[39]Clarke and Jacyna (1987, 115–16).

[40]"We need not, however, be concerned with the details of this somewhat sordid controversy. Again with hindsight, it appears to have been characterized by personal enmities and rivalries, vindictiveness, pettiness, jealousy, and political undertones" (Clarke and Jacyna 1987, 119).

then his contemporaries must have therefore been unduly influenced by his acerbic personality. However, as will be shown in this section, to say that this was the *only* factor that determined Hall's reception is an oversimplification. Certainly his personal style was not conducive to cool reflection, but Hall's notion of reflex function also suffered from perceived shortfalls in his experimental method and theoretical exposition, not just from his ability to raise the ire of his professional colleagues. As will be shown, there was an ever-widening gap between the experiments Hall was conducting and the hypotheses he thought these studies proved. It was these overinflated claims that led Hall into the scientific wilderness, rather than any inherent difficulty with the concept of reflex action.

Marshall Hall first introduced his notion of reflex action to the Zoological Society of London in 1832. The stated aim of this chapter was simple and modest; to present a "series of experiments tending to prove the existence of a source of muscular action distinct from all those hitherto noticed by physiologists: viz. volition, irritation of the motor nerves in some part of their origin or course, that of the muscles themselves."[41] Hall had found that salamander tails, when removed from the body, continued to display some basic nervous function which ceased only if the spinal cord was destroyed. Experimenting with decapitated turtles, Hall noted that "on pinching the eyelid it is forcibly closed; the mouth is opened and the membrane expanded under the lower jaw descends as in respiration."[42] A frog given a solution of opium became tetanic, remained so after decapitation, and only became limp again when the brain and spinal cord were removed. From these experiments, Hall wrote:

> I conclude, then, that there is a property of the sentient and motor system of nerves which is independent of sensation and volition;-a property of the motor nerves independent of immediate irritation:- a property which attaches itself to any part of an animal, the corresponding portion of the brain and spinal marrow of which is entire.[43]

Reflex actions, Hall elaborated, were involved in respiration, coughing, sneezing, vomiting and "the singular effect of tickling." The larynx and anal sphincters were particularly susceptible to disorders of this function, including tetanus and hydrophobia.[44]

From this overview, it is clear that Hall's claims were hardly unpretentious; reflex action supposedly regulated the entire "animal oeconomy." Yet when discussing the nature of reflex actions *per se* he was fairly restrained. At this point Hall only claimed to have found a *characteristic* of the nervous system, a function amongst its many others. Although unaware of the precise mechanisms that allowed reflex actions to operate or the exact pathways that reflex actions took (if any), this was not a cause for concern for either Hall or his audience at the Zoological Society. It is interesting to note, therefore, that Hall's theory was accepted even though it

---

[41] Anon (1832, 190).

[42] Anon (1832, 190).

[43] Anon (1832, 191).

[44] Anon (1832, 191).

contained many elements of vitalistic discourse that were identified earlier in this paper. Hall's discovery was of a temporal process, intangible in itself but known by its effects, a function with the explicit purpose of maintaining the entirety of the animal body. Although not explicitly a vitalist, Hall could use such terminology in ways that satisfied his scientific peers.[45]

Little seems to have been made of these initial experiments. According to a later writer, this was simply because Hall had addressed the wrong audience: "who would look for the 'reflex function' in the Zoological Transactions?"[46] In the following year, Hall gave an expanded version of this presentation to the Royal Society and the British Association for the Advancement of Science, after which he received a small grant so that "the sensibilities of the Nerves of the Brain should be investigated."[47] These later experiments, which included striking a horse with a "poll-axe," were reported the following year.[48] To many present at these meetings, it must have seemed that Hall had made a sound, yet relatively minor, physiological discovery.

By 1836, however, Hall's claims had grown bolder. In an address to the Aldershot School of Medicine and later in his book *Lectures on the Nervous system and its Diseases*, what he had once described as a *function* of the nerves was now elevated to a distinct *structure* within the nervous system. This new anatomy, "one which I [Hall] claim the merit of first pointing out in all its fullness," was seated in the "true" spinal cord, rather than in the brain. This allowed reflex actions to extend throughout the body while bypassing (Hall had hoped) the philosophically charged notions of "sensation" and "volition" that dogged studies of the brain.[49] Its effects, therefore, were widespread. Whereas earlier they were involved in small nervous reactions, now they regulated the entire body. Along with coughing, sneezing and respiration, Hall's schemata now included digestion and various discharges: faeces, urine, semen and foetuses (in other words, reflexes were involved in childbirth). Strange incidents could be reinterpreted and explained by this "excito-motory system." Hall even gave a dramatic example of the power of this system, in which a woman was given a catheter: "certain nerves being excited on introducing this instrument, an action of the muscles has been induced, which has drawn the catheter out of the grasp of the surgeon into the bladder. I show you a catheter which was extracted after such an accident."[50] All spasmodic and convulsive diseases, too, were said to have their origin in reflex actions.[51] In effect, Hall had created a vast physiological system, capable of uniting a disparate array of symptoms into a single framework.

---

[45] Leys seems to equivocate on this point, in one instance calling reflex action "mechanical" (Leys 1990, 240) and "vitalistic" at another (255). This, I believe, stems partly from Hall's changing views on what he was trying to describe.

[46] Anon (1836–1837, 660).

[47] Anon (1834), xxxvii. Cf. Hall (1833, 635–65).

[48] Hall and Broughton (1835, 676–80).

[49] Hall (1836a, 633, b, 25–6).

[50] Hall (1836a, 636).

[51] Hall (1836a, 639).

Perhaps unsurprisingly, many physiologists found this new thesis controversial. When Hall's experiments were conducted in 1834, they completely fulfilled the standards Hall had set for them. By incapacitating the brain, Hall was able to reveal that the medulla oblongata provided a particular function within the animal economy. However, from 1836 on Hall no longer claimed to have found a mere *property* of the "true spinal marrow"; he was advancing an entirely new *structure* within the nervous system. As such, gathering together a number of seemingly connected observations no longer provided the evidence that Hall required – lobotomising horses in 1834 carried a lighter demonstrative burden than in 1836. Hall faced two problems: the first was that he had not conducted any anatomical studies to validate his claims. To Richard Grainger, who employed Hall at his Webb Street Anatomy School in London, this seemed like a massive oversight.[52] While admiring Hall's physiological work, he warned that "it cannot escape the recollection how many theories, none, perhaps, so important, but equally with this the semblance of probability, have ultimately been classed in the number of ingenious but unfounded speculations." If Hall wished to substantiate reflex action – his "great principle of animal economy" – Grainger concluded that Hall "must be satisfied previously to submit his conclusions to the test of anatomy."[53] Luckily, this was exactly the type of experimental support that Grainger aimed to provide.

In his 1836 work, Hall gave some hint of how the excito-motory system could be demonstrated anatomically. He characterised the true spinal cord, the seat of the whole system, as "an axis of excitor and motor nerves."[54] Grainger believed that he had found this axis deep within the spinal nerves. It was little wonder that the small fibres within these nerves had never been satisfactorily examined before, Grainger explained, because these fibres were difficult to see and broke very easily.[55] However after several investigations, Grainger hypothesised that some of these fibres were "lost in the white substance [surrounding the spinal cord], whilst others entering more deeply into the lateral farrows, nearly in a right angle with the spinal cord itself, as far as the grey substance into which they are lost."[56] Grainger believed that this was where the excito-motory system rested and thus were the anatomical foundations that Hall's work conspicuously lacked.

Grainger's enthusiasm, however, was not universally shared. Even if Grainger had discovered some fine nervous fibres around the spinal cord, this did not prove that they operated the excito-motory system. This was the second, more conceptual problem with Hall's doctrine. It was raised by William Carpenter, later a prominent

[52] Hall's familiarity with Grainger, Robert Grant and the British Medical Association has led Desmond to place him within early nineteenth-century political radicalism. Desmond (1989, 130–2, 139–40). More convincing, however, is Diana Manuel's claim that Hall involved himself in politics only when it affected his career prospects. Manuel (1996, 20–1, 138–42).

[53] Grainger (1837, vi).

[54] Hall (1836b, 20).

[55] Grainger refused to use lenses for fear of deception. Grainger (1837, 34–5).

[56] Grainger (1837, 34).

neurologist in his own right. In an influential review of both Hall and Grainger's work on the spinal marrow, Carpenter rejected Hall's claim of a distinct nervous substructure: "this admission we strongly opposed when formerly treating the subject, and we cannot see that the facts brought forward by Mr. Grainger are yet sufficient to warrant it."[57] Reflex action was a physiological theory imbued with a notion of matter in motion. Grainger's experiments, however, were anatomical, where matter was held resolutely still. As such, Carpenter believed that Grainger's studies had not proved anything about reflex actions at all. Rather, he relied on a type of explanatory regress, in which reflex action was *thought* to occur in the (as yet unstudied) white and grey matter of the spinal cord.

In other words, Carpenter hinted towards a conceptual gap between anatomy and experimental physiology. While they may share the same subject matter, their experimental procedures and technical language were inherently and irreconcilably different. When conducting physiological experiments, Hall could talk of bodily functions and teleology, but these were resources that Grainger had not allowed in his purely inert investigations. Instead of trying to amalgamate a number of conceptually different experiments into a single eclectic doctrine, Carpenter reasoned that:

> no enquiry into these functions can be complete which does not include an examination of the vital properties of the organs which it immediately influences: and this examination is particularly desirable at the present time, in order to correct the inaccurate notions which many have entertained regarding the dependence of organic life upon nervous influence; a doctrine which may be regarded as a remnant of the physiology of the Vitalists, which was scarcely less erroneous as a system than the chemical and mechanical theories which it displaced. [58]

What resulted from this confused approach, Carpenter clarified in the following year, was a reified conception of reflex actions. As Carpenter notes, "'property of matter' simply denotes its capability of producing an effect upon the percipient mind … and that it cannot imply any agency distinct or separate from matter."[59] Between their physiology and anatomy, Hall and Grainger had inadvertently stuffed a notion of "vital power" that had ultimately misdirected their investigations. Since both were looking in the wrong areas for experimental proof, their boasts were premature.

Of course, Carpenter found more to criticise in Hall than just his experimental procedures. Hall had placed himself on a scientific pedestal that Carpenter found undeserved for a variety of reasons. The first was the claim that Hall's research was original; an assertion, at best, that needed careful qualification. The second reason related to the gap between what Hall had claimed he had discovered and what he had actually proved. In this way, Hall's manner of presenting his ideas became much more pertinent. As Carpenter appreciated of Hall's work: "we are quite sure that these claims would have received more attention in various quarters had they

---

[57] Carpenter (1838, 500).

[58] Carpenter (1838, 487).

[59] Carpenter (1839, 137).

been more moderate and discriminate."[60] In these circumstances, the later success of the reflex function contrasts sharply with the dismal reception Hall received from his peers. Later physiologists found the concept of reflex function useful, but this was not because Hall was not the scientific genius he thought himself to be.

Here, then, was another theory that did not live up to expectations. Phrenology suffered because it insisted on an isomorphic correlation between cranial bumps and character. Hall, while certainly less ambitious than the phrenologists, came under similar attack when he claimed to have discovered a distinct nervous system. The common problem was attempting to tie physiological notions to anatomical structures: to give motion to matter. But simply crashing these two elements together, as Combe and Hall discovered, was not the simple explanation that physiologists sought. It raised questions of how the mind was constituted by or directed these processes, and it was too crude to withstand anatomical study or charges of atheism. However, this common problem was also a trap of the researchers' own devising; as Carpenter had intimated in his critique of Hall, investigating anatomical structures was not the main purpose of physiological experiments. If a physiological model could be free of any explicit anatomical shackles, then perhaps it could receive the support of the scientific community. Thomas Laycock's "cerebral reflex function" was such a theory. Laycock achieved this by simply ignoring the desire to provide any anatomical foundations to his work. Free from these shackles, cerebral reflex functions could not only be used to describe a broad range of phenomena, it could provide a physiological account of the mind that circumvented the scientific and ideological problems that dogged its predecessors.

4 Cerebral Reflex Function: Thomas Laycock Versus "Vindex"

Thomas Laycock has always had a small but assured place within the history of neurology. As the mentor to John Hughlings Jackson and Professor of Medicine at Edinburgh University, Laycock's work has been seen as an important step towards contemporary understandings of neurophysiology.[61] However, in recent decades this perception has been further developed by investigations into Laycock's philosophical and theological outlook. Roger Smith has studied how Laycock (and Carpenter) were influenced by *Naturphilosophie*, a belief in the unity of nature that allowed them to draw conclusions from comparative anatomy.[62] L.S. Jacyna has extended this notion further, showing how Laycock's belief in a natural hierarchy was directly related to trust in nineteenth-century moral or social hierarchies.[63]

---

[60] Carpenter (1838, 540).

[61] On Laycock's election to the Edinburgh Chair, cf. Barfoot (1995). Barfoot notes that Laycock's neurophysiological work has only been recognised by historians since the 1960s (3).

[62] Smith (1971).

[63] Jacyna (1981, 109–132).

However, belief in the unity of nature did more than just place God, Man and Nature within a firm relational structure. As Smith has pointed out, Laycock's metaphysical garb assured him of the connection between body and mind: "By 'mental forces' Laycock meant the agency of the 'Superior Intelligence' behind nature. In this context, the 'correlation of mental and physical forces' implied that physical forces were correlated with forces executing an immanent purpose."[64] This belief allowed Laycock to develop his notion of a "cerebral reflex function," a doctrine that was firmly based in physiological theory but which freely borrowed terms from traditional mental philosophy. Whereas earlier physicians had felt the need to justify the connections between body and mind experimentally, Laycock's philosophical belief already guaranteed that they were related in some kind of unitary structure. His theoretical boundaries were secure, and he could thus meander peacefully between them.

Laycock first introduced the notion of a "cerebral reflex function" in his work, *A Treatise on the Nervous Diseases of Women*. Public response, however, was negligible. An anonymous reviewer in the *Medico-Chirurgical Review* noted "many novel and hypothetical views are advanced which display considerable research and ingenuity. But as they do not elicit much practical information, we shall pass them over."[65] Laycock had to wait until the 1844 meeting of the British Association for the Advancement of Science for his ideas to receive attention. Summarising the meeting, *The Lancet* said of Laycock's paper, "a most interesting one," but was "of such a nature as not to admit justice being done it by any abstract. A most interesting discussion ensued."[66] Given the unconventional nature of Laycock's scientific claims and that he tended to wear his philosophical convictions on his sleeve, perhaps we should not be surprised.

The following year Laycock's speech was published in the *British and Foreign Medical Review*. He wrote that physiological research was stagnating, driven backwards by an unwarranted belief that the mind was inherently different from the body. Instead he proposed a "law of diffusion," in which all reflexes (emotional, intellectual and excito-motory) were infused through the nervous system:

> I was led to this opinion by the general principle, that the ganglia within the cranium being a continuation of the spinal cord, must necessarily be regulated as to their reaction on external agencies by laws identical with those governing the functions of the spinal ganglia and their analogues in the lower animals.[67]

In short, Laycock was overturning the conventional approach towards the mind and brain. Previous researchers had tried to find an anatomical basis for their physiological theories, but since no composite of anatomical structures could be discerned, physiologists were urged to create their own ways of describing the physiology of the brain.[68] Combe and Hall had tied mental phenomena *down* to a

---

[64] Smith (1971, 258–9).

[65] Anon (1841, 107).

[66] Anon (1844, 57).

[67] Laycock (1845a, 298).

[68] On the language used to describe mental processes during the nineteenth-century, cf. Smith (1992).

physical structure. Laycock argued that physiological explanations should be raised *up* to include the mind. Indeed, compared with the restraints conferred upon phrenology and Hall's reflex action, the freedoms offered by these ideas were great. In an open letter to George Combe, Laycock wrote that:

> And when a visible idea is induced, the concurrent ideagenic change in the sensory grey matter of the brain acts on a substratum, the property of which is to combine the action of the facial and respiratory muscles, so that the sound and muscular movements constituting laughter are produced.[69]

Ironically, now that Laycock had relieved physiologists of the burden of anatomical proof, they could invoke anatomical structures at their leisure.

Laycock's conception of 'reflex', moreover, was considerably more relaxed than that of his contemporaries. Hall's original definition had seen reflexes as unconscious nervous responses. But Laycock could not draw such strong distinctions between conscious and unconscious: "the object of all the purely reflex physiological acts is the conservation of the individual, or of the race."[70] For Laycock the reflexes were goal-directed and purposive; moreover, they could be applied to a range of physiological processes. Hydrophobic convulsions were explained by "ideagenous changes" in the brain produced by the sight of water. The colour red, Laycock believed, could be particularly harmful to the patient with an already distressed nervous system. Smells, too, might create an adverse reaction in those not accustomed to them. "At Rome," he wrote, "delicate odours of flowers and perfumes produce such effects on the nervous system."[71] This grand scheme accounted for sneezing, sighing, gasping, vomiting and dyspnoea, as well as a whole range of emotional reactions.[72]

Laycock had therefore cast a wide net over the type of phenomena to be researched. Moreover, he insisted (like Carpenter) that only physiological experiments could adequately investigate them. This limited studies on humans to pathological case studies and the action of narcotics, but in the animal kingdom the scope for physiological research was potentially limitless. Migratory birds, startled partridges, hard-working bees, and a variety of decapitated wasps, centipedes and worms were included as empirical exemplars. In short, Laycock seemed to offer experimentalists a large number of specimens to study, and a wide array of physiological processes to examine, all within a secure theoretical framework.

However, that sentience and nervous action could be infused in the brain and parts of the nervous system aggrieved some physiologists. One opponent, under the pseudonym "Vindex," is of particular note. Vindex became furious when he read Laycock's letter in *The Lancet*. His reply conveyed his disgust: "The very first statement is an egregious error – 'impressions on the *olfactory* nerves occasion the muscular phenomena of sneezing,' &c. What incipient tyro in his studies would

[69] Laycock (1845b, 347).

[70] Laycock (1845a, 300).

[71] Laycock (1845a, 302).

[72] Laycock (1845a, 306).

write such stuff as this!"[73] As a supporter of Marshall Hall and advocate of the excito-motory system, Vindex believed that Laycock's work risked discrediting a significant physiological discovery by associating it with metaphysical nonsense: "Dr. Laycock's instances are instances of emotion, (I speak in general terms,) and it is most important every way to point out and bear in mind the distinction between this, which is psychical, and the special excito-motor action of the spinal marrow, which is physical." Emotions and thoughts could be *described* as reflex actions, Vindex admitted, but carelessly ignoring this distinction was foolish and potentially dangerous.[74]

Laycock only wrote one letter in reply to Vindex. He admitted worms were perhaps not the best experimental subject with which to prove his claims, but because of their rudimentary nervous system, he considered them to be "brainless."[75] Moreover, he had perused the works of German physiologists, which afforded "the most ample and irrefragable proof of the truth of my proposition, 'that the cerebral ganglia (or even the encephalic) are not necessary to the production of emotional movements.'"[76] However, Laycock's most bitter criticism was reserved for the tone of Vindex's missive:

> Having given this ample response to "Vindex," I can take no further notice of communications like his. They are offences against good manners and sound ethics; they retard the progress of science, and they waste valuable time. Writers on subjects of this kind may be reasonably expected to display so much courage, as to append their names to their communications; and so much good taste, as to write like gentlemen.[77]

Vindex repudiated that a pseudonym was within the realm of "sound ethics": "Surely one may be allowed to pull up weeds with a gloved hand." His real name had been given to the editor of *The Lancet*, Thomas Wakley, which Vindex believed should suffice.[78] Since Laycock (true to his word) did not respond to this letter, it seems that he did not agree. Vindex was quarrelsome and rude, so whatever contribution he may have added to medical discussion could be safely ignored.

To briefly summarise, Laycock's theory had the following attributes: a wide range of phenomena for examination, a mechanism with which to describe these processes ("cerebral reflex function"), and a secure theoretical framework. His schema saw the mind as an assortment of nervous processes related to, yet abstracted from, the physical structure of the brain and spinal cord. His account could survive without recourse to (potentially incorrect) anatomical foundations. Laycock's was an abstract and teleological theory of mind. In other words, he used vitalistic notions to avoid the criticism levelled at his scientific forebears.

This approach to mental phenomena also had political advantages, as William Carpenter discovered. Previously, Carpenter had argued that there was an important

[73] Vindex (1846a, 198).
[74] Vindex (1846a, 198).
[75] Laycock (1846, 424).
[76] Laycock (1846, 425).
[77] Laycock (1846, 425).
[78] Vindex (1846b, 510).

distinction between physiological experiments and anatomical studies. By the 1840s, however, Carpenter had further reason for adopting Laycock's strategy. The publication of his *Principles of General and Comparative Physiology* in 1839 (in which, amongst other things, he had criticised the "vital powers" inherent in reflex action) had embroiled Carpenter in controversy. His work was seen as advocating predestination for mankind and self-government for the universe; in Carpenter's world, many believed, there was no place for God. He only saved his reputation by marshalling together support from the scientific and religious orthodoxy.[79] When he broached the subject again in the 1850s, he employed Laycock's jumble of mental and physical terms:

> The dominant power of the Will, not only over every act of the nervo-muscular system apparatus which is not immediately concerned in the maintenance of the vital functions, but over the course of purely psychical action, is probably the most distinctive attribute of the Human mind in its highest phase of development; and it is that which gives to each individual the freedom of action, which every one is conscious to himself that he is capable of exerting.[80]

These are certainly the words of a physiologist wary of further quarrels. But as Alison Winter has shown, Carpenter's mental physiology also purported to explain a range of mesmeric (and spiritualist) phenomena. There was an element of risk in this, since many considered mesmerism to be just as worthless a practice as phrenology.[81] However, nothing was made of these experimental foundations. The emphasis of Carpenter's work was on his abstract mental schema rather than the physical basis of his ideas. When Carpenter told the Royal Society in 1850 that nervous power "must be regarded as the highest of all forms of vital force," this was not the same vital agent that had earned Hall his criticisms; it referred to the general processes by which the body kept itself alive.[82] By explicitly identifying nervous energy as this force, moreover, Carpenter served two other functions: it drew attention away from his background in radical morphology, and from some of his more dubious sources of evidence. The new approach to mental physiology could afford to disregard these potentially divisive issues.

Laycock's enterprise was physiological yet abstract, scientifically informative yet politically neutral. But this did not mean that the professional community accepted his ideas – they didn't. But significantly, Laycock was able to influence Carpenter, who found that Laycock's theory could explain a range of potentially dubious phenomena with a sense of cool detachment. In fact, so successful was Carpenter's exposition that the theory of "cerebral reflex function" was commonly associated with him, not Laycock. Indeed, when Laycock published his *Mind and Brain* in 1860, he felt it necessary to add an appendix outlining his claim to priority.[83] One of the reasons why the laurel wreath went to Carpenter instead of Laycock,

---

[79] Cf. Desmond (1989, 210–222). Winter (1997).

[80] Carpenter (1855, 649).

[81] Winter (1998, 287–90).

[82] Carpenter (1850, 746).

[83] Laycock (1860, vol. 2, 465–80).

Smith has argued, was because of his "obscure and discursive presentation of the theory of brain reflexes."[84] Compared with Carpenter, already known for his clear and accessible style (his *Principles of Human Physiology*, written as an introductory textbook for anatomy students, would reach six editions), Laycock's philosophical musings must have seemed esoteric, even mystical.[85]

5 Conclusion

Laycock seems to have received as much praise for his theory as Combe and Hall did for theirs – hardly any. While their work did not enjoy the support of the entire scientific community, their failures are nonetheless interesting as attempts to find the right way to describe the mind and to examine its processes. Just as Carpenter found that Combe and Hamilton's imagined courtroom, Hall's obstinate self-promotion, Vindex's melodramatic aggression and Laycock's esotericism were not ideal ways to disseminate scientific ideas, so too did it become clear that a physiological account of the mind required its own methods and terminology. In this regard, it is also interesting that this new field of study – physiological psychology – increasingly adopted vitalistic notions: phrenology had no place for abstract processes, but reflex action was teleological within a particular nervous structure and cerebral reflex function posited that the mind was a schema of intangible mental organs. But while this should not be seen as a deliberate move towards vitalism, especially since each study was an attempt to place the mind within the body rather than asserting its individuality, neither should this be seen as a type of explanatory retreat by physiologists; although more vitalistic ideas were used in response to criticisms levelled at earlier theories, this should not be seen as a flight into obscurity, as traditional histories would have it. Combe, Hall and Laycock may in some ways be considered failures, but only because they were naïve: Combe and Hall because they misunderstood the complexities they were facing, Laycock because his head was in the clouds. To create a successful physiology of mind required a clear understanding of what was known, what could be known, and what could be comprehended by the public. Of course the first sin was graver, but only through this were the limitations of their enterprise uncovered and a new discipline formed; without a clear outline of this field and its new methods, the second sin, few would have been convinced that mental phenomena could be studied in this way. Carpenter recognised all of these aspects of debate, and it was his version of physiological psychology that survived.

From this perspective vitalism was, and is, much more than a specific philosophical outlook or a simplistic explanatory device. To see it as such only leads to the same problems raised at the beginning of this paper: it would lead to a vast array of

[84] Smith (1971, 91).

[85] There is a similarity here with James Hutton's woeful exposition of his theory of the earth. Cf. Playfair (1962, vi–xi).

theories with no other apparent connection except that they were "vitalistic." But thought of as a conceptual tool – one that is, admittedly, created out of the debris of previous theories – it can once again be used for interesting historical analysis. The relationship between experiment and explanation, scientist and audience, and acceptance and outright dismissal become objects for historical study. Although from this perspective, one can no longer discuss a singular history of vitalism in anything other than a deeply philosophical sense, what replaces it is a much more complex and fecund concept: the *histories* of vitalism.

Acknowledgments My thanks to John Forrester, Nick Hopwood, Jim Secord, Charles Wolfe and an anonymous reviewer for their helpful comments on this work.

References

Albury, Randall. 1977. Experiment and explanation in the physiology of Bichat and Magendie. *Studies in the History of Biology* 1: 47–131.

Anon. 1832. A brief account of a particular function of the nervous system. *Proceedings of the committee of Science and Correspondence of the Zoological Society of London* 2: 190–192.

Anon. 1834. *Report of the third meeting of the British Association for the advancement of science; Held at Cambridge in 1833*. London: John Murray.

Anon. 1836–1837. Remarks on the history of some discoveries in medical science. *London Medical Gazette* 20: 657–661.

Anon. 1841. Thomas Laycock, A treatise on the nervous diseases of women. *Medico-Chirurgical Review* 35: 100–108.

Anon. 1844. Meeting of the British Association at York. *Lancet* 44(1101): 56–57.

Barfoot, Michael. 1995. *"To ask the suffrages of the patrons": Thomas Laycock and the Edinburgh Chair of Medicine*. London: Wellcome Institute for the History of Medicine.

Bentley, David. 1998. *English criminal justice in the nineteenth-century*. London: Hambledon.

Benton, Edward. 1974. Vitalism in nineteenth-century scientific thought. *Studies in History and Philosophy of Science* 5: 17–48.

Canguilhem, Georges. 1994. *A vital rationalist: Selected writings of Georges Canguilhem*. Trans. A. Goldhammer. New York: Zone Books.

Cantor, Geoffrey. 1975a. The Edinburgh phrenology debate, 1803–1828. *Annals of Science* 32: 195–218.

Cantor, Geoffrey. 1975b. A critique of Shapin's social interpretation of the Edinburgh phrenology debate. *Annals of Science* 32: 245–256.

Carpenter, William B. 1838. Hall, Grainger, Mayo, on the physiology of the spinal marrow. *British and Foreign Medical Review* 5: 486–540.

Carpenter, William B. 1839. *Principles of general and comparative physiology*. London: John Churchill.

Carpenter, William B. 1850. On the mutual relations of the vital and physical forces. *Philosophical Transactions of the Royal Society of London* 140: 727–757.

Carpenter, William B. 1855. *Principles of human physiology*, 4th ed. London: John Churchill.

Clarke, Edwin, and Leon Stephen Jacyna. 1987. *Nineteenth-century origins of neuroscientific concepts*. Berkeley: University of California Press.

Combe, George. 1826. Letter from George Combe to Francis Jeffrey, Esq. *Phrenological Journal and Miscellany* 4: 1–78.

Combe, George. 1828–1829a. To the editor of the Caledonian Mercury. *Phrenological Journal and Miscellany* 5: 11–14.

Combe, George. 1828–1829b. To the editor of the Caledonian Mercury. *Phrenological Journal and Miscellany* 5: 26–30.
Combe, George. 1828–1829c. Combe to Hamilton. *Phrenological Journal and Miscellany* 5: 56–57.
Combe, George. 1828–1829d. Combe to Hamilton. *Phrenological Journal and Miscellany* 5: 62–65.
Cooter, Roger. 1984. *The cultural meaning of science: Phrenology and the organization of dissent in nineteenth-century Britain*. Cambridge: Cambridge University Press.
Desmond, Adrian. 1989. *The politics of evolution: Morphology, medicine and reform in radical London*. Chicago: University of Chicago Press.
Dixon, Thomas. 2003. *From passions to emotions: The creation of a secular psychological category*. Cambridge: Cambridge University Press.
Driesch, Hans. 1914. *The history and theory of vitalism*. Trans. C.K. Ogden. London: Macmillan & Co.
Fearing, Franklin. 1964. *Reflex action: A study in the history of physiological psychology*. New York: Hafner.
Figlio, Karl. 1975. Theories of perception and the physiology of mind in the late eighteenth-century. *History of Science* 13(3): 177–212.
Gieryn, Thomas F. 1983. Boundary-work and the demarcation of science from non-science: Strains and interests in the professional ideologies of scientists. *American Sociological Review* 48(6): 781–795.
Grainger, Richard Dugard. 1837. *Observations on the structure and function of the spinal cord*. London: Samuel Highley.
Greco, Monica. 2005. On the vitality of vitalism. *Theory, Culture and Society* 22: 15–27.
Hall, Marshall. 1833. On the reflex function of the medulla oblongata and medulla spinalis. *Philosophical Transactions of the Royal Society* 125: 635–665.
Hall, Marshall. 1836a. On the nervous system, being a Lecture delivered at the Aldergate School of Medicine, on Jan. 6, 1836. *London Medical Gazette* 17: 632–641.
Hall, Marshall. 1836b. *Lectures on the nervous system and its diseases*. London: Sherwood, Gilbert & Piper.
Hall, Marshall, and Samuel Daniel Broughton. 1835. *Report on the progress made in the experimental inquiry regarding the sensibilities of the cerebral nerves, Recommended at the last meeting of the association*. Report of the fourth meeting of the British Association for the advancement of science; Held at Edinburgh in 1834. 676–680. London: John Murray.
Hamilton, William. 1828–1829a. Letter to the editor of the Caledonian Mercury. *Phrenological Journal and Miscellany* 5: 1–8.
Hamilton, William. 1828–1829b. Letter to the editor of the Caledonian Mercury. *Phrenological Journal and Miscellany* 5: 14–19.
Hamilton, William. 1828–1829c. Hamilton to Spurzheim. *Phrenological Journal and Miscellany* 5: 38–39.
Hamilton, William. 1828–1829d. Hamilton to Spurzheim. *Phrenological Journal and Miscellany* 5: 42–43.
Hamilton, William. 1828–1829e. Hamilton to Combe. *Phrenological Journal and Miscellany* 5: 52–56.
Hamilton, William. 1828–1829f. Hamilton to Combe. *Phrenological Journal and Miscellany* 5: 58–62.
Hatfield, Gary. 1995. Remaking the science of mind: Psychology as natural science. In *Inventing human science: Eighteenth-century Domains, 184–231*, ed. Christopher Fox, Roy Porter, and Robert Wokler. Berkeley: University of California Press.
Hodge, Clifton F. 1890. A sketch of the history of reflex action. *The American Journal of Psychology* 3(149–67): 343–363.
Jacyna, Leon Stephen. 1981. The physiology of mind, the unity of nature, and the moral order in victorian thought. *British Journal for the History of Science* 14(2): 109–132.
Lawrence, Christopher, and George Weisz (eds.). 1998. *Greater than the parts: Holism in biomedicine, 1920–1950*. Oxford: Oxford University Press.

Laycock, Thomas. 1845a. On the reflex function of the brain. *The British and Foreign Medical Review* 19: 298–311.

Laycock, Thomas. 1845b. Correspondence between Geo. Combe, Esq., Professor Reid, and Dr. Laycock, on the reflex anatomy and physiology of the brain. *Lancet* 46(1152): 347–348.

Laycock, Thomas. 1846. Dr. Laycock on the reflex function of the brain, in Reply to "Vindex". *Lancet* 47(1180): 424–425.

Laycock, Thomas. 1860. *Mind and brain*, vol. 2. Edinburgh: Sutherland & Knox.

Lenoir, Timothy. 1982. *The strategy of life: Teleology and mechanics in nineteenth-century German biology*. Dordrecht/London: Reidel.

Leys, Ruth. 1990. *From sympathy to reflex: Marshall Hall and his opponents*. Harvard dissertations in the History of Science. New York: Garland Publishing.

Manuel, Diana E. 1996. *Marshall Hall (1790–1857): Science and medicine in early Victorian Society*. Amsterdam: Rodopi.

Morrell, Jack, and Arnold Thackray. 1981. *Gentlemen of science: Early years of the British Association for the Advancement of Science*. Oxford: Clarendon.

Playfair, John. 1962. *Illustrations of the Huttonian theory of the earth, with introduction by G. White*. New York: Dover.

Scott, Syme, and Christison. 1828–1829. Proceedings of the arbiters in the reference by Sir William Hamilton and Mr Combe, on the Anatomical Facts of Phrenology. *Phrenological Journal and Miscellany* 5: 34–35.

Shapin, Stephen. 1975. Phrenological knowledge and the social structure of early nineteenth-century Edinburgh. *Annals of Science* 32: 219–243.

Shapin, Stephen. 1979. The politics of observation: Cerebral anatomy and social interests in the Edinburgh phrenology disputes. In *On the margins of science: The social construction of rejected knowledge*, ed. R. Wallis, 139–178. Keele: University of Keele.

Smith, Roger. 1992. *Inhibition: History and meaning in the sciences of mind and brain*. Berkeley: University of California Press.

Smith, Roger. 1971. *Physiological psychology and the philosophy of nature in mid-nineteenth century Britain*. PhD dissertation, Cambridge University.

Spurzheim, Jakob. 1828–1829a. Spurzheim to Hamilton. *Phrenological Journal and Miscellany* 5: 41.

Spurzheim, Jakob. 1828–1829b. Spurzheim to Hamilton. *Phrenological Journal and Miscellany* 5: 47.

Temkin, Owsei. 1946. Materialism in French and German physiology of the early nineteenth century. *Bulletin of the History of Medicine* 20: 322–327.

Van Wyhe, John. 2004. *Phrenology and the origins of victorian scientific naturalism*. Aldershot: Ashgate.

Veitch, John. 1869. *Memoir of Sir William Hamilton*. Edinburgh: W. Blackwood.

Vindex. 1846a. The reflex function. *Lancet* 47(1172): 197–198.

Vindex. 1846b. Dr. Laycock and his reflex function of the brain. *Lancet* 47(1183): 510–511.

Winter, Alison. 1997. The construction of orthodoxies and heterodoxies in the early Victorian life sciences. In *Victorian science in context*, ed. B. Lightman, 24–50. Chicago: University of Chicago Press.

Winter, Alison. 1998. *Mesmerized: Powers of mind in Victorian Britain*. Chicago: University of Chicago Press.

Wolfe, Charles T. 2011. Vitalism. In *Encyclopedia of astrobiology*, ed. M. Gargaud, 1749–1750. Berlin: Springer.

Young, Robert M. 1973. Association of ideas. In *Dictionary of the history of ideas*, vol. 1, ed. P. Wiener, 111–118. New York: Scribners.

Part II
Twentieth-Century Debates on Vitalism in Science and Philosophy

Chapter 6
Vitalism Versus Emergent Materialism

Brian Garrett

> *... and unless ... there is absolute emergence the whole movement and travail of the creation is but a barren shuffling about of the same pieces.*
>
> Arthur Lovejoy (1927)

Abstract During the nineteenth century vitalist theories of life were contrasted with mechanistic and materialist hypotheses regarding the nature of life. Religious, philosophical and empirical reasons were offered for vitalism by numerous thinkers. Mechanistic theories of life appeared problematic, despite their steady empirical success. Emergent evolutionism was thought by some (mostly English-speaking thinkers) to be a compromise position between vitalism and materialism, taking mechanism from the materialists and nonreductionism from the vitalists. The debate was interrupted by World War II and largely forgotten after the discovery of the double helix. In this chapter I introduce some of the thinkers involved and articulate the fundamental tenets and aspirations of vitalists and emergent evolutionists, explaining the philosophical debate (and confusion) over the concepts invoked by each side. I draw on philosophers and biologists from 1900 to 1930 in this study.

Keywords Emergence • Emergent materialism • Reductionism • Vitalism

---

B. Garrett (✉)
McMaster University, 1280 Main St W, Hamilton L8S 4L8, ON, Canada
e-mail: spinozafan2000@yahoo.ca

S. Normandin and C.T. Wolfe (eds.), *Vitalism and the Scientific Image in Post-Enlightenment Life Science, 1800–2010*, History, Philosophy and Theory of the Life Sciences 2, DOI 10.1007/978-94-007-2445-7_6,

1 Introduction

George Santayana famously remarked:

> Progress, far from consisting in change, depends on retentiveness. When change is absolute there remains no being to improve and no direction is set for possible improvement: and when experience is not retained, as among savages, infancy is perpetual. Those that cannot remember the past are condemned to repeat it.[1]

I find myself in sympathy with Santayana, when I reflect upon the philosophy of his time in comparison with my own. Twenty-first century metaphysicians have become "savages" instinctively pursuing their presuppositions – their "intuitions" and semantic preferences – forgetful of the history of their concepts. Twenty-first century metaphysicians of mind have become obsessed with, among other things, the possibility of nonreductive theories of mind, just as philosophers of the 1920s were. Nonreductive materialism has dominated the scene since the 1970s and has been by far the most popular doctrine among (especially) American philosophers. Nonreductive materialism is the view that all events are physical but that some properties of events are not physical. But in the last years of the twentieth century, nonreductivism in its various forms again came under attack from more severe materialists, best exhibited by Jaegwon Kim's resuscitation of the exclusion argument. The exclusion argument holds that there is only one sufficient cause for each effect and so, if all physical effects have physical causes all other putative causes are excluded. Thus no nonphysical mental or vital cause can be operative. The fate of nonreductive materialism is not yet decided, apparently, but if the past is any guide I feel doubtful that the position will survive.[2] After a wave of enthusiasm for emergentism in the 1920s, a generation later the position was entirely abandoned. Perhaps contemporary nonreductive materialism has survived, so far, because those who cannot remember the past are condemned to repeat it?[3] We were warned in the 1920s 'lest we forget', and yet another world war followed, so it is human that we do forget. A *cursory* look at the debates of the early twentieth century offers the impression that little has been learnt and we have been entertaining arguments rehearsed before World War II. Some form of intellectual amnesia, perhaps due to the trauma of World War II and the ascendency of "linguistic" and formal philosophy before and after, appears to have taken hold.

I shall take this cursory commentary somewhat seriously, utilizing it as a dramatic foil to help explicate my historical material with comparisons, admittedly very brief, to current debates over non-reductionism. In what follows I focus on the

---

[1] Santayana (1905).

[2] I have argued in the past, however, that nonreductive materialism has more resources than Kim suggests in offering an "overdetermination" reply to the exclusion argument. I am no longer as sanguine regarding overdetermination as I once was (Garrett 1999, 2000, 2006).

[3] Cf. Blitz (1992).

debate over the autonomy of biology, as it was played out among select biologists and philosophers of the first decades of the twentieth century. The topic is vast, so I focus on the more philosophical disputes between vitalists, emergentists and to a lesser extent, materialists.[4] I briefly examine the vitalist thought of Hans Driesch [1867–1941] and the emergentist theories of biologists J. Arthur Thomson [1861–1933] and H. S Jennings [1868–1947]. I draw on Arthur Lovejoy's [1873–1962] influential discussions of emergence and that of several other philosophers, C. Lloyd Morgan [1852–1936], C.D. Broad [1887–1971] and Samuel Alexander [1859–1938].[5] But given that the latter have received some attention in the historical literature I shall focus on Driesch, Thomson, and Jennings mostly.[6] Biologists Joseph Needham [1900–1995] and Jacques Loeb [1859–1924] represent the materialist mechanist camp, and I shall use D'Arcy Wentworth Thompson's [1860–1948] defense of mechanism in my discussion.[7]

The period involved a wealth of exploration, and other positions, tangentially related to those above, were also recommended.[8] J.S. Haldane's [1860–1936] and J.C. Smuts' [1870–1950] holistic nonreductionism, for example, was intended to be neither vitalist nor reductionist, and so is best seen as an emergentist position, but both were unclearly related to the "organicist" position stated infamously by Alfred Whitehead.[9] Like Haldane and Smuts, Whitehead rejected both reductionism and vitalism, but he tied his concept of organism to his rather obscure notion of "concrescence" and to the "self-creative unity of the universe" that puts its "decisive stamp of creative emphasis upon the determinations of efficient cause."[10]

In this chapter I proceed as follows; first I indicate the *prima facie* case for intellectual amnesia – indicating the broad similarities in debate and doctrine of the two periods. I then turn to selected details and I offer an account of Driesch's vitalism and a longer discussion of the emergentists, Thomson and Jennings. I shall focus on the philosophical discussions offered by these biologists on the autonomy of biology. Issues of determinism, causation and the nature of reducibility will be foremost,

[4]Nouvel (2011).

[5]Alexander (1920).

[6]Burroughs (1912).

[7]'Materialism' is here limited to biology alone. Descartes is thus a mechanist about biology but obviously not a mechanist or materialist regarding psychology. Indeed, many early twentieth-century materialists in biology conceded that the mind was another problem altogether. Second, all the thinkers I discuss hold the conditional "if x is mechanical then x is material or physical". But it should be noted that very diverse views of the physical are consistent with this conditional. Drake held that all physical events have a psychic aspect to them, some of which is revealed in self-consciousness. Thus, although Drake accepts the conditional above he would *not* accept the claim that if x is psychic then x is *not* mechanical. Third, the very concept of "mechanism" was part of the debate. For example, could a mechanism be a self-maintaining entity, as organisms appear to be?

[8]Cf. Boodin (1925), Drake (1925). Cf. Strong (1918) for panpsychic responses to the problem of emergence. For the revival of these arguments cf. Strawson (2006).

[9]Hein (1969).

[10]Whitehead (1929).

along with a discussion of the causal exclusion argument and the causal efficacy of the irreducible (known today as "downward causation").[11]

2 Amnesia Versus Evolution

Why then, might we suspect intellectual amnesia? The main reason for believing in intellectual amnesia is the similarity between doctrines and debates of the two periods along with such a gap of time between them. First there are some broad similarities. In the 1920s the philosophy of biology was taken very seriously and the relevance of evolution to philosophy was debated with much enthusiasm.[12] Biologists were philosophical and philosophers could not ignore evolution and theories of life. In the 1980s the philosophy of biology makes a comeback, having been ignored in the decades after World War II. Evolution is taken up again as a possible resource for the analysis of properties studied in the philosophy of mind. Like the 1920s, the 1980s saw a wave of debate over the naturalization of teleology and its relation to natural selection. But after World War II the *philosophy* of biology all but disappeared, requiring a rebirth.[13] The philosophy of biology in the 1920s was assumed to have some relevance to perennial philosophical issues – for example, the nature of free will and materialism. Emergentists and vitalists were often motivated by the concern for protecting free will (linking free will to "novel" behavior), seeing their non-reductivism as a necessary foundation for their beliefs.

Like then, philosophy of biology has witnessed a renaissance with the work of Elliot Sober, Philip Kitcher, John Dupré, Alex Rosenberg, and William Bechtel, to mention only a few, and its relevance for the philosophy of mind was greatly stimulated by Ruth Millikan's "biological" approach to intentionality and teleology (or "proper function"). Emergentism today makes an appearance in the free will debate and is intended to play a similar role in supporting libertarianism, as it did in the 1920s.[14] After considerable neglect, the relevance of evolution to free will has been re-recognized and developed, not least in the work of Daniel Dennett.

[11]Lloyd Morgan self-consciously traces emergentism to the empiricist philosophy of J.S. Mills [1806–1873] and G.H. Lewes [1817–1878]. Since this is well known I will not touch on it. C.D. Broad and Lloyd Morgan's views are rightly the subject of essays on emergentism – for this reason I examine the less discussed figures J. Arthur Thomson and H.S. Jennings. But the influence of Hegel and Herbert Spencer on the popularity of emergentism is not fully appreciated and is the subject for another paper. Related to this theme I also ignore discussions of cosmic emergence. And finally, I merely touch upon the criticisms of mechanistic evolution leveled by emergent evolutionists. Exactly why they thought mechanistic evolution was incapable of creating "novel" modes of action, or of unity, is somewhat obscure but much was connected to common criticisms of Darwinism, also a topic for another chapter.

[12]Cf. Strong (1918) and Conger (1929).

[13]Cf. Nagel (1961) for the postmortem nail in the coffin.

[14]Cf. O'Connor (2000). Emergentism is not a doctrine that, in itself, necessarily supports libertarian conceptions of freewill, for one might not apply it to consciousness. Although I am not sympathetic, the idea of an irreducible self that produces "novel" unpredictable choices is an idea many find necessary for free will.

Another broad similarity between the periods can be seen in the debate over materialism. The conservation of energy had begun to strangle the immaterial. Nineteenth century materialists (such as Huxley) took the determinism and completeness of physical explanation seriously, recognizing that either the mental or vital are identical to matter or they are epiphenomenal with regard to material change. But arguments for irreducibility, focusing mostly on the alleged irreducibility of teleological phenomena (which includes the intentional), and simple incredulity at the reduction of consciousness, pushed against reductive materialism. We find then, as we do today, the same incredulity regarding the reduction of consciousness and, like then, we have an intense debate over the nature of materialism and the associated concept of mechanism, along with the recognition of the force of the "causal exclusion argument."

The exclusion argument, as it is called today, was a recognized consequence of materialism during the late nineteenth century. If all physical events have physical events as causes, then *irreducibly* psychic or vital events have no effect upon the physical world.[15] William James called it the "automaton theory" and believed it entailed that consciousness is: "the melody [that] floats from the harp-string, but neither checks nor quickens its vibrations; so the shadow runs alongside the pedestrian, but in no way influences his steps."[16] The exclusion argument is still front and centre for any dualist – whether emergentist, vitalist, functionalist or conceptualist[17] – and is the main metaphysical argument against such nonreductionism, as it was back in the 1920s.[18] Although the philosophy of biology appears to have retreated after World War II only to be revived in the 1980s, arguments over materialism continued quietly and the exclusion argument remained a serious concern for postwar thinkers, like J.J.C. Smart, who complained that nonreductionists were indeed committed to nomological danglers, or epiphenomenalism. Nevertheless, ontological discussions of materialism were no longer so popular, the focus having turned to the linguistic or theoretical expressions of ontology.[19] Philosophy

---

[15] Not always made explicit is the further necessary Ockhamist premise that no event has more than one sufficient cause. The argument is this: (1) all physical events have physical causes (2) No event has more than one complete cause (3) biological or mental events are not identical with physical events (non-reductionism), therefore, (4) no biological or mental event causes a physical event: epiphenomenalism.

[16] James (1890, 133).

[17] The conceptualist is one who relies on conceptual thought-experiments for their nonreducibility. So David Chalmers' defense of the irreducibility of consciousness will count here as conceptualist. In the last section I note how things have changed from the 1920s and one of these is the sad retreat away from the empirical as a defense of a non-reductionism.

[18] As we see below Thomson defines a view called "methodological vitalism" which echews ontological considerations for conceptual and methodological considerations. The language and concepts of biology are not mechanistic but neither do they refer to something nonphysical.

[19] See Malaterre (2013) (this volume). I concur with Malaterre's narrative that the debate was mostly ontological in the 1920s and became theoretical or epistemological in the 1950s. However, I don't think the *definition* of emergent phenomena, as those that cannot be deduced from physical theory, changed. Lovejoy uses it. Second, the uncertain ontological implications of the definition were already forefront in the debate during the 1920s since it was a common reply to the vitalist also. But ontology was put aside in the mid-century so there was little left to debate. The loss of interest in philosophy of biology and in ontology during this period is constitutive of what I call here amnesia.

according to Carnap should clarify the logic of science, no more.[20] But despite this debate over nonreductivism familiar from the 1920s, nonreductivism steamed ahead in the late 1960s apparently ignoring the exclusion argument in the excitement over multiple-realizability, the return of holism, and the development of computationalism. By the early 1980s, after a wave of nonreductivist arguments, the exclusion worries were brought back, apparently having been forgotten. They were first leveled against Davidson's nonreductivism and then against nonreductivism generally.

Finally, the characterization of mechanism and reductionism was under some dispute, as today the proper characterization of materialism and reductionism remains controversial. Are there arguments for the irreducible that are not merely arguments *against* mechanism and materialism? Famously, the early twentieth century saw enormous changes in our theories of the physical. The relativity of space-time (and the merging of space and time into space-time) along with the introduction of indeterminism as a possible ontology allowed theorists to doubt the exclusion argument as a symptom of false presuppositions regarding the physical. Perhaps the physical is not so "mechanistic," after-all, goes the still popular refrain.[21]

Further, less significant similarities can be discerned between the periods. The possibility of artificial intelligence and the horror of robots were raised in the debate during the 1920s and 1930s, as it is today. Karel Capek's play *RUR (Russom's Universal Robots)* (first performed in 1921), in which the artificial workers revolt against their human masters, could be invoked for rhetorical support, typically against the reductive materialists.[22]

Thus the concerns and questions of the period are quite similar to those of today: Must we choose between dualism and materialism, or, can nonreductive materialism offer a halfway house, reconciling the arguments and aspirations for irreducibility with arguments for materialism? What sense can be made of "levels of nature" or of the "novel" or "irreducible" in nature (i.e. how are we to understand the autonomy of biology or of Life)? Is "novelty" an epistemic or an ontological feature, according to the nonreductive materialist?[23] Do we have empirical reasons for such nonreductivism and, of course, can we put Humpty Dumpty together again: How can the irreducible properties be causally efficacious properties? What can these properties *do*, given the activities of the

[20] Strawson's (1961) was the exception that proved the rule. At the time it was thought to be a radical departure from Wittgenstein, Austin and the positivist account of philosophy.

[21] Campbell and Bickhard (2011).

[22] Cf. Grossman (1930) who refers to the play "RUR (Rossum's Universal Robots)" by Karel Capek, first performed in 1921. Earlier, R.F. Alfred Hoernle invoked Samuel Butler's *Erewhon*. In both cases the authors mention the dramatic fact that the machines revolt against their makers and both are dubious of the analogy between machines and organisms.

[23] Cf. Emmeche et. al. (1997). Despite their historical overview they still muddy ontology with epistemology, writing: "One of the main characteristics of emergence was the formation of new properties, that is, properties which could not be predicted" (101).

physical properties they emerge from, or supervene upon?[24] Do they interact with the physical, somewhat like a dualist or vitalist would assert, or, are they oddly irreducible yet epiphenomenal? Or is there a third option, that of "downward causation" in which the whole is a cause of its constituent parts? The emergentists always put emphasis on the unpredictability of the emergent "novel" property so they had to ask: How are our current reductive explanatory failures related to predictability and our ontology of the irreducible? Are they signs of human ignorance or of the divisions found in Nature herself?

But a hypothesis of historical amnesia is less certain than a hypothesis of evolution, growth and extinction. After all, as Santayana says, continuity is a necessary presupposition of history. A hypothesis of evolution utilized to explain cursory similarities between debates and doctrines, especially for debates that are not even 100 years apart, is more likely than a hypothesis of amnesia, which requires the arrest and subsequent re-ignition of a debate. Nevertheless, extricating ourselves from the accusation of intellectual amnesia is a worthwhile project. Similarity in argument and doctrine between periods may be due to their close historical proximity and inheritance, but it may also be an effect caused by a theory being all but abandoned, only to be reborn from the ashes of some other theoretical failure.[25]

Below I shall point out what appears to be relatively unchanged in the debate and how that gives rise to the accusation of amnesia. But by the conclusion of our discussion we will be able to see how the debate has indeed been transformed, rather than entirely forgotten and revived. The context – foreground and background – has changed. To use the expressions of Dewey and Santayana, I reckon that nonreductivism today is a "broken-backed, half-hearted" view that continues to struggle with the problems of its full-blooded, empirically proud ancestor.[26] But although the stage has changed and the "props" are more sophisticated, the core problems and arguments, particularly issues of the ontological role of the irreducible, remain pretty much the same. Or so *I* contend.

---

[24] I speak loosely here. The use of the term "supervenient" in the 1920s was just to mean "coming after," not the asymmetric dependence relation familiar from Davidson and later, Kim. However, the emergent properties do seem to possess similar modal properties as those we dub supervenient today, as one can see in Jennings' work.

[25] Is seventeenth-century materialism (i.e. Descartes and Gassendi, who both denied the need for nonphysical explanation in biology) much different from its ancient predecessors (Epicurus and Lucretius)? At one level of (gross) description, it is tempting to say that atomism was forgotten by the Christian West, hence the seventeenth century philosopher had to relearn the arguments for and against such views. We thus see Ralph Cudworth's monumental and encyclopedic discussion of materialism and "atheism" in his *Intellectual System of the World* as offering a solution to their amnesia. Cf. Wolfe (2009) for a more subtle discussion of early modern Epicureanism.

[26] Santayana called Dewey's naturalism "half-hearted" complaining that Dewey was stuck in the foreground (i.e. experience of the world) rather than in the background – the world itself. Dewey replied that Santayana's epiphenomenalist materialism was "broken-backed."

3 Emergentism Cures Vitalism

The development of emergentism as an alternative to substantial dualist interactionism promised to be nonreductivist, meaning by this "nonmechanical," yet also physicalist, meaning that it did not countenance the "indeterminism" involved in Cartesian interactionism.[27] Emergentists wished to retain their allegiance to physical explanations of physical phenomena, maintaining the "continuity of nature" or the causal completeness of the physical (all physical events have physical causes) while allowing that mental and living entities (or "unities") are not entirely mechanistically explicable. Emergentists were keen to avoid the perceived epiphenomenalism of materialism and so conceived of the novel irreducible emerging properties as causally efficacious properties; having an impact on the path and direction of the physical. Today this is called "downward causation" and was identified with the causal efficacy of the irreducible in the 1920s. Emergentism between the wars had fast become the most popular nonreductivist theory of the relationship between mind, life and matter, supplanting Cartesian-style interactionism associated with Bergson's *élan vital* and Hans Driesch's entelechy. It is tempting to say with the waning Anglo-Hegelian[28] spirit of the time, that emergentism was the intended synthesis of anti-materialist and anti-dualist thinking. Thus a three-way debate had emerged between materialists, emergentists and the old-fashioned, but still popular, dualist vitalists.[29]

On one side there were the reductive materialists who thought that eventually a mechanistic interpretation of life and (more rarely) mind would be available. Darwin was often taken as the representative of this nineteenth-century tradition. In biology at the turn of the century Jacques Loeb, among others, advocated for "mechanism," finding the alternatives rather obscure.[30] Given the growing anti-metaphysical tenor of the times, the materialists could ride a wave of methodological criticisms of vitalist positions. Did the vitalists have empirically testable evidence in support of their views, or were they mainly engaged in polemics against their opposition? If merely the latter, then no good scientist would be convinced, lacking positive evidence for the hypothesis.

Opposed to the reductive materialists and vitalists, however, were the emergentists who held that biology (and sometimes psychology) was in some sense autonomous

[27] Materialism then, as now, is defined as a commitment to the causal closure of the physical: that every physical event has a physical cause. As defined, it is neutral with regard to reductionism, allowing for nonreductive and reductive forms of physicalism.

[28] Santayana described Alexander as a Hegelian in 1900.

[29] We cannot describe Bergson or Driesch as supporting substance dualism since the concept of substance had come into some disrepute with both. But the *élan vital* and the entelechy do not emerge from the material properties of the organism despite their problematic relations or need for such properties. There were, of course, Hegelian-inspired idealists who were still offering teleological conceptions of the universe on idealist principles. Bosanquet is the best example. The problem of the individual's relation to the absolute and the nature of the living or mental individual was thought to be related such that two otherwise distinct symposia could be reproduced together in one volume in Clark (1918). Cf. Neal (1916).

[30] Allen (2005).

from the physical sciences, without this implying the violation of the continuity of nature. Non-mechanical features depend upon and arise from the mechanical but cannot be understood as identical or reducible to those features. Genuine novelty emerges out of the mechanical. Then, as today, nonreductivism was the most popular doctrine, especially among the philosophers and the philosophically inclined biologists. But emergentism was plagued by ambiguities and puzzles.

The most prominent puzzle for the thinkers of the time[31] was how to understand the simultaneous novelty and irreducibility of the mental and the living, while also maintaining that all physical change had physical antecedents. How, that is, does emergentism fare better than substantial interactionism when it comes to the fundamental physical laws like that of the conservation of energy?[32]

The second problem was the notion of novelty and the type of prediction utilized. Mental and living "unities" emerge from the physical organization of their bodies and are thus novel entities. But in what does this novelty consist? Emergentists were keen to capture the idea in terms of unpredictability, an epistemic property of theories. The emergent properties were those that could not be predicted from a description of organized physical properties and their laws. As Lloyd Morgan was fond of saying on behalf of Samuel Alexander, the emergence of life from matter must be accepted with "natural piety"; the limits of reductive explanation being met. But the failure of predictability, being an epistemic property, is obscurely related to the ontological issues that motivate the view. When emergentists introduce their position it is often put ontologically: the novelty is not merely unpredictable but the unpredictability is a sign of an ontological novelty – a new "unity." The obscure relation between reductive failure and ontological irreducibility remains in the contemporary debate, and reinforces a feeling of *déjà vu.*[33]

4 Hans Driesch's Vitalism[34]

Hans Driesch and Henri Bergson were the most famous vitalists of the early twentieth century. Both Bergson and Driesch felt their view was new and avoided the problems of past vitalist theories, and both appealed to the existence and nature of teleological processes as indicative of life. Both were keen to distance themselves from the theological

[31] McDougall (1929). McDougall correctly complained, as Ernst Nagel would 30 years later, that deductive failure being epistemic, applied equally for unfamiliar physical conditions as it did for unfamiliar biological entities. Although a good point, McDougall was trying to defend dualism, whereas Nagel uses it in support of reductionism.

[32] The concept of substance was in some disrepute, however, so Bergson and Driesch avoided the term. Although both agreed that material properties were necessary for life, neither thought that life supervened upon the material.

[33] Cf. Stalnaker and Block responding to Jackson and Chalmers, and Dupré 1993. Concepts of irreducible "unity" remain in constitution theories of composition: cf. Baker (2007).

[34] Cf. Freyhofer (1982).

debate over teleology and God, rejecting any inference from the presence of teleology in natural phenomena to the existence of a conscious purpose-giver, or God. Bergson's speculations were *a priori* and philosophical, and made sparse and dubious use of scientific inquiry. His contemporaries did not stop reminding their readers of Bergson's selective use of empirical research. Driesch, however, was a scientist turned philosopher, so his views were nominally based on his empirical research into sea urchins.

Driesch had studied with August Weismann and Ernst Haeckel but differed with his teachers over the autonomy of biology. In 1891 Driesch performed experiments upon the blastomeres of sea urchins. When he cut blastomeres having two cells in half, each cell would still develop into a smaller but fully formed sea urchin. This was a remarkable phenomenon that Driesch argued could not be mechanical and led credence to the belief in an entelechy. Driesch's career quickly took him from empirical research into philosophy. In 1907 he was made chair of natural theology at the university of Aberdeen. He popularized his scientific results for a receptive English-speaking world through the Gifford lectures (1908) and his *History of Vitalism* (1918).[35] In contrast to Bergson, who had little *practical* experience in empirical matters, Driesch brought the authority of science to his theories, much as did emergentist physiologist J. S. Haldane. It is hard, however, to disagree with historian of science Georges Canguilhem who writes of Driesch (and we could include J.S. Haldane here too):

> The vitalist biologist who turns philosopher of biology thinks he brings a certain capital with him to philosophy, but in reality he brings to it only rent, which continually decreases in the market of scientific values – for the simple reason that empirical research, in which he no longer participates, continues to move forward.[36]

There is little doubt that Driesch's scientific capital was losing value as the twentieth century matured. The proposal of an immaterial entelechy was rarely urgent for the working biologist and appeared mostly when the biologist reflected upon his work, not so much while engaged in it. By 1930 Joseph Needham[37] could confidently remark:

> I cannot refrain from adding that a famous series of Gifford lectures was entirely devoted to introducing a certain biological phantom into the (usually) polite society of theology. I felt it my duty to devote a part of the present lecture to withdrawing and cancelling such credentials as Entelechy has had. There is no future for it in theoretical biology, and theology is the home of a quite sufficient number of lost causes already.[38]

5 Teleology and Mechanism

According to Driesch, older teleological views were static rather than dynamic and thus could accommodate the "mechanistic" image of the world as God's artifact. But mechanism was false: the *dynamic* construction of an organism during development

---

[35] Driesch (1908, 1914, 1894, 1918).

[36] Canguilhem (2008, 68).

[37] Needham (1930).

[38] Ibid., 5.

could not be explained mechanistically. Driesch grants that machines are teleological but this is entirely derivative of the acts of mankind, which are indubitably teleological. Static teleology is thus distinct from dynamic teleology. We may see the purpose or function in the parts of the machine but we do not see the machine assembling itself, the parts moving themselves into functionally effective positions. The teleological phenomenon that supports vitalism is dynamic teleology: processes of self-construction, self-maintenance and self-reproduction. Driesch limits the domain of the phenomena to beings that can reproduce themselves, although he offers no deep explanation for the connection between teleology in general and the specific goal of reproduction.[39]

Up front, Driesch admits to a limitation to his argumentative strategies. He notes that most of his proofs of vitalism are indirect; they can only show that "mechanical or singular" causality is insufficient to explain the phenomena.[40] That makes his arguments all the more reliant on a clear understanding of both the vital properties and those essential to mechanisms. No in-between conditions are granted possibility – either something is a machine, or it is subject to an entelechy – but there is no third option and, importantly, no matters of degree between the two.

Driesch offers three "proofs" for the entelechy. Two are based on his theoretical considerations of his experimental studies in biology and the third "the direct argument" is based on his reflections upon human psychology.[41] Driesch defines "harmonious equipotential systems" as those systems in which the whole is formed by the harmonious cooperation of its parts. The best examples of such systems are those of restitution. When a plant restores a part that has been damaged, the parts of the organism need to come together for the purposes of re-growth. But this dynamic reconstitution is a dynamic teleological property *par excellence* and is not easily assimilated into the notion of a machine (which has, at best, static teleological features). The dynamic cooperation of parts, required to restore a severed limb, cannot be given a clear mechanistic interpretation, he thought.

The second phenomenon that lent itself to vitalist interpretation was the existence of complex equipotential systems. Driesch considered sea urchins to be an example. A complex equipotential system is one in which each element of a whole is capable of forming the totality of an organism. Each element is "equally" able to produce the organism. Complex equipotential systems are indeed remarkable and were the focus of Driesch's empirical studies.

But why could complex equipotential systems and harmonious equipotential systems not be mechanical? The argument Driesch offered was a venerable argument from division. If the development of sea urchins from the blastomeres were mechanical then the division of the blastomere into two or more parts would prevent that development. If you take a machine and cut it "arbitrarily" into sections then the mechanism will be dysfunctional and the machine destroyed. But with artificial cell division this did not happen. Instead Driesch found that each blastomere would

[39] Driesch (1908, 133).

[40] Driesch (1914, 208).

[41] For want of space I shall ignore the particularly obscure "direct" argument.

develop into a smaller but fully functioning sea urchin. A change in quantity had occurred without a change in quality (i.e. without change in functional unity).

Driesch emphasized that he could section the cells in any number of overlapping ways and still the whole organism could grow. No matter where we cut, sea urchins would still develop. But if the area cut overlapped with another area cut (in a distinct experiment) then each volume must contain a complete machine capable of producing the sea urchin. And yet those machines somehow overlap. Further, reasoned Driesch, since each volume removed still allowed for the growth of the sea urchin, each volume must contain all the same parts as any other volume (i.e. a complete machine):

> Indeed there do exist almost indefinitely many V_n all of which can perform the whole morphogenesis, and all of which ought to possess the machine….the different volumes V_n overlap each other…but what then about our machines?… For a machine, typical with regard to the three chief dimensions of space, cannot remain itself if you remove parts of it or if you rearrange its parts at will.[42]

6 How Does Entelechy Work?

Driesch could not ignore the enormous developments of nineteenth-century physics. The conservation of energy and the law of entropy delivered a serious challenge to the modern vitalist and had given impetus to the materialist movement of the nineteenth century. Driesch admitted the conservation of energy as an *a priori principle,* so it could not be rejected. The issue thus became just how to reconcile the entelechy's activities with the conservation of energy. If the entelechy is not a form of energy, then the problem dissolves immediately. Not being a state of energy, naturally it does not add or subtract energy from the system. Driesch takes this path, holding that the entelechy could not be "energetic" because energy is a quantitative phenomenon and the entelechy is an "intensive manifold." The entelechy is responsible for the order of relations between elements, a qualitative feature of the system, as Driesch thinks of it.[43] But the entelechy can be understood as a cause of change in the organism since causation is a very general concept and importantly neutral with regard to the kind of properties and events it applies to. The entelechy acts differently from matter so we may say it has a distinct kind of causality, acting as it does by suspension of mechanical and chemical reactions, not by any transference of energy. It might also act by way of directing the physical forces and energy without, however, utilizing energy to do so.

> But entelechy is able, as far as we know from the facts concerned in restitution and adaptation, to suspend for as long period as it wants any one of all the reactions which are possible with such compounds as are present, and which would happen without entelechy. And entelechy may regulate this suspending of reactions now in one direction and now in the other, suspending and permitting possible becoming whenever required for its purposes.[44]

---

[42] Driesch (1908, 140–141).

[43] Ibid., 168.

[44] Ibid., 180.

The actions of an entelechy do not entail some form of unacceptable indeterminism. Driesch supports what he takes to be a further *a priori* principle of sufficient reason: that the very same conditions will have the very same effects. Thus the presence of the entelechy in some material compound will always have the same effect if conditions of the entelechy and compound are repeated.[45] Importantly, Driesch does not think determinism distinguishes entelechies from mechanisms. The presence of an entelechy in physical circumstances would have the same result if repeated in that environment.[46] This was an astute point, for it was commonly thought that a successful science would discern laws and regularities, leaving the indeterminate (or the "free") to be anomalous.[47] But a science without laws of nature is not much of a science, so a vitalism that rejects determinism is unscientific.[48] The mechanical effects of a system are determined just as much as effects from the combination of mechanical and non-mechanical principles is determined, allowing us to formulate the laws of such transitions, at least in theory.

The "indeterminism" that vitalism is accused of is a matter of limited perspective. Such reactions are not indeterministic *absolutely*, rather they are due to the action of the non-perceivable entelechy in that circumstance. But an entelechy could be in a different qualitative state at different times yet be associated with the same physical state, giving rise to the appearance of indeterminism. Such an appearance was dubbed by Jennings "experimental indeterminism" and Driesch accepts this without holding to "absolute" or ontological indeterminism. Driesch's ontological assurance was not sufficient for Jennings, however, who complained that such "experimental indeterminism" violated good scientific method.[49]

But there are no inexplicable or uncaused phenomena in the vitalist world-view, thinks Driesch. The principle of sufficient reason is thus a broader concept than that of causation which is either mechanical or not. The cause of an effect is identified, empiricist-style, as the immediate change before the effect[50] hence both a change in entelechy or in something mechanical may be considered as causes and neither would violate the principle of sufficient reason or the conservation of energy. But Driesch's metaphor is somewhat mixed: "suspension in one direction or another" offers a teleological sounding phrase – one direction or another – but in fact describes a retardation of processes that are *already* "going somewhere," due to their mechanical features. Stopping the flow of blood with a clot so that it takes one path rather than another seems very different from directing the blood towards that path. But Driesch would reply that the *occasion* of such suspension is prescient and does indeed bring about a goal – maintaining a steady state within the organism.[51]

---

[45] Ibid., 153.

[46] Driesch (1914, 155).

[47] Compare Davidson (1993) on mental events and his appeal to anomalous psychology to preserve free will.

[48] This is the main point of Elkus (1911). Cf. Jennings (1918) who takes *experimental* indeterminism to be a problem.

[49] Jennings (1911, 1912).

[50] Driesch (1908, 158).

[51] Thanks to one of the anonymous reviewers for this point.

7 Some Responses to Driesch's Vitalism

Pretty much everything in Driesch's system was contested. It was not clear whether one could generalize results from the few harmonious and complex equipotential systems he studied to life in general. There were the broad concerns regarding entropy and the second law of thermodynamics, but also concerns regarding the reducibility of teleological phenomena. And the action of the entelechy looked pretty mysterious. But the most damaging claim, made by geneticist H.S. Jennings (and others) was that Driesch's empirical research had been superseded. *The discovery of chromosomes and their role in reproduction allowed Jennings to reply that indeed there were tiny machines in each segment cut.* Driesch's argument was very weak, for it merely demanded that the mechanist believe in the possibility of very, very small machines, like chromosomes. Indeed, the sharp distinction between organic and inorganic looked impossible to maintain as P. Chalmers Mitchell noted in his response to J. S. Haldane's notion of biological autonomy:

> ...many organic compounds have been made in the laboratory from inorganic materials... Many of the observed phenomena of living protoplasm have been copied by artificial non-living preparations. The fertilization of the egg-cell, which seemed a supreme case of the action of life upon life, has been achieved by the action of an inorganic salt on the ovum.[52]

But the vitalist interpretation of biology was also criticized by nonreductivists. J.B.S. Haldane did not sympathize with Driesch's work due to its *atomistic* stance. Haldane saw the clue to nonreductivism in holism, in contrast with the atomism of mechanistic science. The organism could not be conceived as distinct from its environment. But the entelechy of the vitalist is an individual entity localized within the organism.[53] Haldane's holism was distinct from Hobhouse's and Smuts' and could be dubbed "environmental holism" or even "wide holism." In the 1970s Putnam and Burge utilized relational properties to deny local reduction. Significantly, Burge invoked Hegel at the start of his famous paper on wide content (Burge 1979, 73f.).

8 The Emergentists

Philosophical and biological emergentists were numerous in the 1920s, although not all of them embraced the same terminology. In biology, J. Arthur Thomson and H.S. Jennings were prominent, along with J.S. Haldane in physiology. Among the philosophers, the most noteworthy were Samuel Alexander, C.D. Broad and Arthur Lovejoy, but there were many others, such as J.T. Hobhouse and J.C. Smuts. In biology the ideas of Thomson and Jennings were closest to the philosophical view expressed by C. Lloyd Morgan who characterized emergentism thus:

---

[52]Clark (1918, 57).

[53]Haldane (1931, 1932).

> Under what I call emergent evolution stress is laid on this incoming of the new. Salient examples are afforded in the advent of life, in the advent of mind, and in the advent of reflective thought. But in the physical world emergence is no less exemplified in the advent of each new kind of atom, and of each new kind of molecule. It is beyond the wit of man to number the instances of emergence. But if nothing new emerges – if there be only regrouping of pre-existing events and nothing more – then there is no emergent evolution.[54]

Morgan self-consciously identifies the tradition of emergentist thinking with Lewes' and Mills' distinction between heteropathic and homopathic laws. Heteropathic laws governed phenomena that violated the principle of the composition of causes. The effects of some composite causes could not be predicted from the laws governing their parts acting alone. Lewes coined the terms "emergent" and "resultant" to capture Mill's point that the activities of atoms alone are different from when they are together in special circumstances (the context of an organism).

9 J. Arthur Thomson on the Autonomy of Biology

Thomson published his *The System of Animate Nature* in 1920 as his contribution to the Gifford lectures.[55] Thomson begins his discussion with a few preliminary points. He notes that there are skeptics regarding the closure of the physical but that this question cannot be decided by a biologist, but by a physicist. The question for Thomson then becomes whether we can offer a mechanical interpretation of biological properties. Thomson notes and approves of Bates' remark that everyone is currently a vitalist since at this moment in the history of science no mechanical interpretation of vital processes has been given. Vitalism therefore is the default epistemic position. But Thomson realizes that such an argument had limited force for it contains no guarantee that future research will or will not offer mechanical interpretations of the phenomena.

Thomson distinguishes negative vitalism, which is simply the claim that current theory has not provided a mechanical interpretation, from positive vitalism that offers an explanation for the autonomy of biology in ontological terms. The positive vitalist posits a special force or entelechy to explain the difference between living and nonliving entities. The difference between the views is that the latter offers a hypothesis or explanation in addition to the claim of irreducibility. In the "scientific spirit" Thomson shies away from any commitment to hypotheses that are speculative due to insufficient evidence. His caution leads him to frame his methodological or emergentist conception of vitalism.

Thomson also notes that the problem of animation and the problem of vitalism are distinct. Some have thought that all life is mentalistic in some sense – that the difference between the living and the nonliving is the presence of a rudimentary consciousness, a phenomena that is full-blown only in humans. Bergson seemed to

---

[54] Morgan (1923, 1–2).

[55] Thomson (1920).

speculate this in *Creative Evolution*. Thomson sensibly puts aside the hypothesis as being too uncertain for acceptance since we have no clear knowledge of the existence of minds in plants. Thus the vitalist position is not necessarily a theory of animation, he concludes.

Three possible vitalist positions can now be discerned. One is epistemic: that the configurations in the living are so different from that of the nonorganic that no prediction of the living can be made from nonliving systems or models. But the only difference between the living and the nonliving is organization. The second vitalist view is one Thomson associates with medieval philosophy. There is a special kind of *physical* energy at work in the living, according to this view. As a special physical force this force should be measurable or "perceivable." The third species of vitalism, which Thomson takes to be the most popular, holds that there is some kind of "transcendent" immaterial vital force at work, hence one that is not measurable or "perceivable."

Taking the last position as the typical one Thomson asks: must we choose between mechanism and the idea of a nonphysical force? He replies: no, neither is correct. A third option is available, which he calls *methodological vitalism.*[56] Against the first idea he notes that it is, ontologically, a mechanistic position, stating merely that we cannot predict the activities of the organism from "lower-level" mechanistic considerations. The organism is only a complex mechanism on this view:

> This is a vitalistic view in so far as it recognizes the apartness of living creatures from things in general, but it does not admit that the problem of the Amoeba on the hunt is more than a very difficult problem in dynamics.[57]

Crucial to Thomson's argument about mechanistic epistemic vitalism is that it does not require new *concepts* for the description of biological properties, and hence is not an account of the autonomy of biology after-all. Thomson's complaint appears to be that mechanists utilizing only the concepts and vocabulary of physicists overlook the distinctly biological concepts that are currently required to describe the phenomena of life adequately. Thus, this concept of vitalism is not going to respect the contemporary state of the science.

Thomson invokes with approval Arthur Lovejoy's conception of the unity of science as the idea that special scientific phenomena and laws can be derived or deduced from the laws of physics. But writing in 1919, Thomson recognizes that no such derivation was available to determine the teleological properties of the living from the physical. Without such a derivation, the epistemic materialist view appears inadequate.

Thomson's criticism on this point could be seen as a symptom of not recognizing, as Ernst Nagel later did, that bridge laws are needed to connect the predicates of biology to the predicates of physics. For the failure to deduce the *concepts* of biology from the concepts and theories of physics could be a problem in notation or

---

[56] Ibid., 159.

[57] Ibid., 147.

translation. If separate theories are analogous to separate languages, then some "translation manual" will be required for communication. Overlooking this requirement will leave us with the impression that nothing much can be deduced from physical theory since the concepts and laws of physics are very different from those of biology or chemistry. Physics alone, without bridge laws connecting its vocabulary to the vocabulary of the special science, will be silent regarding the description of macrophysical phenomena.[58]

Thomson's criticism of this view is somewhat disappointing. He had already noted that the current inability to provide mechanistic interpretations underdetermined which of the three accounts should be accepted, yet he goes on to reject this mechanistic account based solely on the current state of research. Mechanism has not yet provided an answer "to any biological question" and this is sufficient to reject this option, despite possible future discoveries. On the other hand, Thomson is keen to remain scientific – to accept only theories that he has pretty good current evidence for. The problem, however, is that he ends up merely describing the problem without offering any guidance or prediction to what the proper ontological solution is.

Thomson thinks the second option also fails to account for the autonomy of biology. Since it posits a verifiable measurable physical force, whatever arguments we have for the inadequacy of a mechanical reduction will simply reappear for this special physical force.[59] Furthermore, distinguishing this alleged force from other forces seems quite obscure.

Thomson finally turns to the vitalism of Driesch, noting that Driesch holds that the entelechy is not a result of prior physical conditions, that it introduces a kind of indeterminism in the physical continuum on which it acts and thirdly that it be counted as a genuine "agent counting for something 'at work'." Thomson's objections to this view are expected yet disappointing: it contradicts the fundamental law of the conservation of energy. While admitting the problem, he reverts to future empirical discoveries: they might show that the law of conservation is limited and does not apply to the organism after-all! So he is not convinced that Driesch's entelechy violates the law of conservation because that law might not apply, and he avoids Driesch's own assertion that the suspension of physical activity by the entelechy does not itself require any change in energy. Thomson's reasoning therefore seems rather desperate, especially given Driesch's view that the total amount of energy of a living being may be maintained while the configuration of the particles of the organisms is changed.

Thomson notes that Driesch's arguments always contrast machines with entelechies. But as before, he notes that we cannot canvass all possible machines but only

[58]Cf. Hoernlé (1918) who argues that teleological concepts dominate biological theories while being consistent with mechanism. Hoernlé is the epistemic materialist that Thomson seems to attack in favor of emergentism. Hoernlé also appeals to Lovejoy's notion of reduction as deduction and prediction from physical theory.

[59]Thomson speaks of perceptual versus non-perceptual forces by which he means experimentally observable and mathematically measurable.

those we are currently familiar with. Thus the negative argument that no known machine can account for biological properties is a weak one, for we need an account that would show no *possible* machine could do the job.[60] But then Thomson assumes that we cannot utilize human-made machines in our discussion since they bear the mark of human intelligence and are therefore not "fair samples of the inorganic world" but are elaborated tools that have "inside of it, so to speak, a human thought."[61] Thomson had in mind to deny the materialist reply found in the debate between J. S. Haldane and D'Arcy Thompson.

D'Arcy Thompson was one idiosyncratic representative of the materialist camp, despite his own eclectic views on growth and form.[62] But like so many biologists of his time, Thompson's materialism did not apply to psychology. In the introduction of *On Growth and Form* (1917), Thompson made it clear that biological phenomena were physical: "the construction and growth and the working of the body, as of all that is of the earth earthy, physical science, is my humble opinion, our only teacher and guide."[63] Thompson had little interest in positing nonphysical forces in biology, but philosophically he speculated on teleological explanations for reality as a whole. "Like warp and woof, teleology and mechanism are interwoven together, and we must not cleave to one and despise the other."[64] Such a teleology would not license the belief that there were any discontinuities in nature, as countenanced by Driesch. D'Arcy Thompson did not agree with Haldane or Driesch's notion of the autonomy of life and represented the mechanist camp in a roundtable discussion with Haldane.[65] He responded to J.S. Haldane's claim that self-production, self-maintenance and self-regulation were not mechanistically explicable by pointing out that there existed mechanisms that perform the task. D'Arcy Thompson cheekily speculates on J.B.S. Haldane's own area of expertise, the physiology of respiration, suggesting that unknown mechanical features might well be discovered to accord with the "mechanical laws of diffusion." Haldane's own description of the process, claimed D'Arcy Thompson, seems presumptive given our ignorance. But worse, machines do have the features Haldane and others think are irreducible to mechanistic concepts. D'Arcy Thompson writes:

[60]Thomson is not as subtle as he could be. Driesch denies that the entelechy is a force at all, but does agree to its directive ability, and Driesch also holds that he has captured the essential nature of machines and is not merely arguing from current examples. The objection might be re-tooled as this: future research will reveal a different *essence* to machines, allowing them to explain Driesch's phenomena of restitution, etc.

[61]Thomson (1920, 157).

[62]D'Arcy Thompson's broad sympathy for Oken's *Naturphilosophie,* following in Hegel's footsteps, did not prevent him from holding to the continuity of the physical. D'Arcy Thompson, it is tempting to say, was a Hegelian realist or materialist, somewhat as Samuel Alexander was. He was Hegelian in finding that a philosophical immanent account of teleology is required to supplement the mechanist account available from science.

[63]Thompson (1917, 8).

[64]Ibid., 5.

[65]Clark (1918).

> Many a machine is constructed to oil itself the more copiously when it works the faster, as the printing press, as we urge it to put out more newspapers on the one side, pulls in more blank paper on the other.[66]

In a clever comment worthy of a nineteenth-century natural theologian, D'Arcy Thompson reminds us that human machines are rather crude compared to natural self-maintaining, self-sustaining systems:

> But in nature herself, if we look at her larger handiwork, self-regulation and self-maintenance becomes paramount attributes and characteristics of her machines. The solar system, *qua* mechanism, is the perfect specimen, the very type and norm, of a self-maintaining self-regulating mechanism...[67]

Thomson's complaint is that D'Arcy Thompson can "reduce" the biological to the mechanical simply because the properties that make an organism irreducible, its teleological properties, are presupposed in man-made mechanisms. But then the reduction of teleology to the physical fails. In some ways, Thomson is expressing Driesch's distinction between static and dynamic teleology. Static teleology is applicable to machines. Their parts have purposes to serve given to them by an external source. But in a dynamic teleological system the parts have purposes that are internal to them, independent of the attitudes or plans of external agents.

Finally, J.H. Thomson invokes Haldane and Jennings against the vital principle. Their objection, one that was awfully old, was that it is very hard, if not impossible, to understand how the entelechy could do the job of coordination unless it *knew* about the physical particles it arranged. Indeed, Pierre Bayle made this very complaint in response to the vitalist speculations of Nehemiah Grew.[68] How could the entelechy know what chemical and physical processes to suspend and when to do it? Having just rejected the machine analogy for relying implicitly on the intelligence of the designer to explain the organization of its parts, the intelligence of a designer appears to be smuggled into the entelechy itself. Thomson admits this argument is powerful, and so having surveyed the problems of each rejects all three conceptions for his own "methodological or descriptive vitalism."

A different order of explanation, Thomson believes, utilizing concepts not found in mechanistic science but not necessarily inconsistent with mechanical science, captures the autonomy of biology. Thomson takes Arthur Lovejoy's account of the unification of science seriously. Lovejoy argues that a unified science would deduce the phenomena from a simple law or set of laws. When laws cannot be deduced from some other laws then we have a case of the disunity of science. Thomson takes it that biological laws cannot be so deduced and novel concepts are required for the explanation of biological phenomena. He picks on the old favorites: the "self-maintaining, self-preserving, and purposiveness of

---

[66] Clark (1918, 37).

[67] Thomson (1920, 37).

[68] Cf. Garrett (2003).

organisms" and their ability for "organic retention, the capacity of enregistering experience, garnering the past, captilizing on gains" and finally, variability, "the capacity and habit of giving origin to the new." [69]

Exactly how Thomson's methodological view differs from the views he discusses is not so clear. Like J.S. Haldane and J.C. Smuts[70] he asserts that new categories are required and it is not just because there is some special complexity of matter. Thus he opposes the epistemic mechanistic definition of vitalism. Thomson takes it that his view differs from the second view that interprets the *élan vital* as a physically measurable special force, thus the properties of living organisms are not captured by complexity or by a special physical force. His view appears to be that there is more than an epistemic difference between organisms and mere complex mechanisms, but why this is not to assert an ontological position, like that of Driesch, is somewhat puzzling.

Thomson's methodological view holds to a distinction between living and nonliving but makes no commitment to the existence of any special force, or special principle, material or immaterial. He cannot appeal to an immaterial force, for this would deny the central tenant of evolution: the continuity of nature. Thomson thus finds himself holding to an emergentist position similar to Morgan and Alexander. There are no gaps in the nonorganic mechanistic world, yet there are clearly nonmechanistic features emerging out of such processes that cannot be accounted for without a new, irreducible vocabulary.

10 Arthur Lovejoy on the Disunity of Science

As noted above, Thomson appeals to Lovejoy's conception of reductionism:

> It would consist in showing that a given organic process, A, can be subsumed under and deduced from a given generalization, B, of the more "fundamental" science. The proof of the autonomy of biology, on the other hand, would consist in showing that there are modes of action characteristic of matter when organized into a living body which can never be deduced from any law that describes any modes of action of inorganic matter. But here an explanation about deducibility is needful, since the notion has been somewhat confused in some recent discussions.[71]

Lovejoy takes reductionism to involve deducing the laws of one science from more general laws of another science. Materialism would be the complete reduction of all special scientific laws to the laws of physics. Failure to reduce the laws of one science to another leaves us with the disunity of science or rather, the autonomy of the sciences. Lovejoy recognizes, however, that one cannot get very far deducing anything from just a law-like statement. He notes that we need to know some of the variables the law covers in order to make predictions. Nevertheless, even with such

[69] Thomson (1920, 165).

[70] Smuts (1926).

[71] Lovejoy (1911, 611; See also Lovejoy 1912).

physical information we are unable to deduce the laws of biology from the laws of chemistry or the laws of chemistry from those of physics. Lovejoy takes it that such a nonreductive view of biology could well be termed a "vitalist" view but he is clear to distinguish it from Driesch's and Bergson's. Lovejoy does not wish to countenance distinct modes of causation or the violation of the continuity of Nature, as they seem to do. Lovejoy's concerns are more philosophical than scientific and he identifies as one of the main objections to his emergentism assumptions regarding causation. Utilizing the traditional terminology from the debates over biological reproduction he explains the "preformationist" view of causation inherited from tradition. The preformationist view treats

> …natural events as combinations or re-arrangements of relatively simple, pre-existent entities, of which the total number or quantity remains invariant, and of each of which the qualities and laws of action remain the same through all the combinations into which it may enter. By this mechanistic conception of causation there is nothing substantive in the consequent, which was not in the antecedent, and the supposed paradox of epigenesis is thus avoided.[72]

This conception of causation is crucial to the *objections* against emergentism. If one cannot get "more" from a cause than what is found within the cause then no distinct "novelty" or "levels of reality" will be possible, for if some ontological novelty really were produced from the material, its novelty would constitute something emerging from nothing. By denying this traditional concept of causation it is possible, argues Lovejoy, for the irreducible and novel to emerge from the physical. The novel need not be "preformed" or "contained" in its cause. But what then do we make of the irreducible "unity's" causal powers? What work is there to do for the emergent properties if all physical change has physical causes?

11 Jennings on Downward Causation

H.S. Jennings argued for emergent evolutionism in his 1930 book *The Biological Basis of Human Nature* but had also been vigorously involved in debate a decade earlier.[73] Jennings defends emergent evolution over mechanistic evolution, but his arguments for emergentism tended to be arguments against mechanism, dependent as all such arguments were, on the adequacy of the inferences drawn from materialism. Jennings argues that the mechanist evolutionist holds that every change in the world is calculable in advance from knowledge of the material atoms and mechanistic laws, but he is not as careful as Thomson who worries about the relation between the epistemology and ontology of emergence. With the standard definition in hand, Jennings complains that no novel properties could come about if mechanism is true:

---

[72] Ibid.

[73] Cf. Jennings (1913, 1927, 1918, 1930), Gurwitsch (1915), and Hoernlé (1918). Mackenzie (1926), Spaulding (1906), Warren (1916a, b, c, 1918).

> We are not agents; we are merely stages in the working out of the world-formula. Furthermore, since the laws of nature are discoverable at any early epoch, and are immutable, the world must continue to act in the future as it has in the past. Nothing new in principle can appear. No new methods of action can come into operation.[74]

The determinism and materialism implied in mechanistic evolution will rule out the creativity and "new methods of action" required to make sense of human agency and free will. But it would also rule out the novel methods of action found in plants and animals, these seemingly quite different from the "actions" of physical particles. By contrast, the emergent evolutionist holds:

> …that in the course of evolution there have emerged things that are new, of a different kind from anything that has gone before; and that are not predictable from a knowledge of the preexisting things, from a knowledge of the preexisting particles, their arrangements and motions. … New laws of motion, new methods of action, have appeared, as new arrangements of particles occur, and particular mental states arise. And this is continuing as evolution proceeds, at present as in the past.[75]

But why would mechanical evolution have the consequence that no new methods of action can come about? Jennings claims that the materialist thinks there would be no change in the distribution of physical properties were mental items absent. Physical determinism implies that the irreducible mental or vital realm have no causal effect. Jennings applies a form of what Jaegwon Kim belatedly dubbed the "exclusion argument."[76] If physical events are caused only by further physical events (and hence their effects are calculable merely from such knowledge of the laws and antecedent physical conditions), then irreducible mental events would be epiphenomenal, having no effect whatsoever on the course of those physical events. This "exclusion argument" was familiar from James, was accepted by Santayana, and presupposed the irreducibility of the mental.

Jennings offers an interesting argument against "mechanist evolution." As he characterized the position, mind must be epiphenomenal since only mechanism is responsible for change. But if mind is epiphenomenal and has no effect on the physical world then, were mind absent the world would continue on as it were (presumably remaining a series of changes in atoms alone.) What mechanistic evolutionists are worried about is that their calculations of matter will be affected if mental states have causal efficacy.[77] So mechanists deny that irreducible or novel mental states have any relevance to physical change.

But if this is the mechanistic evolutionary perspective *then it is empirically un-testable*. Thus, using the budding positivist terminology of the day, Jennings

[74] Jennings (1930, 363).

[75] Ibid., 369.

[76] Norman Malcolm utilized a curious version of the exclusion argument in his 1969 paper "The Conceivability of Mechanism." Malcolm inverts the argument to claim that mechanism is in fact incomplete. Malcolm's argument depends on a dubious Wittgensteinian premise that psychological laws are not empirical, but *a priori*, hence only empirical physical laws can be up for denial or revision.

[77] Jennings (1930, 366).

describes the hypothesis as "metaphysical, unscientific, without meaning."[78] Mechanistic evolution is untestable, argues Jennings, because we cannot find circumstances in which we can separate mental and physical phenomena. To test the mechanical evolutionist hypothesis we would need to see whether there could be a change in matter *without a change in consciousness*, and whether there could be *change in consciousness without a change in matter*, for these are the alleged consequences of (nonreductive epiphenomenal) materialism! But neither of these options is experimentally verifiable. Mechanistic evolution is not experimentally verifiable and thus is not a legitimate scientific position! We cannot offer an experiment that would subtract the mental from the physical to see whether the physical processes would continue in the very same manner as they do in the presence of the mental.

Put in contemporary philosophical terms, since the mental is supervenient on the physical there is no possibility of the mental property being instantiated without some physical properties instantiated and no possibility of these exact physical properties being instantiated without the emergent property instantiated. This complicated dependence notion leads Jennings, like some today, to slip and slide between identity and difference.[79] The tight dependence guaranteed by emergence or supervenience leans us towards an identity of mental, biological and physical properties, at least as a *practical* matter:

> And since they are inseparable, identified, the specification of either implies the other. They are two names for the same situation. It follows that either may be employed in describing the results of experimentation. The outside observer, experimenting by physical and chemical methods...may without error attribute the results of his experimentation to the physical and chemical conditions, provided that he refrains from asserting that the mental state has no role. He may do this because the mental state has its characteristic accompanying physical situation, different for every diverse mental state; so that there never lack diverse physico-chemical conditions for every difference in results.[80]

Jennings cannot quite make up his mind. On the one hand the emergent is a novelty and brings with it a novel manner of acting, so it cannot be identified with the physico-chemical conditions from which it emerges. On the other hand, one cannot ever separate out the physico-chemical property from any instance of the supervenient or emergent property, thus there is no experimental or observable conditions of their separation. But if there is no such experimental possibility then they cannot be identified from the point of view of empirical science. So mechanistic evolution is an untestable hypothesis – we cannot experimentally prove the epiphenomenalism it implies.

The emergentist hypothesis, however, has some serious implications. The presence of such emergent properties must make a difference to physical change, if those properties are not conceived as epiphenomenal. Jennings thus takes emergent materialism to have implications for what today is dubbed "downward causation":

---

[78] Ibid., 368.

[79] Jaegwon Kim and David Armstrong (along with some Australian metaphysicians) think of logical supervenience as a reductive thesis.

[80] Jennings (1930, 368).

> It affirms . . . that the same is true for the steps from electrons to atoms, from atoms to molecules, from molecules to crystals. It holds that the properties of atoms do indeed depend on those which the electrons have when they are in the atom; the properties of molecules on those which the atoms have when they are in the molecules. It holds too that the properties of living things depend on those of their physical constituents when the latter are in living things; the activities of thinking beings on the action of their physiological constituents when the latter are part of a thinking being; the activities of societies on those of their unit individuals when these individuals form part of the society. . . . But it contends that the constituents of each grade acquire new properties, new modes of action, in becoming part of the "emergent" thing of "higher" grade. *It holds that the physics of atoms, of molecules, is not fully known till these are studied in the living as well as in the non-living.*[81]

Since the emergent is not an epiphenomenon it makes a difference to physical change. Thus the physical will act one way in the presence of the emergent and another way in its absence. Physical laws and physical action are distinct in distinct contexts: the biological matrix gives birth to physical novelty. The incredible implication is that a scientist cannot do physics completely and adequately without doing biology and, ultimately, psychology! The biological under Jennings' interpretation is not only autonomous; it is colonial. Physics is not, according to this view, the most general of sciences, being in a sense context-free. Rather, how the physical behaves depends on whether the physical processes occur within an organism or not, thus biology and psychology are required to complete physics.

> The emergent evolutionist is free to believe that when specific mental states occur, the rules of motion of the parts are different from those prevailing when no such mental state occurs.[82]

Jennings' argument against mechanistic evolution is interesting and resonates with recent debates over the significance of the supervenience of irreducible mental properties upon the physical. If the supervenient is to be causally efficacious then it "makes a difference" to the physical, to use Davidson's phrase.[83] But if it does make a difference then it would appear that a physicist had better know that he is studying the particles or forces within the "unity" of a living being. If the physicist did not know that the mental or vital supervened upon his particles or forces then he would be misled and his theory would not be universal. It seems difficult to deny that under the non-epiphenomenal emergentist position there must be a "biological" or "psychological physics" distinct from inorganic physics.

12 Conclusions

I began this discussion with a wild accusation of intellectual amnesia. One suspects the debates of the last 30 years over nonreductivism could have been cut short or moved faster had we remembered the arguments of the 1920s. Kim's "exclusion

[81] Jennings (1930, 369). Emphasis mine.

[82] Ibid., 379.

[83] Davidson (1993).

argument" was well known in the nineteenth century and the resulting epiphenomenalism for nonreductivists about life or consciousness was also well known. Santayana was famous for biting the epiphenomenal bullet, as David Chalmers today is famous for a similar circus act. The fight over nonreductivism involved, as it does today, confusions and disputes over the relation between epistemology and ontology. As in the 1920s, the epistemic and ontological conceptions of nonreductivism are often confused and their relations made obscure. By the 1930s it was well-known that emergentism had not got its epistemology distinct from its ontology: the nondeducibility of the emergent looked no different, really, from nondeducibility in physics due to our ignorance of the laws or details of the physical. We also find in the 1920s a common assumption, kept popular by Arthur Lovejoy, that reduction consists in deduction from physical laws, a view that is championed today by Frank Jackson and David Chalmers, albeit with different whistles and bells. But even if the failure of such a deduction were the case, its ontological significance is unclear. As their debate with Stalnaker and Block reveal, the old worry that nonreductivists draw ontological nonreduction from theoretical nonreduction applies to Chalmers and Jackson as much as it did for Jennings and Thomson back in the 1920s.[84]

Among scientists, emergentism waned due to its empirical ambitions being thwarted. As Brian McLaughlin notes, empirical developments in chemistry and biology sounded the death knoll for emergentism.[85] In philosophy, ontology bowed to the philosophy of language, which remained stubbornly *a priori* and nonnaturalist. Indeed, philosophy of language, as it is still practiced by twenty-first century philosophers shows little sign of recognizing that evolution ever occurred in syntax or semantics. The blurry and merely pragmatic lines found between species finds no counterpart in contemporary philosophical semantics where the concepts of intension and extension appear to be well-regulated and precise, handed down as they are from Friar Frege and Brother Russell. Vagueness is treated as a *special* case. If the ascendancy of philosophy of language was one reason why the ontological and evolutionary disputes over materialism and non-reductionism languished, then it is perhaps no surprise that this field ignores the relevance of evolution to their theories of meaning and reference. There is, of course, some contemporary interest in reviving the concept of emergence. The revival is plagued with the puzzles inherited from the 1920s. Is emergence an epistemic concept or an ontological concept and what is the relation between these two?[86] The concepts of dependence and novelty still remain unclear and their implications uncertain.[87] But criticism of the current literature is beyond the scope of this discussion.

So what is new in the debate over life, mind and reductionism in the early years of the twenty-first century? Back in 1912 Burroughs described the dilemma facing philosophers of *his* time:

---

[84] Block and Stalnaker (1999).

[85] McLaughlin (2003). See also Beckermann (1992).

[86] Bedau (1991, 1992).

[87] Silberstein and McGeer (1999).

> He must either admit of a break in the course of nature and the introduction of a new principle, the vital principle, which, if he is a man of science, he finds it hard to do; or he must accept the theory of the physico-chemical origin of life, which, as a being with a soul, he finds it equally hard to do. In other words, he must either draw an arbitrary line between the inorganic and the organic when he knows that arbitrary lines in nature, and fencing off one part from another, is an unscientific procedure, and one that often leads to bewildering contradictions; or he must look upon himself with all his high thoughts and aspirations, and upon all other manifestations of life, as merely a chance product of the blind mechanical and chemical action and interaction of the inorganic forces. Either conclusion is distasteful.[88]

Today the vitalist and the Cartesian style immaterialist no longer hold the sway they used to. The context, background and foreground, have changed. Cartesian dualism is roundly dismissed, no sharp division between the mental and physical can be found. But emergentism in the 1920s was intended to be a synthesis of anti-materialist and antidualist thought guided by good empirical evidence. The sharp division of nature it implied could not be maintained in the face of empirical advances. Although there are now some excellent attempts to justify nonreductionism with careful, empirically sensitive studies in the philosophy of biology,[89] nonreductionism tends to have a philosophical *a priori* inclination towards arguments that scientists find irrelevant (e.g. arguments from multiple realizability and considerations of logical [not empirical] possibility). Contemporary emergentism is not vitalistic, anymore than the emergentism of the 1920s was, but all forms of irreducibility, if they do not merely reflect our epistemic limitations, must face the puzzles stemming from the bifurcation of nature into 'levels'. Given the force of the exclusion argument, the mysterious implications of downward causation and the general lack of scientific support, I doubt the fate of contemporary nonreductivism will be much different from that of emergent evolution and of vitalism.

References

Alexander, Samuel. 1920. *Space, time, and deity*, vol. 2. London: Macmillan.

Allen, Garland E. 2005. Mechanism, vitalism and organicism in late nineteenth- and twentieth-century biology: The importance of historical context. *Studies in History and Philosophy of Biological and Biomedical Sciences* 36: 261–283.

Baker, Lynne. 2007. *The metaphysics of everyday life*. Cambridge: Cambridge University Press.

Beckermann, Ansgar, Hans Flohr, and Jaegwon Kim (eds.). 1992. *Emergence or reduction? Essays on the prospects of nonreductive physicalism*. New York: Walter de Gruyter.

Bedau, Mark. 1991. Can biological teleology be naturalized? *Journal of Philosophy* 88: 647–657.

Bedau, Mark. 1992. Weak emergence. *Philosophical Perspectives* 11: 375–399.

Blitz, David. 1992. *Qualitative novelty and the levels of reality*. Dordrecht: Kluwer.

Block, Ned, and Robert Stalnaker. 1999. Conceptual analysis, dualism and the explanatory gap. *Philosophical Review* 108: 1–46.

Boodin, John. 1925. *Cosmic evolution*. London: MacMillan and Co.

Burge, Tyler. 1979. Individualism and the mental. *Midwest Studies in Philosophy* 4: 73–121.

[88] Burroughs (1912, 759).

[89] Dupré (1993).

Burroughs, John. 1912. The new vitalism. *The North American Review* 196(685): 759–771.
Campbell, Richard J., and Mark A. Bickhard. 2011. Physicalism, emergence and downward causation. *Axiomathes* 21(1): 33–56.
Canguilhem, Georges. 2008. *Knowledge of life*, eds. Paola Marrati and Todd Meyers. Trans. Stefanos Geroulanos and Daniela Ginsburg. New York: Fordham University Press.
Clark, W. (ed.). 1918. *Life and finite individuality: Two symposia*. London: Williams and Norgate.
Conger, George P. 1929. *New views of evolution*. New York: Macmillan.
Davidson, Donald. 1993. Thinking causes. In *Mental causation*, ed. John Heil, 3–17. Oxford: Clarendon.
Drake, Durant. 1925. *Mind and its place in nature*. London: Macmillan.
Driesch, Hans A. 1894. *Analytische Theorie der organischen Entwicklung*. Leipzig: W. Engelmann.
Driesch, Hans A. 1907 [1908]. *The science and philosophy of the organism: The Gifford lectures delivered before the University of Aberdeen in the year 1907 [and 1908]*. London: Adams and Charles Black.
Driesch, Hans A. 1914. *The history and theory of vitalism*. London: Macmillan.
Dupré, John. 1993. *The disorder of things*. Cambridge: Harvard University Press.
Elkus, Savilla Alice. 1911. Mechanism and vitalism. *The Journal of Philosophy, Psychology and Scientific Methods* 8(13): 355–358.
Emmeche, Claus, S. Køppe, and F. Stjernfelt. 1997. Explaining emergence: Towards an ontology of levels. *Journal for General Philosophy of Science/Zeitschrift für allgemeine Wissenschaftstheorie* 28(1): 83–119.
Freyhofer, Horst H. 1982. *The vitalism of Hans Driesch*. Frankfurt: Peter Lang.
Garrett, Brian. 1999. Pluralism, causation and overdetermination. *Synthese* 116(3): 355–378.
Garrett, Brian. 2000. Defending non-epiphenomenal event dualism. *Southern Journal of Philosophy* 38(3): 393–412.
Garrett, Brian. 2003. Teleology and vitalism in the natural philosophy of Nehemiah Grew (1641–1712). *British Journal for the History of Science* 36(1): 63–81.
Garrett, Brian. 2006. What the history of vitalism teaches us about consciousness and the hard problem. *Philosophy and Phenomenological Research* 72(3): 616–628.
Grossman, Edgar F. 1930. Something about vitalism and mechanism. *The Scientific Monthly* 30(6): 541–546.
Gurwitsch, Aron. 1915. On practical vitalism. *The American Naturalist* 49(588): 763–770.
Haldane, John Scott. 1931. *The philosophical basis of biology*. London: Hodder and Stoughton.
Haldane, John Scott. 1932. *Materialism*. London: Hodder and Stoughton.
Hein, Hilde. 1969. Molecular Biology vs. Organicism: The enduring dispute between mechanism and vitalism. *Synthese* 20(2): 238–253.
Hoernlé, R.F.Alfred. 1918. Mechanism and vitalism. *Philosophical Review* 27(6): 628–645.
James, William. 1890. *The principles of psychology*. Reprint, 1950. New York: Dover Reprints.
Jennings, H.S. 1911. Vitalism and experimental investigation. *Science,* New Series 33(859): 927–932.
Jennings, H.S. 1912. Driesch's vitalism and experimental indeterminism. *Science,* New Series 36 (927): 434–435.
Jennings, H.S. 1913. Doctrines held as vitalism. *The American Naturalist* 47(559): 385–417.
Jennings, H.S. 1918. Mechanism and vitalism. *Philosophical Review* 27(6): 577–596.
Jennings, H.S. 1927. Diverse doctrines of evolution, their relation to the practice of science and of life. *Science,* New Series 65(1672): 19–25.
Jennings, H.S. 1930. *The biological basis of human nature*. London: Faber and Faber.
Lovejoy, Arthur O. 1911. The meaning of vitalism. *Science,* New Series 33(851): 610–614.
Lovejoy, Arthur O. 1912. The unity of science. *University of Missouri Bulletin* i(1): 1–35.
Lovejoy, Arthur O. 1927. The meanings of 'emergence' and its modes. *Journal of Philosophical Studies* 2(6): 167–181.
Mackenzie, W. Leslie. 1926. Methods of analysis. *Aristotelian Society, Supplementary Volume* 6: 56–62.

Malaterre, Christophe. 2013. Life as an emergent phenomenon: From an alternative to vitalism to an alternative to reductionism. In *Vitalism and the scientific image in post-Enlightenment life science, 1800–2010*, ed. Sebastian Normandin and Charles T. Wolfe, 155–178. Dordrecht: Springer.

McDougall, William. 1929. *Modern materialism and emergent evolution*. New York: D. Van Nostrand.

McLaughlin, Brian. 2003. Vitalism and emergence. In *The Cambridge history of philosophy: 1870–1945*, ed. Thomas Baldwin, 631–639. Cambridge: Cambridge University Press.

Morgan, C.Lloyd. 1923. *Emergent evolution*. London: Williams and Norgate.

Nagel, Ernest. 1961. *The structure of science*. New York: Harcourt.

Neal, H. V. 1916. The basis of individuality in organisms. A defense of vitalism. *Science*, New Series 44(1125): 82–97.

Needham, Joseph. 1930. *Order and life*. Cambridge, MA: M.I.T. Press.

Nouvel, Pascal (ed.). 2011. *Repenser le vitalisme. Histoire et philosophie du vitalisme*. Paris: PUF.

O'Connor, Timothy. 2000. *Persons and causes: The metaphysics of free will*. Oxford: Oxford University Press.

Santayana, George. 1905. *The life of reason*, vol. 1. London: Constable.

Silberstein, Michael, and John McGeever. 1999. The search for ontological emergence. *The Philosophical Quarterly* 49: 182–200.

Smuts, Jan Christian. 1926. *Holism and evolution*. London: Macmillan.

Spaulding, Edward G. 1906. Driesch's theory of vitalism. *Philosophical Review* 15(5): 518–527.

Strawson, Peter Frederick. 1961. *Individuals: An essay in descriptive metaphysics*. London: Methuen.

Strawson, Galen. 2006. *Consciousness and its place in nature*. Exeter: Imprint Academic.

Strong, Charles A. 1918. *The origin of consciousness: An attempt to conceive of the mind as a product of evolution*. London: Macmillan.

Thompson, D'Arcy Wentworth. 1917. *On growth and form*. Cambridge: The University Press.

Thomson, J.Arthur. 1920. *The system of animate nature*. New York: Henry Holt and Company.

Warren, Howard C. 1916a. A study of purpose. *Journal of Philosophy, Psychology and Scientific Methods* 13(1): 5–26.

Warren, Howard C. 1916b. A study of purpose II: Purposive activity in organisms. *Journal of Philosophy, Psychology and Scientific Methods* 13(2): 29–49.

Warren, Howard C. 1916c. A study of purpose III: The role of purpose in nature. *Journal of Philosophy, Psychology and Scientific Methods* 13(3): 57–72.

Warren, Howard C. 1918. Mechanism versus vitalism in the domain of psychology. *Philosophical Review* 27(6): 597–615.

Whitehead, Alfred. 1929. *Process and reality*. Cambridge: Cambridge University Press.

Wolfe, Charles T. 2009. A happiness fit for organic bodies: La Mettrie's medical Epicureanism. In *Epicurus in the enlightenment*, ed. Neven Leddy and Avi Lifschitz, 69–83. Oxford: Voltaire Foundation.

Chapter 7
Life as an Emergent Phenomenon: From an Alternative to Vitalism to an Alternative to Reductionism

Christophe Malaterre

Abstract In this contribution, I investigate the changes of focus in the philosophical concept of emergence in the nineteenth and twentieth century period, especially in connection with the problem of characterizing life and its origins. Since its early philosophical formulation in the nineteenth century, "emergence" has been applied to vital phenomena, but also to chemical compounds and mental states. In each case, the whole is said to be more than the sum of its parts: a higher level of organization appears to exhibit properties that are claimed to be non-deducible, non-predictable or unexplainable on the basis of the properties of its lower level components. In the early twentieth century, the concept of emergence was strongly stimulated by the wish to formulate a philosophical alternative to both vitalism and mechanism. The concept experienced a golden age that proved to be short lived as it encountered several scientific and philosophical setbacks in the mid-twentieth century. The concept somehow re-emerged in the late twentieth century, especially as it became a central topic in philosophy of mind, and as it also received the unexpected support of the science of complex systems. In the first decade of the twenty-first century, benefiting from a growing awareness of the complexity of biological phenomena, the concept of emergence re-emerges as a way of characterizing life and its origin, not so much as an alternative to vitalism, but as an alternative to reductive explanations of life. Its relevance remains a debated topic.

Keywords Emergence • Mechanism • Origins of life • Reductive explanation • Vitalism

---

C. Malaterre (✉)
Département de philosophie, Université du Québec à Montréal, Case postale 8888, Succursale Centre-Ville, Montréal, QC, H3C 3P8, Canada
e-mail: malaterre.christophe@uqam.ca

S. Normandin and C.T. Wolfe (eds.), *Vitalism and the Scientific Image in Post-Enlightenment Life Science, 1800–2010*, History, Philosophy and Theory of the Life Sciences 2, DOI 10.1007/978-94-007-2445-7_7,

1 Introduction

At the end of the nineteenth century, despite the foundational work of John Stuart Mill (1843) and of George Henry Lewes (1875), the philosophical concept of emergence was still fairly imprecise and led to different interpretations. This does not come as much of a surprise, since many properties – even simple ones – of matter and of living organisms were still at that time unexplained by science on the basis of an analytic study of their constituents: this was typically the case of chemical elements but also of living organisms and of the mind. At that time, the concept of emergence was construed with a view to applying it to unexplained qualitative gaps between parts and wholes. In the early twentieth century, the concept was forged into a philosophical stance: "emergentism" gained momentum among philosophers, most notably as an alternative to the new spiritualist and metaphysical vitalism popular in Europe at that time.[1] In the "golden age" philosophical writings of Alexander, Morgan, and Broad, the concept of emergence was actually construed not only as an alternative to vitalism and its dualist ontology, but also as an alternative to mechanism and its constraining determinism.[2] In the twentieth century, two phenomena occurred: on the one hand, the concept of emergence gradually gained in formalism and precision in philosophy while, on the other, the domains to which it was deemed applicable shrink or grow depending on scientific advances and theories. For instance, the discovery of quantum mechanics and its application to chemistry in the 1930s tended to shift the concept of emergence away from chemistry;[3] and similarly in the 1950s, the discovery of the structure of DNA[4] as well as the first prebiotic experiments[5] tended to push the concept of emergence away from the phenomenon of life. In parallel, logical positivists such as Hempel and Nagel turned emergence into a purely epistemic concept.[6]

Nonetheless, at the end of the twentieth century, the debate on emergence regained unexpected vigor in philosophy, first in the tracks of the philosophy of mind debate, but also in the new field of complex systems and computational modeling.[7] Interestingly, emergence percolated back into biology, in particular because of the bewildering complexity of living organisms that was then discovered through research studies such as the Human Genome Project. The interesting side effect of this re-emergence of emergence in science and in philosophy is its claimed relevance to the phenomenon of life. In this contribution, I show how, more than 50 years later and after much scientific and philosophical misfortune, the notion of emergence is again construed by some as a key concept to account for the specificity of the phenomenon of life yet this time as an alternative to tentative reductionist explanations

---

[1] Driesch (1908). See also Garrett, (2013) this volume.

[2] Alexander (1920), Morgan (1923), and Broad (1925).

[3] James and Coolidge (1933).

[4] Watson and Crick (1953).

[5] Miller (1953).

[6] Hempel (1948) and Nagel (1961).

[7] Cf. Bedau (1997).

of life.[8] Of course, this recent popularity of emergence does not imply that the notion is empirically adequate: my only aim here is to document the return of the notion as a key discussion topic among scientists and philosophers when it comes to characterizing life. I have discussed this empirical relevance of the notion of emergence with regards to life elsewhere, defending a pragmatic epistemic construal of it.[9] In this contribution, my account starts in the mid-nineteenth century with the first formal philosophical discussions on emergence (Sect. 2) and continues into the early twentieth century when the concept gained momentum as an alternative to vitalism and to mechanism (Sect. 3). I describe the mishaps, both scientific (Sect. 4) and philosophical (Sect. 5) that the concept encountered in the mid-twentieth century. I then show how the concept later on gained a strong foothold in philosophy of mind and received support from anti-reductionist debates (Sect. 6). Further, I explain how the concept was found to be of much relevance for complex systems (Sect. 7). I then describe the novel percolation of emergence into the life sciences (Sect. 8) and into origins of life studies (Sect. 9) due, in particular, to questions related to those of complex systems science.

2 Life as an Emergent Phenomenon: A Nineteenth-Century Legacy

The paternity of the concept of emergence is often attributed to the great English philosopher John Stuart Mill.[10] In his extensive 1843 work, *A System of Logic,* Mill contrasts two modes of causal composition as exemplified by mechanics and chemistry: in mechanics, such quantities as forces or momenta can be added somehow linearly (or vectorially as one would say today); in chemistry, however, the properties of the chemical reagents cannot be used in any composition whatsoever to establish the properties of the resulting chemical compounds. These two modes of causal composition, which Mill calls the "homopathic" and the "heteropathic" modes respectively, are taken to reveal a fundamental distinction in nature: on the one hand, certain causes can be composed linearly into their effects; on the other, other causes do not lend themselves to such composing principles. It is this latter class of relationships that is also characteristic of biology and of living organisms:

> All organized bodies are composed of parts similar to those composing inorganic nature, and which have even themselves existed in an inorganic state; but the phenomena of life, which result from the juxtaposition of those parts in a certain manner, bear no analogy to any of the effects which would be produced by the action of the component substances considered as mere physical agents. To whatever degree we might imagine our knowledge of the properties of the several ingredients of a living body to be extended and perfected, it is certain that no mere summing up of the separate actions of those elements will ever amount to the action of the living body itself.[11]

---

[8] Luisi (2002), Hazen (2005), and Deamer (2007).

[9] Malaterre (2010).

[10] McLaughlin (1992) and Fagot-Largeault (2002).

[11] Mill [1843] (1866), vol.1, Book III, Ch.6, § 1, 407–408.

Strictly speaking, Mill does not use the word "emergent," but the concept is lurking behind his heteropathic mode of causal composition. The privilege of introducing the word "emergent" in a technical sense in philosophy is often attributed to George Henry Lewes,[12] who, on the basis of the conceptual distinction made by Mill, classifies the effects of causal composition either as "resultants" or as "emergents":

> Although each effect is the resultant of its components, the product of its factors, we cannot always trace the steps of the process, so as to see in the product the mode of operation of each factor. In this latter case, I propose to call the effect an *emergent*. … The emergent is unlike its components in so far as these are incommensurable, and it cannot be reduced either to their sum or their difference.[13]

Aside from the major problem of characterizing the autonomy of mind, Lewes is strongly concerned with characterizing the phenomenon of life, and what differentiates living organisms from machines. He is fiercely opposed to vitalism, yet at the same time also thinks that the doctrine of mechanism is insufficient:

> A theory which reduces vital activities to purely physical processes is self-condemned. Not that we are to admit the agency of any extra-organic principle, such as the hypothesis of Vitalism assumes …; but only the agency of an intra-organic principle, or the abstract symbol of *all* the co-operant conditions – the special combination of forces which result in organisation.[14]

Of course, the development of the concept of emergence with a technical construal in philosophy deserves to be seen in a broader European intellectual context. Indeed, reconciling mechanical determinism with life and with personal freedom was an issue in many intellectual circles at that time. And dualist vitalism was far from being unanimous. No wonder then that ideas very similar to that of emergence appeared in several places across Europe, even if the word "emergent" itself might have been lacking.

In Germany, for instance, several philosophers and psychologists contributed background ideas that were strongly related to the concept of emergence:[15] Wundt spoke of "psychic resultants" to describe the effects of a "principle of creative synthesis" in his book *Grundriss der Psychologie* of 1896; he was also quoted by a leading English emergentist, Morgan, for his "principle of creative resultants"; another example is Sigwart, who discussed these concepts in his 1873 *Logik*.[16]

In France, even if the notion of emergence was not defended by any particular group, several philosophers formulated ideas in the nineteenth century that were subsequently found at the very heart of the emergentist debate in the United Kingdom in the early twentieth century. Anne Fagot-Largeault (2002) proposes to group a certain number of spiritualist positivist philosophers, such as Ravaisson, Lachelier, Boutroux or even Bergson, including also a scientist such as Boussinesq, into a "French school of emergence." Above all, these intellectuals were seeking to

---

[12] Stephan (1992) and Fagot-Largeault (2002).

[13] Lewes (1875), vol. 2, Problem V, Ch. III, 412–413, my italics.

[14] Lewes (1877), Ch. 2, §17, 324.

[15] Stephan (1992).

[16] Ibid., 25.

highlight the limits of mechanism in order to restore some meaning to the notion of human freedom.[17] A "second French school of emergence" can also be identified among those scientists studying life and medicine at that time, in particular in several medical schools or universities.[18] As a matter of fact, the evolution of the concept of vitalism in France in the nineteenth century, since its initial definition by the founders of the Montpellier medical school to its criticism by the Paris medical school, and to Claude Bernard's "neo-vitalist" position, clearly shows interesting similarities with the foundational ideas of the notion of emergence: a simultaneous opposition to dualist vitalism and to mechanism; a holistic approach to life; as well as the assertion of the irreducibility of living organisms to the properties of their physico-chemical components.

3 Emergence as an Alternative to Vitalism and Mechanism

It was in response to the resurgence of a spiritualist stream of vitalism in the late nineteenth and early twentieth centuries that the concept of emergence went through one of its major developments. This resurgence of vitalism was Europe-wide and mainly directed against materialism and its implications for human freedom.[19] Such vitalism was clearly visible in some parts of biology under the form of a resurgence of Stahl's animism (among others) or through the dissemination of such novel ideas as Driesch's entelechy. It was also visible in some parts of metaphysics with, for instance, Bergson's *élan vital* or Nietzsche's vital affirmation.[20] It was this late vitalism, inspired by a spiritualist metaphysics, which flourished in the late nineteenth and early twentieth centuries.[21] And it is partly in response to such vitalism that the notion of emergence developed. The debate between mechanists and vitalists also happened to be fueled by biochemists trying to define their own discipline as distinct from both chemistry and physiology.[22]

The specificity of living organisms thus came to occupy a very central place in the thoughts of emergentist philosophers: the aim was to propose a non-mechanistic alternative to the position of dualist vitalism, that is to say an alternative that would be compatible with some form of monist materialism. In the 1920s, several philosophers developed alternative positions to both vitalism, with its ontological dualism and its reliance on a non-material non-identifiable entity, and mechanism, with its determinism that was found to be too restricting for human freedom. The concept of emergence was thereby conceived in very close relationship with the problem of life and also of its material origin on Earth. But this concept was intended to have a much

[17] Fagot-Largeault (2002, 954).
[18] Malaterre (2007).
[19] Rey (2000, 18).
[20] Vanderlinden (1989).
[21] Duchesneau (2000).
[22] Weber (2007).

broader range of relevance as well: in particular, it sought to apply to both the characterization of novel properties of chemical compounds and to the relationship between mind and brain. More generally, it aimed to extend its scope to all levels of nature where new properties occur during major transitions between these levels, these transitions being regarded as resulting from a general evolutionary process of matter: transition from the physical to the chemical, from the chemical to the biological, from the biological to the mental. Most of the emergentist debate took place in the United Kingdom and three great names of this "golden age of British emergentism" stand out: Samuel Alexander, C. Lloyd Morgan and Charlie Dunbar Broad.

In *Space, Time, and Deity* (1920), Samuel Alexander envisions a theory of emergent evolution that accounts for the dynamic appearance of complex entities, of new types of structures that would thereby engender higher levels of organization and exhibit properties that would be novel when compared to the properties of their constituents. He borrows the term "emergent" from Lewes and puts it into a cosmological evolutionary context with a view to describing as "emergent qualities" the novel properties that appear as a result of the aggregation of components into ever more complex "wholes" belonging to higher levels of organization. For him, life definitely is one of these emergent qualities:

> Physical and chemical processes of a certain complexity have the quality of life. The new quality life emerges with this constellation of such processes, and therefore life is at once a physico-chemical complex and is not merely physical and chemical, for these terms do not sufficiently characterise the new complex which in the course and order of time has been generated out of them. … The higher quality emerges from the lower level of existence and has its roots therein, but it emerges therefrom, and it does not belong to that lower level, but constitutes its possessor a new order of existent with its special laws of behaviour.[23]

The detailed interpretation of Alexander's construal of emergence is somehow challenging: Is it a question of irreducibility, of unpredictability or of lack of explanation? Alexander remains quite imprecise as to what exactly emergence means, yet he is convinced that it exists and that it corresponds to a remarkable trait of nature that must be acknowledged. To borrow Alexander's own words, emergence seems to be a brute fact of nature that must be accepted with "the 'natural piety' of the investigator".[24] Nevertheless, the idea is there, and the phenomenon of life is found to play a critical role in this respect.

In *Emergent Evolution* (1923) – a philosophical work that was conceived at the end of his career – the ethologist C. Lloyd Morgan wishes to demonstrate that the evolution of the natural world – an "orderly sequence of events" (1923, 1) – gives rise to emergent and unpredictable phenomena, that are due, in part, to the complexity of this evolution. Combining the concept of emergence with a cosmology

[23] Alexander [1920] 1927, vol. 2, 46.

[24] For Alexander, "the existence of emergent qualities thus described is something to be noted, as some would say, under the compulsion of brute empirical fact, or, as I should prefer to say in less harsh terms, to be accepted with the 'natural piety' of the investigator. It admits no explanation" ([1920] 1927, vol. 2, 46–47). Alexander will be severely criticized for his 'natural piety', including by fellow emergentist philosophers (e.g. Broad 1925).

inspired by the Darwinian theory of evolution, Morgan develops a theory of emergence whose main objective is to describe a world in which evolution proceeds not by continuous shifts but by an orderly series of steps or jumps. With the process of emergent evolution, novel relationships amongst compound systems generate new emergent properties at each step, and life is but one of them:

> What is supervenient at any emergent stage of evolutionary progress is a new kind of relatedness, new terms in new relations hitherto not in being. In virtue of such new kinds of relatedness, not only have natural entities new qualities within their own proper being, but new properties in relation to other entities. The higher entities are not only different in themselves; but they act and react differently in presence of others. . . . There is more in the events that occur in the living organism than can adequately be interpreted in terms of physics and chemistry, though physico-chemical events are always involved. Changes occur in the organism when vital relatedness is present the like of which do not occur when life is absent. This relatedness is therefore effective.[25]

The problem of the emergence of life is at the heart of Morgan's questioning: he seeks to highlight the limits of mechanism while also rejecting dualistic vitalism. For him, there is only one "order of nature" and yet physico-chemical theories are not sufficient to account for the phenomenon of life.[26]

The concept of emergence is developed further, and in a more technical fashion, by the philosopher Charlie Dunbar Broad in his major book *The Mind and Its Place in Nature* (1925). The debates on the phenomenon of life and on the identification of an alternative to both vitalism and mechanism occupy a central position in his emergentist thoughts. Broad also believes, like other emergentist philosophers, that the problem is wider in scope, encompassing all levels of organization in nature. Although Broad follows Driesch (1908) when it comes to criticizing mechanism, he markedly does not endorse the proposal of a non-material vital principle – an entelechy – that would be responsible for life (1925, 57). As a matter of fact, he refuses the presence of such a non-isolable explanatory entity and opts for a monist metaphysical type of physicalism. Yet, like Driesch, Broad argues that mechanism is not acceptable, as can be seen, for instance, in the difficulties faced by mechanism when it comes to accounting for the sense qualities (1925, 52) or in its incapacity to take into account the non-deductibility of the behavior of a whole from the behavior of its parts (1925, 59). As an alternative, Broad argues in favor of an emergentist stance. He construes emergence as a philosophical theory that claims the existence of systemic properties that are *not deducible* from the properties of the components and their organization. He thereby formulates one of the very first formal philosophical accounts of the concept of emergence, based on this notion of non-deducibility:

> Put in abstract terms, the emergent theory asserts that there are certain wholes, composed (say) of constituents A, B, and C in a relation R to each other; that all wholes composed of constituents of the same kind as A, B, and C in relations of the same kind as R have certain characteristic properties; that A, B, and C are capable of occurring in other kinds of complex where the relation is not of the same kind as R; and that the characteristic properties of the whole R(A, B, C) cannot, even in theory, be *deduced* from the most

[25] Morgan (1923, 19–20).

[26] Morgan (1923, 5–6; 1925, 73).

> complete knowledge of the properties of A, B, and C in isolation or in other wholes which are not of the form R(A, B, C). The mechanistic theory rejects the last clause of this assertion.[27]

In this context, if life is an emergent property of living organisms, then any analytical knowledge obtained by studying the parts of such living organisms separately from one another will not be sufficient to deduce the characteristically vital property that is exhibited by the organism as a whole. Broad thereby leaned towards a holistic approach to living organisms; this approach was, for him, the only way to account for life. And he defined his alternative to mechanism and to dualist vitalism as the philosophical theory of "emergent vitalism" (1925, 58).

In the years following the publication of these three major emergentist works, the concept of emergence sparked quite a debate in philosophy of science – even a "little philosophical fury" according to some.[28] In 1926, at the Sixth International Congress of Philosophy in Boston, many papers were presented on the concept of emergence: Lovejoy spoke of "the meanings of emergence and its modes"; Wheeler discussed the concept of "emergent evolution in the social sciences"; even Hans Driesch, in person, gave a talk on "emergent evolution." That same year, the Aristotelian Society organized a symposium specifically dedicated to "The concept of emergence," attended by Russell, Morris, Mackenzie and many others. And it was also in 1926 that Stephen Pepper published his article "Emergence."[29] In parallel, the debate moved to the science arena too, especially in England. For instance, the biologist J. S. Haldane argued that the phenomena of biology and of life in particular can only be understood by adopting a holistic approach to living organisms:

> The parts of the actively maintained whole which constitutes a unit of life do not exist independently of one another and their environment. They are not things which can be separated without essential change of properties. ... To attribute the maintenance and origin of specific structure to specific structure itself is only to reason in an evident circle – to substitute mere words for ideas. It is impossible to maintain that the physical and chemical structure of a living organism accounts for its life.[30, 31]

Meanwhile, and following a totally opposite stance, many scientists also shared the thoughts of the zoologist Hogben who saw absolutely no reason to abandon the reductionist methodology of biochemistry.[32] The emergence debate had a golden age, but it did not last long. For McLaughlin, *The Mind and Its Place in Nature* is really the last great work of the emergentist tradition (1997, 34): the quantum revolution broke out just after its publication, and, in the years that followed the reductive explanation of the chemical bond by quantum physics, the concept of emergence experienced a series of setbacks both in science and in philosophy.

---

[27] Broad 1925, 61, italics mine.

[28] Ablowitz (1939).

[29] For more historical details, see for instance Stephan (1992) and McLaughlin (1992).

[30] This is the physiologist and biologist John Scott Haldane, father of John Burdon Sanderson Haldane (also known as J.B.S. Haldane).

[31] Haldane 1926, Part I, Lecture I.

[32] Hogben (1930).

4 Scientific Setbacks to Emergence

After its golden age in the 1930s, the concept of emergence encountered two significant setbacks due to two major scientific advances: the advent of quantum mechanics, and the simultaneous rise of molecular biology and prebiotic chemistry.

The theory of quantum mechanics was progressively formulated in the 1920s and early 1930s. To mention a few key dates: in 1924 de Broglie postulated the wave nature of the electron; in 1925, Heisenberg developed his "matrix mechanics"; in January 1926, Schrödinger proposed his "wave mechanics"; and in February 1926, both matrix and wave mechanics were unified with Schrödinger's proof of the mathematical equivalence of both formalisms. Quantum mechanics was born. Also in 1926, Born proposed his probabilistic interpretation of the wave function. And in 1927, Heisenberg developed his famous "uncertainty relations." Many experiments were carried out that revealed both the corpuscular and the wave nature of light, and quantum mechanics received numerous experimental corroborations: it accounted for the Stark and Zeeman effect, for the photoelectric effect, as well as for a number of dispersion phenomena of elementary particles that had remained hitherto unexplained.[33]

Very rapidly, quantum mechanics was applied to the resolution of problems that were traditionally rooted in chemistry: this new physics opened up new explanatory avenues and, in particular, the possibility of deducing certain chemical properties of molecular compounds from the physical properties of their atomic constituents and their organization. And this is precisely when the concept of emergence encountered its first scientific setback. In 1927, Heitler and London proposed a quantum model of the hydrogen molecule, which is the simplest of all molecules to study since it is composed of only two hydrogen atoms: they predicted the energetic stabilization of the molecule and explained the formation of a bond between the two hydrogen atoms forming the molecule.[34] A few years later, James and Collidge carried out the first quantum estimation of the energy of the chemical bond from a quantum model. The calculations were certainly complex and it took a year for the two scientists to check them up to the thirteenth digit of the wave function, but their efforts were rewarded.[35] At the end of the 1930s, there was no doubt that quantum mechanics could be used to predict the chemical properties of molecular compounds on the basis of the physical properties of their atomic constituents.

While these predictions were, in reality, limited to relatively simple cases because of the complexity of the calculations involved, they constituted evidence against the theory of emergence: contrary to claims made by emergentist philosophers, the chemical properties of molecules could indeed be deduced from the physical properties of atoms. From then on, the application of quantum mechanics to chemistry even gave rise to a new discipline: the discipline of "quantum chemistry." The calculation

[33] For a historical account of quantum mechanics, cf. for instance Jammer (1974).

[34] Heitler and London (1927).

[35] James and Collidge (1933).

methods were further developed and refined. The approach of Heitler and London was, for instance, extended by the chemists Slater and Pauling, and became known as the "Valence-Bond (VB) method" or the "Heitler-London-Slater-Pauling (HLSP) method." In parallel, another calculation approach called the "molecular orbital method" was developed by Friedrich Hund and Robert S. Mulliken. The calculations of quantum chemistry being extremely complex and tedious, one had to wait for the onset of computers in the 1950s to extend quantum calculations to slightly more complex molecular systems, including systems possessing up to two dozen atoms. These calculations – called "*ab initio*" since they were carried out exclusively on the basis of the formalism of quantum mechanics – were soon complemented by "*semi-empirical*" calculations, the latter being based on a mixture of purely quantum calculations and of empirical measurements from experimental chemistry.[36] The relevance of the concept of emergence in chemistry thereby appeared seriously compromised. The rise of quantum chemistry in the mid-twentieth century can even be said to constitute a "snub" to the emergentist claims of unpredictability and non-deducibility of chemical properties from physical properties.

Emergentism encountered a second scientific setback in the mid-twentieth century due to the simultaneous rise of molecular biology and of prebiotic chemistry. It is a historical coincidence that the year 1953 saw the publication of two major scientific articles of extreme relevance to the chemical basis of life: the first one in *Nature* by Watson and Crick on the structure of DNA; the second in *Science* by Miller on the abiotic synthesis of amino acids. The discovery of the double helix structure of DNA marks the beginning of molecular biology. If this structure is now quite familiar, the mechanisms of genetic information encoding and transmission were then totally unexplained. The discovery of the double helix structure, with the complementary pairing of nucleotides immediately suggested a possible mechanism of replication. The principle of heredity – the "secret of life" for many – seemed within reach of elucidation, and with it the possibility to understand the functioning of living organisms and their reproduction in molecular chemical terms. With the identification of the three-dimensional structure of the DNA molecule and with the elucidation of the mechanism of replication and of transmission of genetic information, a whole new area of research opened up to biologists.[37] The analysis of the molecular compounds that lie at the heart of living organisms provided a glimpse of possible explanations of life's most fundamental properties, such as reproduction and heredity. With the birth of molecular biology, the explanatory ideal of the phenomena of life became that of explanations formulated at the molecular – hence chemical – level. In other words, the philosophical notion of emergence, according to which biological properties could not be deduced from physico-chemical ones, became obsolete. Molecular biology offered instead the promise of a physico-chemical explanation of life and of its mechanisms.

Furthermore, it became clear by the mid-twentieth century that the key molecular components of living organisms could be synthesized in prebiotically plausible

[36] For more details on the application of quantum mechanics to chemistry cf. Atkins and Friedman (1999).

[37] For a historical overview of molecular biology, cf. Morange (1994).

chemical conditions. Of course, since the early nineteenth century, it was known that some "organic" substances such as urea could also be synthesized in total absence of living organisms, that is to say on the basis of totally inorganic compounds and following the laws of what will come to be known as "synthetic chemistry."[38] Yet, it was only in the mid-twentieth century that one discovered that a broad range of major organic substances, such as amino acids, could indeed be created altogether from some of the most simple and abundant molecular species, such as water, hydrogen and ammonia, simply by adding energy.[39] In other words, it was shown that organic molecules could spontaneously form under conditions similar to those prevailing on the primitive Earth some 4 billion years ago, before life appeared. As such, the work of Stanley Miller in the 1950s marked the birth of a new scientific discipline: the discipline of prebiotic chemistry. Under the supervision of his research director and Nobel laureate Harold Urey, Miller developed an experimental device consisting of glass flasks connected by tubing in a closed circuit: a first flask was filled with boiling water and simulated the primeval ocean and its steam; a second flask was filled with a gas mixture of methane, hydrogen, ammonia and equipped with electrodes that sparked electric discharges into the gas, thereby simulating the primitive atmosphere, its storms and lightning. After several days of running the fluids and gases in circuit from one flask to the other, the device was stopped, and the liquids were analyzed: the results showed the astonishing presence of many organic compounds, some of which, such as amino-acids, were of major biological interest. Miller's work has had a far-reaching impact.[40] Although the composition of the primitive atmosphere is still subject to much debate,[41] Miller's experiment showed that simple chemical reactions in abiotic conditions could be at the origin of relatively sophisticated organic molecules.

From the mid-twentieth century onward, this experiment became the cornerstone of the new discipline of prebiotic chemistry, thereby casting new light on the plausibility of different scenarios of the origins of life, such as those proposed in the 1920s independently by Alexander Oparin (1924) and John B. S. Haldane (1929): such scenarios that were previously taken for extremely speculative became serious candidates for research programs. With these new lines of investigation that took shape in the 1950s, the phenomena of life and of its origins were rooted even more strongly into the physico-chemical sciences. Indeed, if it was possible to synthesize some of the most fundamental constituents of living organisms in prebiotically-plausible conditions, why not imagine being able to synthesize them all, including the most central molecule of life: DNA? The outlook that prebiotic chemistry offered on the question of the origins of life brought support to a mechanistic explanation of life and to a physico-chemical anchoring of its

[38] Wöhler (1828).

[39] Miller (1953). For a historical overview of origins of life studies, cf. Maurel (1994) and Raulin-Cerceau (2009).

[40] Bada and Lazcano (2003).

[41] Cf. Kasting (1993, 2005).

appearance. And clearly, such an explanation ran contrary to an emergentist construal of life. As a result, in the mid-twentieth century, the concept of emergence no longer seemed relevant to the characterization of life and its origins.

5 Philosophical Setbacks to Emergence

While emergence suffered two major scientific setbacks in physics and in the life sciences, the concept also underwent much criticism in philosophy of science in the wake of logical positivism. In their famous 1948 article on the logic of explanation, Hempel and Oppenheim propose to construe the concept of emergence as purely epistemic: they criticize the early twentieth-century views according to which there would exist "emergent phenomena" that would be "new" not just in the psychological sense of "being unexpected" but also in a much stronger sense of being intrinsically "inexplicable" or "unpredictable." Hempel and Oppenheim advocate a formal definition of the concept of emergence that they spell out in terms of the notion of "explanation": a phenomenon, at a given level of organization, is said to be emergent if it cannot be explained by means of the scientific theories that account for the properties of its constituents. And because Hempel and Oppenheim construe "explanation" in the deductive-nomological way, they end up with a definition of emergence in terms of deductive impossibility, which is quite reminiscent of Broad's definition (quoted above). Yet, unlike what Broad seemed to mean and unlike many other emergentists for whom emergence was a fact of nature, Hempel and Oppenheim construe "emergence" as epistemic and relative to a given theory:

> The occurrence of a characteristic *W* in an object *w* is emergent relative to a theory *T*, a part relation *Pt*, and a class *G* of attributes if that occurrence cannot be deduced by means of *T* from a characterization of the *Pt*-parts of *w* with respect to all the attributes in *G*.[42]

Such a definition makes "emergence" relative to a given scientific background: the set of theories that are available at any given point in time. As a result, the use of the concept of emergence is a perfectly legitimate one, yet only when one specifies the specific scientific background against which this emergence is claimed. In short, one should not say: "a property *P* is emergent," but rather: "a property *P* is emergent *relative to a theory T*." So construed, the concept of "emergence" is an epistemic concept (that characterizes our knowledge of Nature), and no longer a metaphysical or ontological one (that would say something about the way Nature really is). And this construal of emergence is meant to apply to all classes of phenomena, be they chemical, biological or mental. In particular, Hempel and Oppenheim propose to reformulate the assertion of the emergent character of life as follows:

> The emergentist assertion that the phenomena of life are emergent may now be construed, roughly, as an elliptic formulation of the following statement: Certain specifiable biological phenomena cannot be explained, by means of contemporary physico-chemical theories, on

---

[42] Hempel and Oppenheim (1948, 151).

the basis of data concerning the physical and chemical characteristics of the atomic and molecular constituents of organisms.[43]

A similar epistemic formalization of the concept of emergence is proposed by Nagel in *The Structure of Science* (1961). For Nagel, to say that a property of a whole is emergent is to assert that one cannot logically *deduce* statements about the occurrence of this property from theoretical statements pertaining to the constituents of this whole. In short, the concept of emergence "must be understood as stating certain *logical* facts about formal relations between statements" (1961, 369). This construal of emergence applies, among others, to the statements of biology. And, given the complexity of living organisms and the state of biological theorizing in the 1950s–1960s, it is not surprising that some biological statements were not deducible from available physico-chemical statements. Yet this impossibility should be understood as relative to the set of scientific knowledge at the time and, in no case, as a "metaphysical" fact about some allegedly "inherent" property of nature:

> It is an elementary blunder to claim that, because some physico-chemical theory (or some class of such theories) is not competent to explain certain vital phenomena, it is *in principle* impossible to construct and establish a mechanistic theory that can do so.[44]

This epistemic reading of emergence in the wake of logical empiricism weakened the reach of the concept: whereas philosophers of the "golden age of British emergentism" and their followers in the first decades of the twentieth century construed emergence as a metaphysical thesis revealing an inherent property of Nature, the reading of emergence that became the new orthodoxy in the mid-twentieth century is an epistemic one: "emergence" did not characterize raw facts of Nature any longer, but at most, the impossibility of a logical deduction of certain statements from others at a certain point in time, given the state of scientific knowledge at that time.

6 The Special Sciences and the Criticism of Logical Empiricism Regarding the Rescue of "Emergence"

Despite the serious setbacks it encountered in the physical and biological sciences and in philosophy, the concept of emergence remained in use in some domains of the special sciences in the second half of the twentieth century.[45] As a matter of fact, the concept was used by several psychologists and philosophers to characterize the relationship between mind and brain, between the mental states of a subject and the neuro-physiological states of his/her brain: mental states like "being hungry" or "seeing the color red" were claimed to be emergent from neuro-physiological phenomena. This emergence was generally taken in the sense of "not being reducible to," yet at the same time it was not taken as an argument in favor of a doctrine of spiritual

---

[43] Hempel and Oppenheim (1948, 151).

[44] Nagel (1961, 438).

[45] See also Garrett, (2013) this volume.

dualism: rather, physical monism was stressed, and yet at the same time, mental phenomena appeared to have some degree of autonomy from their physical basis. The motivations behind such an idea were somehow reminiscent of what had motivated the French spiritual positivist philosophers of the late nineteenth century mentioned above (see Sect. 2), namely the wish to account for human mental freedom, despite a biological body that would remain under the governance of mechanistic laws. As such, emergentist claims with regards to mental states arose as an alternative to both a dualist metaphysical stance and a physico-chemical mechanistic one.

In support of such emergentism, Donald Davidson proposed the thesis of the "anomalism of the mental" in the early 1970s: according to him, mental events (or states or phenomena) do not obey any strict or specific law. The mind is "anomalous" in that there are no laws connecting mental events to other mental events, and also in that there are no laws connecting mental events to physical events.[46] The mind therefore is not subject to any nomological connection at all. As a consequence, it is emergent from physico-chemical phenomena in a much stronger ontological or metaphysical sense than the epistemic sense championed by Hempel and Oppenheim (1948) or Nagel (1961) mentioned above. At about the same time, Hilary Putnam proposed the thesis of "multiple realizability."[47] According to this thesis, a mental state like "being hungry" can be realized in countless different ways at the neurological level. In other words, many different neurophysiological states can engender the very same mental state, in different persons at the same time, or in the same person at different times. For instance, my mind was in the "being hungry" state just before lunch today, and again just before dinner tonight, yet the neurological state of my brain is likely to be quite different tonight, after the fatigue of a day's work, from what it was at noon. Putnam's thesis was interpreted as an argument in favor of a construal of the concept of emergence that is much stronger than the epistemic one proposed by the logical positivists. The debate on the emergent status of the mind with respect to the brain was very active in the last decades of the twentieth century. Emergentist philosophers such as Bunge (1977), Popper and Eccles (1977) or Sperry (1980) made strong claims about the emergent status of mental states, the mind being governed by properties and laws of its own that do not exist at the lower level of neurophysiology and that are irreducible to them. And such positions triggered much debate, with sustained opposition from philosophers such as Kim (1993, 1996).

In the background of this philosophy of mind debate, two other philosophical disputes took place that contributed to a renewed interest in emergence. The first one concerned the very foundations of logical positivism. To be brief, the dispute concerned the criteria chosen by logical positivists to anchor science in logic and to demarcate it from non-science or metaphysical speculation. The debate somehow began in the early 1950s, spanned three decades and involved numerous philosophers. I will illustrate the arguments briefly by referring to Quine, Popper and

---

[46] Davidson (1970).

[47] Putnam (1967).

Hempel. In 1951, Quine published a now often-cited article entitled "Two Dogmas of Empiricism" in which he criticizes the soundness of two assumptions that are taken to be foundational for logical empiricism: one is the analytic-synthetic distinction between analytic and synthetic truths, explicated by Quine respectively as truths grounded only in meanings and as truths grounded in facts; the other is verificationism, the theory that each meaningful statement receives its meaning from some logical constructions of terms that refer exclusively to immediate experience. This article, and subsequent criticisms, started to crack the logical-positivist edifice in the 1950s. Adding to this debate, Popper criticized the positivist claim that science might be demarcated from non-science on the basis of the verification of scientific statements, contrary to non-scientific ones. For Popper, it is not "verification" but "falsifiability" that ought to be taken as demarcation criteria for science, that is to say the possibility for a theory to be proven false by observation (1959). The critique of logical positivism can also be seen in the changes that occurred in the philosophical views of Carl Hempel from the 1940s until the late 1980s.[48] The point here is not to develop an historical account of the questioning of logical empiricism, but rather to show that the weakening of logical-positivist theses contributed, by the same token, to a weakening of the logical-positivist construal of emergence (in an epistemic sense), and thereby to a strengthening of alternative (and more metaphysical) construals of emergence in the last decades of the twentieth century.

The second philosophical debate that contributed to a renewed interest in the concept of emergence at that time is the debate on reductionism and on its relevance in the special sciences. Whereas positivists, such as Nagel (1961, 345–357), had proposed formal accounts of reductionism as the logical deduction of a theory (the "reduced theory") from another theory (the "reducing theory") together with a set of "connecting principles" (that logically connect terms of the reduced theory to terms of the reducing theory), philosophers voiced concerns over such connecting principles (also called "bridge laws" or "correspondence laws"). Fodor, for instance, criticized the possibility of formulating connecting principles in the social sciences and, in particular, in economics, arguing that such laws would take the form of countless disjunctive statements (1974). Concerning more specifically the biological sciences, the critique of reductionism was very active with regards to two major issues: the reduction of classical Mendelian genetics to molecular biology, and the autonomy of biology from physics and chemistry. Hull also called into question the relevance of the model of inter-theoretical reduction for the discipline of biology in which theories do not adopt the clean nomological axiomatized form theories of the physico-chemical sciences may have.[49] Ruse, however, while agreeing with Hull on a number of issues, believed that the empiricist model of reduction provided relevant insights into the development of population genetics.[50] In the heart of the debate, Schaffner sought to improve on the positivist model of inter-theoretical

[48]Cf. Fetzer (2001).

[49]Hull (1972).

[50]Ruse (1976).

reduction by introducing a relationship of "close similarity" between successive theories; Schaffner thereby defended the relevance of reductionism in biology, arguing in particular that molecular biology does offer partial reductions of biology to physics and chemistry (1967, 1976). Yet, Kitcher and many other philosophers of biology, argued that, despite its explanatory role with regards to classical genetics, molecular biology would never succeed in reducing to classical genetics since, owing to the complexity of living organisms, such a reduction would require too large an amount of data (1984). Again, my aim here is not to provide any detailed historical account of such a debate, which is furthermore far from being settled. Nevertheless, I argue that the criticism of the concept of reductionism in the 1970s and 1980s led to a new anti-reductionist orthodoxy that, in turn, lent support to emergentist stances.

Taken together, the debates in the philosophy of mind and more generally in the special sciences, as well as the questioning of logical empiricism and of its construal of reductionism, led to a much-renewed philosophical debate on emergence in the 1980s and 1990s, maybe up to the point of creating – if I may borrow from Ablowitz (1939) – "a [second] small philosophical fury."

7 Unexpected Support from the Physical Sciences: Complex-Systems Studies and Artificial Life

It is remarkable that in the 1990s the concept of emergence also made a significant "comeback" within a scientific milieu that had earlier provided some of the first empirical arguments against it, namely the physical sciences. As a matter of fact, a new field of research appeared in the late twentieth century that focused on "complex systems." Complex systems are physical systems whose characteristic features are said to result in our "inability to discriminate the fundamental constituents of the system or to describe their interrelations in a concise way."[51] Paradigmatic examples include fluids that, under certain conditions, give rise to instability or turbulence phenomena like the appearance of Rayleigh-Bénard thermal convection cells in a heated fluid or similar phenomena in nonlinear optical systems, including lasers. Other examples include chemical systems that give rise to surprising phenomena like the creation of reaction–diffusion wave fronts, as is the case in the Belousov-Zhabotinsky two-dimensional chemical phenomenon. Such phenomena are indeed very puzzling: Why do circular convection cells suddenly appear in a fluid that was previously characterized by linear heat transfer movements? Why do wave fronts appear in the Belousov-Zhabotinsky reaction whereas none occur in more traditional chemical reactions? Can such properties be explained, predicted on the basis of the properties of the constituents of the systems under study? Are there common models, laws behind all such intriguing, unpredictable complex systems?

[51] Badii and Politi (1997, xi).

Because the concept of emergence is meant to apply to systems that display a range of properties that are novel and unexpected when compared to properties of their constituents, it perfectly characterizes the surprising properties of complex systems. And, as a matter of fact, the new field of complex systems studies makes abundant use of the concept of emergence.[52] While more and more sophisticated computational models are being developed with a view to accounting for such sudden appearances of properties and phase transitions, many researchers label these phenomena "emergent." This is very much so in the case of complex formal networks,[53] of specific chaotic systems governed by strange attractors[54] or of particular features of cellular automata, such as "gliders" in Conway's "game of life" or bird flocks in Reynold's "boids."[55]

Related to the field of complex systems studies, especially through computational perspectives, is the field of "artificial life" (also called "ALife"). ALife definitely is another area where the concept of emergence has been put to much use. As a matter of fact, this concept is central to this field of research as the life-like properties that are exhibited by computer programs, machines or even physico-chemical systems are very often labeled "emergent." The underlying idea of ALife is that the study of key properties of life and of their appearance within systems that are initially devoid of life can be pursued regardless of the carbon-based substrate that forms the basis for biological life on Earth.[56] Hence one of the foundational objectives of this field: the identification of the universal laws of life (and not just – so to speak – the terrestrial laws of life). ALife exists at the intersection of several streams of research, including studies of self-assembly and self-organization, growth and development, evolutionary and ecological dynamics, and even social and cultural evolution. Such research streams rely, to a large extent, on models and simulations, some of which happen to display quite unexpected and surprising properties: appearance of life-like properties (including reproduction with variation, evolution, growth, development), appearance of amazing structural forms, display of sudden changes akin to phase transition phenomena, and many others. The concept of emergence appears thereby ideally suited to describe the novelty, the unexpectedness, and the unpredictability of such properties:

> The key concept in Artificial Life is *emergent behavior.* Natural life emerges out of the organized interactions of a great number of nonliving molecules, with no global controller responsible for the behavior of every part. Rather, every part is a behav*or* itself, and life is the behav*ior* that emerges from out of all of the local interactions among individual behav*ors*.[57]

The concept of emergence is indeed found abundantly in artificial life research and has become one of its cornerstone concepts, even up to the point of becoming

[52]Cf. Bonabeau and Dessales (1997).

[53]Cf. Kauffman (1993) and Barabasi and Reka (1999).

[54]Cf. Newman (1996).

[55]Cf. Bedau (1997) and Holland (1998).

[56]Langton (1989).

[57]Langton (1989, 2–3).

an object of study in itself. Emergence is everywhere, in the appearance of multicellular organisms,[58] in the surprising properties of small-scale societies[59], in the formal features of complex systems theories[60] or even in unexpected simulation results.[61]

8 The Re-emergence of Emergence in the Life Sciences

In the wake of its abundant use in complex systems studies and in artificial life, the concept of emergence has also been put to use in different areas of biology, and has somehow "re-emerged" after half a century of disgrace. In the early 2000s indeed, scientists started to realize – quite concretely – how complex genomes could be. The human genome project, for instance, quickly resulted in a wealth of data, the analysis of which raised overwhelming problems. Whereas it was initially expected that we could easily identify numerous and useful "genetic laws" that would explicitly connect sets of genes to set of phenotypes, the analysis of the data produced proved to be much more difficult than anticipated.[62] As more and more gene data was gathered, it became clearer and clearer that genetic laws were far from obvious. Rather than being the expression of some sort of straightforward isomorphism, the relationships between genotypes and phenotypes appeared to result from extremely complex networks of interactions and regulations, in addition to being modulated by newly discovered sets of epigenetic factors (including the action, for instance, of methyl groups and chromatin).

In response to this new and abundant genetic complexity, a new discipline of biology has recently appeared: the discipline of "systems biology."[63] By combining powerful mathematical modeling and computational analysis, systems biology has taken it upon itself to integrate the wealth of available genetic information into meaningful causal webs of interactions. Yet, owing to the complexity of the models at hand, some of the tools developed by the science of complex systems happened to make much sense in this new epistemic context and were quickly put to use. No wonder then that some of the key concepts pertaining to the study of complex systems – such as that of emergence – happened to be used, in turn, in this new area of biology that arose specifically to deal with the tantamount complexity of genetic expression.

It is worth noting that the percolation of the concept of emergence into the life sciences is not limited to genetics, but extends to several other domains of biology. It concerns, for instance, some of the new organicist approaches that have been

[58] Cf. Furusawa and Kaneko (1998).

[59] Cf. Read (2003).

[60] Cf. Kubik (2003).

[61] Cf. Ronald et al. (1999).

[62] Keller (2005) and Morange (2005).

[63] O'Malley and Dupré (2005).

proposed to explain the development of organisms in developmental biology.[64] It is also of central importance to some "systemic" theories of cancer that locate the causes of carcinogenesis in the disruption of the emergent properties of tissues rather then in a malfunction of the genetic and molecular machinery of cells.[65] Still others see in animal behavior, such as that of ants interacting with each other, emergent properties of the same type as those highlighted by some models of complex systems.[66] In the 2000s, emergence has thereby re-entered into the life sciences quite at large, and from there, has ended up well positioned to percolate back into questions related to life and its origin.

9 Emergence, Life and the Origin of Life

A century or so after having played a key role as an alternative to both vitalist and mechanist theories of life, the concept of emergence has recently re-entered the very debate of the characterization of life, from a synchronic point of view, as well as the more diachronic question of its origin and of the transition from non-living matter to living matter.[67] Many of the aforementioned factors contribute to this return: the discovery of the complexity of living systems and in particular of the intricate network-like functioning of their genomes; the percolation of emergence in the life sciences through concepts and methods imported from complex-systems science; or even philosophical stances with regards to anti-reductionist theses in the special sciences, including biology.

For some scientists, what is characteristic of life is a set of emergent properties that initially appeared in some particular complex "prebiotic" molecular systems on primitive Earth. Although such properties would solely result from molecular interactions, they are claimed by some to be irreducible to physico-chemical properties. Living organisms are taken to be physico-chemical systems that, of course, display the most significant properties of life – reproduction, variation, metabolism – but that are also characterized by unexpected, irreducible "emergent" phenomena:

> Complex systems display properties, often called "emergent properties," that are not demonstrated by their individual parts and cannot be predicted even with full understanding of the parts alone. For example, understanding the properties of hydrogen and oxygen does not allow us to predict the properties of water. *Life is an example of an emergent property.* It is not inherent in DNA, RNA, proteins, carbohydrates, or lipids but is a consequence of their actions and interactions. A comprehensive understanding of such emergent properties requires systems-level perspectives and cannot be gleaned from simple reductionist approaches.[68]

[64] Gilbert and Sarkar (2000).

[65] Sonnenschein and Soto (1999).

[66] Solé and Goodwin (2000).

[67] Cf. Morowitz (2002), Luisi (2002), Hazen (2005), and Deamer (2007).

[68] Aderem (2005, 511), italics mine.

Such a characterization of life as an emergent phenomenon – or as the result of emergent properties of living organisms – is also present in the field of origins of life studies. For instance, for the geologist and biologist Robert Hazen whose work has been much focused on origins of life questions, emergence is a typical trait of life, and life is indeed one of the most significant embodiments of emergence:

> The science of emergence seeks to understand complex systems – systems that display novel collective behaviors that arise from the interactions of many simple components .... From the chemical interactions of individual ants emerge the extraordinarily complex social behavior of ant colonies. From the electrical interactions of individual neurons in your brain emerge thought and self-awareness. Emergence is nature's most powerful tool for making the universe a complex, patterned, entertaining place to live. *Life itself is arguably the most remarkable of all emergent systems*. Many origin-of-life experts adopt the view that life began as an inexorable sequence of emergent events, each of which was an inevitable consequence of interactions among versatile carbon-based molecules. Each emergent episode added layers of chemical and structural complexity to the existing environment.[69]

Another example is the biochemist Pier-Luigi Luisi whose research focuses on the spontaneous formation of lipid vesicles, especially in prebiotically plausible environments. Luisi describes the appearance of life on Earth as the result of emergence, while clearly discarding both mechano-reductionist and vitalist theories:

> Life can be seen as a particular kind of *emergent property*. The single components, such as DNA, proteins, sugars, vitamins, lipids, etc., or even the cellular organelles such as vesicles, mitochondria, Golgi islands etc., each per se are inanimate substances. From this multitude of non-living structures, life arises once a given space/time organization of these non-living components is given. *Life itself is indeed the most dramatic outcome of emergence*. .... No vitalistic principle, no mysterious force, is invoked to explain life by modern biochemistry or molecular biology, just molecules and their relatedness as determined by the long history of evolution. Also the field of origin of life is based on this view. As it is well known, the modern view of the origin of life assumes that life on Earth originated from the inanimate matter throughout a spontaneous increase of molecular complexity (the so-called prebiotic molecular evolution). Clearly, at each level of growing complexity, novel properties arose (binding, catalysis, self-reproduction, etc.) up to the complexity of the first protocells. The analysis of this historical progression of molecular complexity and of the corresponding emergent properties is indeed a fascinating field of inquiry.[70]

There seems to be something irreducibly complex in the phenomenon of life that makes living systems possess properties that appear emergent from their molecular constituents. Such views are not associated with a return of vitalism but rather with anti-reductionist conceptions of the irreducibility of life to physico-chemical phenomena. One may wonder, however, whether the interdisciplinary research stream on the origins of life – "origins of life studies" – that has considerably developed since the mid-twentieth century, will indeed bring continued support to the concept of emergence, or whether new developments will once more reduce its scope as I have argued elsewhere.[71] Most importantly, one may wonder to what extent the formulation of a "theory of life" will impact the emergentist construal of life or will comfort some more reductionist views.

---

[69] Hazen (2005, xiv–xv), my italics.

[70] Luisi (2002, 197), my italics.

[71] Malaterre (2010).

10 Conclusion

The recent "re-emergence" of emergence in regards to characterizing life and its origin may seem somehow surprising as this nineteenth-century concept faced such serious setbacks throughout the twentieth century, both from scientific and philosophical standpoints. In particular, it would have seemed that the quantum explanation of chemical properties on the one hand and of the molecular explanation of genes on the other would have refuted emergentism as applied respectively to chemistry and life. In addition, the field of research on the origins of life has grown considerably since the 1950s. Whereas origins of life questions were considered as nothing other than speculations at the time when Oparin and Haldane published the first prebiotic soup hypotheses,[72] the prebiotic chemical experiments of Miller (1953) turned these questions into respectable and tractable scientific ones. And, since then, the field of prebiotic chemistry has developed strongly, together with a set of disciplines that address the question of the origins of life from different complementary angles: biochemistry, phylogeny, molecular biology, systems chemistry, synthetic biology and even astrobiology. It would seem that such massive research efforts would have brought about additional setbacks to the concept of emergence. Yet, the concept has reentered the arena through the back door of the sciences of complex systems and of artificial life. While this re-emergence of emergence is not associated with a return of vitalism, emergence is nonetheless used, by some, to depict a key characteristic of life. Are we back to the "golden age of British emergentism" of the 1920s, when publications flourished on emergence and its application to life as an alternative to both mechanism and vitalism? In a sense, yes, yet the meaning of the concept of emergence has evolved since then. More precisely, different construals of the term have been formalized, some with strong ontological or metaphysical theses even to the point of granting downward causal powers to emergent wholes onto their own constituents, others with weaker epistemological viewpoints that turn emergence into a question of relationships between knowledge statements.[73] Nonetheless, while Ablowitz was claiming in the 1930s that "the scientific transcendence of the distinction between life and matter would be a death-blow to the theory of emergence" (Ablowitz 1939), it seems that the deathblow has yet to come. Of course, this may have nothing to do with science or with an explanation of the origins of life, but simply with the fact that theories, even philosophical ones, are never totally refuted.[74]

Acknowledgments An earlier version of this paper was presented at the 2006 HOPOS conference in Paris, and I thank the audience for valuable insights. I also thank Charles Wolfe and Sebastian Normandin for their editorial contribution.

---

[72] Oparin (1924) and Haldane (1929).

[73] Stephan (1999).

[74] Duhem (1906).

References

Ablowitz, Reuben. 1939. The theory of emergence. *Philosophy of Science* 6: 1–16.
Aderem, Alan. 2005. Systems biology: Its practice and challenges. *Cell* 121: 511–513.
Alexander, Samuel. [1920] 1927. *Space, time, and deity*. London: Macmillan.
Atkins, P.W., and R.S. Friedman. 1999. *Molecular quantum mechanics*. Oxford: Oxford University Press.
Bada, Jeffrey L., and Antonio Lazcano. 2003. Prebiotic soup – Revisiting the Miller experiment. *Science* 300: 745–746.
Badii, Remo, and Antonio Politi. 1997. *Complexity. Hierarchical structures and scaling in physics*. Cambridge: Cambridge University Press.
Barabasi, Albert-Laszlo, and Albert Reka. 1999. Emergence of scaling in random networks. *Science* 286(5439): 509–512.
Bedau, Mark. 1997. Weak emergence. *Philosophical Perspectives* 11: 375–399.
Bonabeau, Eric, and Jean-Louis Dessales. 1997. Detection and emergence. *Intellectica* 25(2): 85–94.
Broad, Charlie Dunbar. 1925. *The mind and its place in nature*. London: Kegan Paul, Trench, Trübner & Co.
Bunge, Mario. 1977. Emergence and the mind. *Neuroscience* 2: 501–509.
Davidson, Donald. 1970. Mental events. In *Experience and theory*, ed. L. Foster and J.W. Swanson, 79–101. Amherst: University of Massachusetts Press.
Deamer, David. 2007. Emergent phenomena in biology: The origin of cellular life. In *Planetary systems and the origins of life*, ed. Ralph E. Pudritz, Paul G. Higgs, and J.R. Stone, 89–109. Cambridge: Cambridge University Press.
Driesch, Hans [1908] 1921. *The science and philosophy of the organism* vol. I. Aberdeen: University of Aberdeen.
Duchesneau, François. 2000. Préface. In R. Rey, *Naissance et développement du vitalisme en France de la deuxième moitié du 18e siècle à la fin du Premier Empire*, ix-xii. Oxford: Voltaire Foundation.
Duhem, Pierre. [1906] 1954. *The aim and structure of physical theory*. Reprint. Tran. P.P. Wiener. Originally published as *La théorie physique: son objet, et sa structure* (Paris: Marcel Rivière & Cie). Princeton: Princeton University Press.
Fagot-Largeault, Anne. 2002. L'émergence. In *Philosophie des sciences II*, ed. D. Andler, A. Fagot-Largeault, and B. Saint-Sernin, 951–1048. Paris: Gallimard.
Fetzer, James (ed.). 2001. *The philosophy of Carl G. Hempel*. Oxford: Oxford University Press.
Furusawa, Chikara, and Kunihiko Kaneko. 1998. Emergence of multicellular organisms with dynamic differentiation and spatial pattern. *Artificial Life* 4(1): 79–93.
Garrett, Brian. 2013. Vitalism versus emergent materialism. In *Vitalism and the scientific image in post-Enlightenment life science, 1800–2010*, ed. Sebastian Normandin and Charles T. Wolfe, 127–154. Dordrecht: Springer.
Gilbert, Scott F., and Sahotra Sarkar. 2000. Embracing complexity: organicism for the 21st century. *Developmental Dynamics* 219: 1–9.
Haldane, John Scott. 1926. *The sciences and philosophy*. Gifford lectures. Accesible on-line at: http://www.giffordlectures.org/Browse.asp?PubID=TPSAPH&Cover=TRUE.
Haldane, John Burdon Sanderson. [1929] 1967. The origin of life. *The Rationalist Annual* 148: 3–10. (Reprinted in Bernal J.D. 1967. *The origin of life*, 242–249. London: Weidenfeld and Nicolson.)
Hazen, Robert M. 2005. *Genesis. The scientific quest for life's origin*. Washington, DC: Joseph Henry Press.
Heitler, Walter, and F. London. 1927. Wechselwirkung neutraler Atome und homoopolare Bindung nach der Quantenmechanik. *Zeitschrift für Physik* 44: 455–472.
Hempel, Carl G., and Paul Oppenheim. 1948. Studies in the logic of explanation. *Philosophy of Science* 15(2): 135–175.
Hogben, Lancelot. 1930. *The nature of living matter*. London: Kegan Paul, Trench, Trübner.

Holland, John. 1998. *Emergence: From Chaos to order*. Cambridge, MA: Perseus Books.
Hull, David. 1972. Reduction in genetics – Biology or philosophy. *Philosophy of Science* 39: 491–499.
James, Hubert M., and Albert S. Coolidge. 1933. The ground state of the hydrogen molecule. *Journal of Chemical Physics* 1: 825–834.
Jammer, Max. 1974. *The philosophy of quantum mechanics. The interpretations of quantum mechanics in historical perspective*. New York: Wiley.
Kasting, James F. 1993. Earth's early atmosphere. *Science* 259(5097): 920–926.
Kasting, James F. 2005. Methane and climate during the Precambrian era. *Precambrian Research* 137(3–4): 119–129.
Kauffman, Stuart A. 1993. *The origins of order: Self-organization and selection in evolution*. Oxford: Oxford University Press.
Keller, Evelyn Fox. 2005. The century beyond the gene. *Journal of Biosciences* 30(1): 3–10.
Kim, Jaegwon. 1993. *Supervenience and mind*. Cambridge: Cambridge University Press.
Kim, Jaegwon. 1996. *Philosophy of mind*. Boulder: Westview Press.
Kitcher, Philip. 1984. 1953 and all that. A tale of two sciences. *Philosophical Review* 93(3): 335–373.
Kubik, Ales. 2003. Toward a formalization of emergence. *Artificial Life* 9(1): 41–65.
Langton, Christopher G. 1989. Artificial life. In *Artificial life: Proceedings of an interdisciplinary workshop on the synthesis and simulation of living systems, Los Alamos – 1987*, ed. C. Langton, 1–47. Reading, MA: Addison-Wesley.
Lewes, George Henry. 1875. *Problems of life and mind*, vol. II. London: Trübner and Co.
Lewes, George Henry. 1877. *The physical basis of mind: Second series of problems of life and mind*. London: Trübner and Co.
Luisi, Pier Luigi. 2002. Emergence in chemistry: Chemistry as the embodiment of emergence. *Foundations of Chemistry* 4: 183–200.
Malaterre, Christophe. 2007. Le néo-vitalisme au XIXe siècle: une seconde école française de l'émergence? *Bulletin d'Histoire et d'Épistémologie des Sciences de Vie* 14(1): 25–44.
Malaterre, Christophe. 2010. *Les origines de la vie: émergence ou explication réductive?* Paris: Hermann.
Maurel, Marie-Christine. 1994. *Les origines de la vie*. Paris: Syros.
McLaughlin, Brian. 1992. The rise and fall of British emergentism. In *Emergence or reduction? Essays on the prospects of nonreductive physicalism*, ed. A. Beckermann et al., 49–93. Berlin: Walter de Gruyter.
McLaughlin, Brian. 1997. Emergence and supervenience. *Intellectica* 2: 25–43.
Mill, John Stuart. [1843] 1866. *System of logic*. London: J.W. Parker.
Miller, Stanley. 1953. A production of amino acids under possible primitive Earth conditions. *Science* 117: 528–529.
Morange, Michel. 1994. *Histoire de la Biologie Moléculaire*. Paris: La Découverte.
Morange, Michel. 2005. *Les secrets du vivant. Contre la pensée unique en biologie*. Paris: La Découverte.
Morgan, C.Lloyd. 1923. *Emergent evolution*. London: Williams and Norgate.
Morgan, C.Lloyd. 1925. Emergent Evolution. *Mind* 34(133): 70–74.
Morowitz, Harold. 2002. *The emergence of everything: How the world became complex*. New York: Oxford University Press.
Nagel, Ernest. 1961. *The structure of science: Problems in the logic of scientific explanation*. New York: Harcourt Brace and World.
Newman, David V. 1996. Emergence and strange attractors. *Philosophy of Science* 63(2): 245–261.
O'Malley, Maureen A., and John Dupré. 2005. Fundamental issues in systems biology. *BioEssays* 27: 1270–1276.
Oparin, Aleksander I. [1924] 1967. *The origin of life*, Trad. A. Synge. (Reprinted in Bernal J.D. *The origin of life*, 199–234. London: Weidenfeld and Nicolson.)
Popper, Karl R. 1959. *The logic of scientific discovery*. New York: Basic Books.

Popper, Karl R., and J.C. Eccles. 1977. *The self and its brain*. New York: Springer International.
Putnam, Hilary. 1967. Psychological predicates. In *Art, mind, and religion*, ed. W.H. Capitan and D.D. Merrill, 37–48. Pittsburgh: University of Pittsburgh Press.
Quine, Willard V. 1951. Two dogmas of empiricism. *Philosophical Review* 60: 20–43.
Raulin-Cerceau, Florence. 2009. *Les origines de la vie: histoire des idées*. Paris: Ellipses.
Read, Dwight W. 2003. Emergent properties in small-scale societies. *Artificial Life* 9(4): 419–434.
Rey, Roselyne. 2000. *Naissance et développement du vitalisme en France de la deuxième moitié du 18e siècle à la fin du Premier Empire*. Oxford: Voltaire Foundation.
Ronald, Edmund M.A., Moshe Sipper, and Mathieu Capcarrère. 1999. Design, observation, surprise! a test of emergence. *Artificial Life* 5(3): 225–239.
Ruse, Michael. 1976. Reduction in genetics. In *PSA Proceedings 1974*, ed. R.S. Cohen et al., 32:633–651. Boston: D. Reidel.
Schaffner, Kenneth. 1967. Approaches to reduction. *Philosophy of Science* 34(2): 137–147.
Schaffner, Kenneth. 1976. Reductionism in biology: prospects and problems. In *PSA proceedings 1974*, ed. R.S. Cohen et al., 32: 613–632. Boston: Reidel.
Solé, Ricard, and Brian Goodwin. 2000. *Signs of life: How complexity pervades biology*. New York: Basic.
Sonnenschein, Carlos, and Ana Soto. 1999. *The society of cells: Cancer and control of cell proliferation*. New York: Springer.
Sperry, Roger W. 1980. Mind-brain interaction: Mentalism, yes; Dualism, no. *Neuroscience* 5: 195–206.
Stephan, Achim. 1992. Emergence – A systematic view on its historical aspects. In *Emergence or reduction? Essays on the prospects of nonreductive physicalism*, ed. A. Beckermann, H. Flohr, and J. Kim, 25–47. Berlin: Walter de Gruyter.
Stephan, Achim. 1999. Varieties of emergentism. *Evolution and Cognition* 5: 49–59.
Vanderlinden, Sonja. 1989. Le vitalisme dans la littérature néerlandaise. In *Aspects du vitalisme*, ed. S. Vanderlingen and G. Jacques, 190–209. Bruxelles: Editions Nauwelaerts.
Watson, James D., and Francis H.C. Crick. 1953. A structure for desoxyribose nucleic acid. *Nature* 171: 737–738.
Weber, Bruce H. 2007. Emergence of life. *Zygon* 42(4): 837–856.
Wöhler, Friedrich. 1828. Ueber künstliche Bildung des Harnstoffs. *Annalen der Physik und Chemie* 88(2): 253–256.

Chapter 8
Wilhelm Reich: Vitalism and Its Discontents

Sebastian Normandin

Abstract In order to understand the work of the radical psychiatrist Wilhelm Reich in wider context, I ask whether or not he can be considered a vitalist. Beginning with some brief discussion of the idea of vitalism, this chapter moves to look at Reich's seminal writings and the details of his life in an effort to properly situate him in the context of the vitalist tradition. I argue that Reich encountered deep resistance to, and criticism of, his ideas, and his role as an outsider in relationship to mainstream psychological paradigms was coveted but also unavoidable. Not only does Reich clearly fit into the vitalist paradigm – understanding him in these terms actually helps explain why he struggled for legitimacy and recognition in the psychiatric field throughout his career. Regardless of this peripheral role, or perhaps because of it, much interest remains in Reich's thought, and his insights can still give us a more nuanced sense of sexuality and its relationship to life, spirituality, society and politics.

Keywords Animism • Counter culture • Sigmund Freud • Orgone • Outsiders • Psychiatry • Wilhelm Reich • Vital force • Vitalism

Was the radical psychiatrist Wilhelm Reich (1897–1957) a vitalist? This question requires elaboration and the exploration of historical complexities, which are found both in the details of Reich's life and work, and in the history of the idea of vitalism. Using vitalism's history as scaffolding, I fit the building blocks of Reich's biography and research tangents into a larger context.

Reich's impact has been significant. A lack of recognition regarding his role is a result, I believe, of confrontational encounters with the society around him, particularly in the later part of life. That this pioneering psychiatrist and sexologist, once a

S. Normandin (✉)
Lyman Briggs College, Michigan State University, 919 E. Shaw Lane, East Lansing, MI, 48825, USA
e-mail: vitaliste@hotmail.com

S. Normandin and C.T. Wolfe (eds.), *Vitalism and the Scientific Image in Post-Enlightenment Life Science, 1800–2010*, History, Philosophy and Theory of the Life Sciences 2, DOI 10.1007/978-94-007-2445-7_8,

favorite of Freud, died in a cell in Lewisburg Penitentiary in Pennsylvania in 1957, a mere 60 years young, requires explanation. Only by investigating the particulars of Reich's life and thought will their tragic trajectory be mapped.

1 Reich and the History of Vitalism

Currently, vitalism gets a lot of "bad press" in scientific circles.[1] This derives from a misunderstanding. To its opponents, it is synonymous with superstition and spiritualism. In fact, vitalism is deeply entrenched in the history of thought, medicine and psychiatry. The word, coined in the late eighteenth century, became narrowly defined as the resistance to a strictly physico-chemical interpretation of life, but this only grazes over its wider meaning and broader nuance. To begin with, vitalism concerns itself with the idea of a "life force." The "life force" is widespread in many healing traditions – one could call it an archetype; whether it is *qi* in Chinese medicine, *prana* in Indian Ayurvedic practice or *mana* in the thought of many Polynesian cultures, there is no denying its entrenched ubiquity.

In his promulgation of a tripartite soul, or *anima*, in all living things, Aristotle gives life to a general idea of vitalism in the Western mind.[2] One also finds clear elements of a vitalist perspective in the Hippocratic and Galenic medical traditions, especially in the importance of the Stoically inspired conception of *pneuma* (or breath).[3] This focus on breath is taken up in the work of Reich. Vitalism also has echoes in the long-held belief in phenomena like abiogenesis or spontaneous generation.[4] Aristotle, for example, thought aphids formed organically from the morning dew on flowers. Classically, others saw flies spawning spontaneously from rotting meat or carrion. Not until the seventeenth century does spontaneous generation or the vitalistic perspective face any real criticism or challenge.[5] Even though challenged, it persists in many forms; one sees vitalism in the Montpellier school[6] of the late eighteenth century and the Romantic medical thought of the early nineteenth century in a thinker like Blumenbach.[7] The rise of mechanistic views during the

[1] As Canguilhem (2008, 60) says: "This term has served as the label for so many extravagances that, at a moment when the practice of science has imposed a style of research and, so to speak, a code and deontology of scientific life, vitalism carries a pejorative value even for those biologists least inclined to align their object with that of physicists and chemists. There are few biologists who, classified as vitalists by critics, willingly accept this label. In France, at least, it is not exactly a compliment to invoke the names and fame of Paracelsus or Jan Baptist van Helmont."

[2] Aristotle (1956).

[3] For the Stoics, *pneuma* was a "vital spark," the source of life. Galen was influenced by this perspective and in a sense partly materialized it in his thoughts on physiology.

[4] Geison (1974), Farley (1977), and Strick (2000).

[5] Redi (1964).

[6] Williams (2003).

[7] For a deeper understanding of Blumenbach's vitalism cf. Chap. 3 by Steigerwald in this volume.

Scientific Revolution (perhaps best embodied in Descartes) begins the challenge to the idea and the materialism of the biochemical laboratory in the nineteenth century further erodes the influence of vitalism.

And yet there were more recent manifestations, whether in the *Odic force* of speculative scientist Karl Reichenbach (1788–1869) in the mid-nineteenth century or the *élan vital* of French philosopher Henri Bergson (1859–1941) in the early twentieth century.[8] Early twentieth-century biology also featured vitalist elements, in the notion of *entelechy* proposed by the German Hans Driesch (1867–1941), a term borrowed from Aristotle and referring to a kind of internalized sense of purpose, a drive or structure that seemed to transcend the material, again long considered characteristic of living things.[9]

Reich was driven by the same broad concerns as these vitalist predecessors. In *The Function of the Orgasm* (1942) he says: "The question, '*What is life?*' lay behind everything I learned. Life seemed to be characterized by a peculiar reasonableness and purposefulness of instinctive involuntary action."[10] Reich claimed he was well aware that the human race had long known about the existence of a universal energy related to life, but his desire, different from most earlier thinkers, consisted in making this energy usable. As such, there was always a tension in his work between science and philosophy: "Reich was not so much a misguided natural scientist as a life philosopher whose observations were governed by his vitalist and holistic beliefs."[11]

Reich was first inspired to look at the question of vitalism while reading about research into the rational organization of ants and, in his investigations into contemporary examples of the idea, fell upon the work of Driesch:

> Between 1919 and 1921 I became acquainted with Driesch's [*Science and Philosophy of the Organism*] and [*Theory of Order*]. The former I understood, but not the latter. It became clear that the mechanistic concept of life, which dominated our study of medicine at the time, was unsatisfactory. There could be no quarrel with Driesch's contention that, whereas in the living organism the whole could be formed out of a part, one could not make a whole machine out of a screw. However, his explanation of living functioning by means of the concept of 'entelechy' was unconvincing. It gave me the feeling that a gigantic problem was being evaded by way of a word.[12]

Driesch's ideas, derived from experiments on living development he performed with sea-urchin embryos (blastomeres), prompted reflection from Reich, but his materialist inclinations made him resist the embryologist's overall conclusion: "I ruminated a good deal about Driesch's three proofs of the specific difference between the organic and the inorganic. They seemed to be sound, but the metaphysical quality of the life principle did not seem quite right to me."[13] Reich's vitalist

---

[8]Reichenbach (1968) and Bergson (1911).

[9]Cf. Driesch (1908, 1914). On teleology and vitalism cf. Garrett (2003) and Chap. 6 in this volume.

[10]Reich (1968, 45).

[11]Pietikainen (2007, 154).

[12]Reich (1968, 45).

[13]Ibid.

thinking, as we will see, was less mystical, and avoided superadded elements; in this sense there is a parallel between Reich's vitalism and the "physical vitalism" proposed by Claude Bernard.[14] Though Reich's "orgone biophysics" was rooted in the physical sciences, there was always a suspicion that understandings of phenomena like electricity or magnetism were insufficient models.[15] In one paper, Reich says "the basic question under discussion is whether the orgone energy is electricity or whether it represents a *primordial form of energy* of which electricity and magnetism are only specific functions."[16] Still, the German Driesch was ever an inspiration: "I … had Driesch's concepts in mind when I thought of vitalism." Reich's suspicions about Driesch's ontological orientations, however, eventually proved correct: "My vague feeling of the irrational nature of his assumptions proved to be true. He later found refuge among the spiritists."[17]

Disenchanted with Driesch, Reich moved to Bergson, whose *élan vital* was the most cogent, convincing expression of vitalism in the early twentieth century. Bergson caused a sensation with *Creative Evolution* (1907 [1911]), a book with wide-ranging appeal. While he repudiated vitalism *per se*, many still saw him as a vitalist. In the immediate pre-war period, Bergson was one of the world's most recognized philosophers, and it is no exaggeration to speak of his thinking as a kind of *bergsonisme*.[18] Reich credits Bergson's work as inspiring and importantly influential in his own development:

> I had better luck with Bergson. I studied his writing very thoroughly, especially his *Essai sur les données immédiates de la conscience, L'Evolution créatrice,* and *Matière et Mémoire*. I felt instinctively the validity of his effort to reject a mechanistic materialism as well as a finalism. His explanation of the perception of *time-duration* in mental life and of the unity of the self only confirmed my inner perceptions of the non-mechanistic nature of the organism. All this was quite dark and vague, a feeling rather than a knowledge. My present theory of psychophysical identity and unity originated from thoughts of Bergson, though it became a new *functional psychosomatic theory*.[19]

Reich struggled to remake Bergson and Driesch into more concrete form, seeking to balance his sympathy with vitalism and his penchant for a mechanistic understanding of science:

> For some time, I was taken for a 'crazy Bergsonian', because I agreed with him in principle, without, however, being able to state exactly where his theory left a gap. His *élan vital* was highly reminiscent of Driesch's 'entelechy'. There was no denying the principle of a

[14] Normandin (2007).

[15] Reich's concepts of "orgone biophysics" and "orgone energy" remind us that he was not seeking to transcend all physico-chemical understandings of life, rather he saw the mechanical metaphors upon which these understandings relied as insufficient. Like many vitalist thinkers, I believe he felt "life" was a variable – arguably *the* variable – that could not be left out of the equation.

[16] Manuscripts, box 11, "Orgone Biophysics, Mechanistic Science and 'Atomic' Energy," Wilhelm Reich Archives, Countway Library of Medicine, Boston, MA, hereafter WR Archives.

[17] Reich (1968, 46). For Hans Driesch's spiritual turn cf. Driesch (1933).

[18] Grogin (1988).

[19] Reich (1968, 46). Canguilhem also describes vitalism as more of a requirement than a theory. Cf. Canguilhem (2008, 62).

creative power governing life; only it was not satisfactory as long as it was not tangible, as long as it could not be described or practically handled. For, rightly, this was considered the supreme goal of natural science. The vitalists seemed to come closer to an understanding of the life principle than the mechanists who dissected life before trying to understand it. On the other hand, the concept of the organism working like a machine was more appealing to the intellect; one could think in terms of what one had learned in physics.[20]

Reich grapples here with dualism, a specter haunting the West and typically associated with Cartesianism. Was Reich immersed in this tradition or eclipsing it? Like good dualists, commentators on Reich are divided over this issue. A recent portrayal fits him directly in this tradition:

This dualistic thought pattern can be seen in the fundamental opposition of body and soul, psyche and soma, heaven and hell, as well as the medical concept of manic-depression and the cultural infatuation with Jekyll-and-Hyde doubles. The cosmic struggle of opposites in Reich's Orgonomic Functionalism can be situated in this traditional Western mythology of polarities.[21]

Another biographer argues the opposite:

...Reich had a basic philosophical distrust of any idea that would split nature into two contrasting *primary* principles, and he appealed to the thought of the (then highly influential) French philosopher Henri Bergson to shore up his vitalistic and monistic arguments. Bergson insisted that all of nature was linked by a surging primal energy that could not be quantified at its source: the *élan vital*, or vitalistic component, which remained central to Reich. The monistic component worked against any mind/brain dualism that would either impose a mysterious psychophysical parallelism or derive the conscious and unconscious mind from the brain in a reductive manner.[22]

There is some confusion about Reich's philosophical inclinations here. Seeing him as a vitalist helps clarify this problem. There are strong arguments to be made that vitalism represents a "middle way" between the extremes of materialism and spiritualism,[23] and while under the unavoidable influence of the dualisms of language (something thinkers have struggled with since Heraclitus), there is an attempt among vitalists to bring about a deep synthesis beyond this superficial duality. Reich's later work, *Ether, God and Devil: Cosmic Superimposition* (1949) clearly fits into this struggle. In *Ether*, he actually seeks to transcend the longstanding struggles in Western thought: "Orgonomic functionalism ... stands outside the framework of mechanistic-mystical civilization."[24] Reich also delves into history, finding particular inspiration from *animism*, a conception of vitality he distinguishes from mysticism:

The primitive view of emotional life was not mystical, as is our view today; neither was it spiritualistic or metaphysical. It was *animistic*. Nature was regarded as "animated," but this animation was derived from man's real sensations and experiences. The spirits had human

[20] Ibid. One senses here Reich is trying to "revitalize" a traditional mechanistic-materialism.

[21] Pietikainen (2007, 157–8). Cf. Wright and Potter (2000).

[22] Corrington (2003, 43).

[23] This was first articulated by Dumas in the early 1800s when the word vitalism was coined. Cf. Wolfe and Terada (2008).

[24] Reich (1973a [1949], 11).

> form, the sun and the stars acted like real, living people. The souls of the dead continued to live in real animals.[25]

It is beginning with this animistic sense that one connects Reichian vitalism to holistic thinking and healing, and to the origins of the word "holism."[26] There is also a way in which Reich's orgone is an attempt to "vitalize" or "animate" a universe conventionally seen as made up of largely inert, dead matter. We get an idea of Reich's animism realizing that he saw Johannes Kepler's *vis animalis* in this light.

Reich eventually moved to a kind of "holistic" early modern viewpoint of the harmony between microcosm and macrocosm, suggesting in *Ether*: "The same energy that guides the movements of animals and the growth of all living substance indeed also guides the stars."[27] His manuscripts include a series of quotes from Kepler about the sun and planets being moved by the power of a "soul," and he even underlined all the explicit references to "souls" and their relation to movement.[28] Reich's interest in Italian philosopher of nature Giordano Bruno, who proposed the idea of a "plurality of worlds," can also be viewed through this animistic lens; he "saw Bruno," burned to death by the Inquisition in 1600, "as his true predecessor."[29]

And yet, there was also a practical, pragmatic bent in Reich. In his studies of orgasm described in *The Function of the Orgasm*, Reich writes that he happened upon a particular relationship between fluid and electric discharge. This was a unique phenomena that drove him to "the inevitable conclusion … that the *particular combination of mechanical and electric functions was the specific characteristics of living functioning*." Reich "was now," he says "in a position to make an essential contribution to the age-old difference between living and non-living substance." He elaborates:

> To make living function comprehensible, [vitalists] always adduced some metaphysical principle, such as 'entelechy'. The mechanists, on the other hand, contended that living matter, physically and chemically, differed in no way from non-living matter; that it simply was not yet sufficiently investigated. That is, the mechanists denied a fundamental difference between living and non-living matter. The formula of tension and charge showed both schools to be right, though not in the way they had thought.

Reich suggests his findings unified two disparate ideas into a fully harmonious whole:

> As a matter of fact, living matter does function on the basis of the same physical laws as non-living matter, as is contended by the mechanists. It is, at the same time, fundamentally different from non-living matter, as is contended by the vitalists. For, in living matter, *the functions of mechanics* (tension-relaxation) *and those of electricity* (charge-discharge) *are combined in a specific manner which does not occur in non-living matter*. This difference of living matter, however, is not ascribed – as the vitalists believe – to some metaphysical

---

[25]Ibid., 87.

[26]For the origins of holism cf. Smuts (1926). Also cf. Ash (1995).

[27]Reich (1973a, 94).

[28]Manuscripts, box 17, WR Archives. The folder with the Kepler quotes includes a wide-range of material on astronomy and physics.

[29]Pietikainen (2007, 156). On Bruno cf. Rowland (2009).

> principle beyond matter and energy. Rather, it is itself to be understood on the basis of laws of matter and energy. *The living is in its function* at one and the same time *identical with the non-living and different from it.*[30]

The process Reich was explaining, remember, is the orgasm, the *sine qua non* of his system. Reich would later say that his investigations of orgastic potency "represented the coastal stretch from which everything else has developed."[31] To see its central connection to his theories, we turn to look at orgone energy and the libido.

2 Orgone Energy: A "Vital Force"?

Reich's orgone energy is an idea inspired by vitalism, and becomes central to his work in the 1930s and 1940s. While it has significant parallels to earlier ideas of vital energy, orgone germinates as a variant of Freud's theory of the libido. One author sees orgone as "libido unbound," and notes that Reich essentially "conceptualizes energy as entirely sexual, and pursues a quantitative approach to the libido."[32] In *Civilization and its Discontents* (1930) Freud argues that healthy biological urges (the libido) are suppressed or sublimated to the demands of the social order (with bourgeois morality at its core). Seeing the libido as a kind of "life force," Reich responded critically to Freud's conception:

> When Freud published *Civilization and Its Discontents* … Reich was appalled. In his hundred-page monograph Freud presented his pessimistic evaluation of the eternal conflicts among the libido, ego, and superego within the self and between the self and its community. Eros and death remained the same major players they were nine years earlier (in *Beyond the Pleasure Principle*), and the aggressive drive assumed center stage. Needless to say, none of this sat well with the antidualist Reich, who thought that there was no death drive and that the so-called drive for aggression was the result of bodily armoring rather than an innate piece of nature.[33]

In the early 1920s the two thinkers were in closer concert, but quickly diverged as Reich developed a pathological vision of modern society: "Indeed, at its most extreme, orgonomy turned against the Freudian virtues of sublimation, strength of character, and self-knowledge, abominating them as toxic substances, literally carcinogens."[34] Reich explored repression that marked people in deep, physically

[30]Reich (1968, 357). Reich relating the orgone to the function of electricity (hence his focus on functionalism) – the importance of a build-up of charge and need for this charge to be released – is a crucial metaphor. One could say it is an epistemic model. The focus on electricity is further interesting in light of the history of vitalism. Cf. Steigerwald's Chap. 3 in this volume for the way in which new understandings of electricity by Galvani and Volta inspired a reconceptualizing of ideas of "life" in philosophy. This occurred even more famously in literature.

[31]Sharaf (1983, 86).

[32]Seelow (2005, 50–51).

[33]Corrington (2003, 45).

[34]Shechner (1985, 104).

manifested ways. In *Character-Analysis* (1933), he talked of "character armor" – a kind of bodily transformation that externally demonstrated the profound internal struggles for release and sexual expression he felt challenged so many.[35] He proposed a therapeutic solution to this – "vegetotherapy" – a process involving breathing and relaxation techniques, and even direct and aggressive manipulation by the therapist, designed to break through the build-up of tension and "body armoring." There was a kind of primitivist, vital tone to this approach: "Reich's 'vegetotherapy', with its subliminal suggestion that healthy people aspired to be vegetables, indicated a Romantic primitivism so radical as to take the psyche out of psychoanalysis."[36] Of course, the orgasm was one of the key ways to release this natural build-up of psychic and physical energy, ensuring health. His energetic and fluidic view also resonates with the traditions of alternative medicine, and is further reliant on vitalist notions.

A deep chasm formed between Reich and Freud, rooted in divergent conceptions of psychoanalysis *and* reality:

> For Reich, the optimistic romantic, the libido remained the core of psychoanalysis, a kind of Niagara of vital energy that had to find its outlet if it was not to become completely destructive. If Reich was correct about the libido, then Freud must be wrong about the death-wish and the superego; there was hardly room for both. It really was a basic clash of temperaments: romantic optimist versus realistic pessimist, and the pessimist was reacting to the optimist with increasing antipathy.[37]

In fact, their souring relationship in the late 1920s is traceable through their correspondence.[38] Reich once boasted that *Civilization and Its Discontents* was written in response to one of his lectures in 1929, but he moved away from Vienna in 1930, and from the introspective vision of Freud, preferring to understand neuroses as a more patently overt and physical phenomena:

> Under the Reichian dispensation, self-inquiry became just another layer of suppressive armor, a clinically fashionable way of blocking the flow of natural vegetative juices. If the hero of Freud's old age was Moses, that of Reich was the segmented earthworm. The modern therapeutic offshoots of the Reichian ethos such as EST have maintained this hostility to reasoned self-interrogation which, according to them, merely reinforces the inhibitions that afflict the neurotic.[39]

Reich expands on the libido, seeing it as a key concept in understanding repression as the source of much mental illness and neurosis, and beyond this as a fundamental force in life:

> What excited him so much was his feeling that … libido was nothing less than the vital force itself. If so, this concept could unify all the 'vital sciences' – biology, psychology, zoology – as Newton's concept of gravity unified the physical sciences.… Reich compared

[35] Reich (1961).

[36] Fuchs (2011, 44).

[37] Wilson (1981, 89).

[38] Correspondence, box 2, Freud to Reich, WR Archives.

[39] Shechner (1985, 104).

> the libido to electricity, which can never be observed directly, but only through its manifestations – light, heat, and so on. [It] should be possible to measure the libido as directly as we can measure electric current with a voltmeter.[40]

This immanent vital force, a libido made manifest in even the simplest living systems,[41] anticipates Reich's research tangent, and his development of orgone.

While Reich's orgone energy fits organically into the vitalist tradition, he had a fairly material conception of "vital force":

> In 1945, he wrote that, while Freudian *Id*, Aristotelian and Drieschian *entelechy*, Bergsonian *élan vital*, and his own Orgone describe the same thing … his concept fundamentally differed from those other concepts in that it was not merely an expression of 'human *intuitions* of the existence of such an energy', but '*a visible, measurable, and applicable energy of a cosmic nature*'.[42]

Reich imagined himself a "scientist"; and though he integrated the vitalism of Bergson and Driesch, and even dabbled in the occult through the "anthroposophy" of Theosophist Rudolf Steiner (1861–1925),[43] he saw orgone energy in more concrete, physicalist terms.

What, then, was orgone energy? In the *Function of the Orgasm* Reich says he "discovered" orgone sometime between 1936 and 1940.[44] He described it as "primordial cosmic energy" that was "demonstrable visually, thermically, electroscopically and by means of Geiger-Mueller counters."[45]

Escaping Germany in 1933, Reich ended up in Denmark, and briefly spent time in Sweden – he eventually settled in Oslo in 1934. Disenchanted with radical politics, Reich began experiments to "seek the origins of life." He started observing single-celled protozoa, growing cultures in which he noted glowing blue vesicles that gave off observable energy. Reich named these vesicles "bions" (from the Greek word "life"):

> The *bions* are microscopic vesicles charged with orgone energy ('energy vesicles'). They can be produced from organic and inorganic material by a process of disintegration and swelling up. They propagate like bacteria. They also develop spontaneously in the soil, or, as in cancer, from disintegrating tissues. My book, '*Die Bione*' (1938), shows the significance which the formula of tension and charge assumed for the experimental investigation of the natural organization of living substance out of non-living substance.[46]

In *The Bion Experiments* he discusses spontaneous generation, suggesting that the distinctions between plant and animal life, and the organic and inorganic, were too rigid.[47] This moves him away from an Enlightenment-era vitalism focused on

[40] Wilson (1981, 34).

[41] "Through the work of two Germans, the biologist Max Hartmann and a zoologist, Ludwig Rhumbler, Reich was able to relate his two basic directions of energy flow – 'toward the world' in pleasure and 'away from the world' in anxiety – to the movements of the amoeba" (Sharaf 1983, 208).

[42] Pietikainen (2007, 156).

[43] Wilson (1981, 35).

[44] Reich (1968, 270).

[45] Ibid.

[46] Ibid., 359.

[47] Reich (1979, 144–5).

the categorical, but brings him closer to ancient animisms and contemporary holisms. In an interesting interaction Reich describes with a local he had befriended and started to work with after he moved to northern Maine, we can get a sense of this perspective:

> We had come close to each other when I told him about the nature of bions. This simple man disclosed a spontaneously acquired knowledge of the living with which no academic biology or physics can compete. I asked him whether he wanted to see the life energy under the microscope. I was flabbergasted when my friend, even before looking in the microscope, gave me a correct description of the bions. For decades, he had been observing the growth of seeds and the character of the humus with the unerring instinct of somebody who always lived close to nature. There are, he said, very small vesicles ("bubbles") everywhere. From these, everything develops that is "life." They are so small, he said, that they could not be seen with the naked eye. Yet, the moss on rocks developed from them: the rock, always exposed to the weather, "softens up" on the surface and forms these life bubbles. He said he had often tried to talk about this with academic tourists, but had only met with a peculiar smile. Nevertheless, he said, he was sure that he was right. I had to admit that he was right, for how could moss "germs" "strike root" in the rock?[48]

With orgone energy, Reich believed he "had stumbled upon the bioenergetic foundation of humankind," and "moreover," found "a foundation that was an object not of metaphysical speculation but of natural scientific observation."[49] Experiments in 1936 with what he called T-bacilli (the T standing for *Tod*, the German word for death) also showed that this energy could twist, decay and have darker manifestations, leading to pathological forms – he wrote about these discoveries in *The Cancer Biopathy* (1973 [1936]). Throughout *The Cancer Biopathy* Reich remains focused on the slow, steady, insidious nature of cancer and speculates on the importance of an endogenous decline in orgone energy as the source of the initial cellular transformation. His programmatic would be familiar to the contemporary alternative or holistic healer – a focus on prevention rather than treatment after the fact; an acknowledgement of the important emotional and attitudinal dimension of apparently physiological problems; and suggestions for treatment which incorporated important social transformations.[50]

In *The Cancer Biopathy*, Reich further credits the vitalists for being the first group of thinkers to at least acknowledge the existence of a heretofore unknown form of energy essential in organic function, what he calls the orgone:

> [There] are major contradictions which are impossible to resolve within the framework of known forms of energy. They have been well known to biology and natural philosophy for a long time. Attempting to bridge the gap, some people have put forward concepts that were intended to make the specific life function comprehensible. Most of these concepts were advanced by the opponents of mechanistic materialism, the vitalists. Driesch suggested an "entelechy," a life force inherent in all living matter and governing it. But, since it was neither measurable nor tangible, it ended up a contribution to metaphysics. Bergson's *élan vital* attempted to take account of the incompatibility between known forms of energy and

[48] Sharaf (1983, 341).

[49] Pietikainen (2007, 152).

[50] Reich (1973b).

> living functioning. His *force créatrice* represents an explosive function of matter which manifests itself more clearly in the way life functions. Bergson's hypothesis was directed against both mechanistic materialism and teleological finalism. In theory, it grasped correctly the basic *functional* character of the life process, but it lacked empirical validation. The force in question was not measurable, tangible, or controllable.[51]

Reich believed he solved some of these intractable dilemmas with the orgone. Through orgone energy Reich had found a unifying principle, a variation on historical "life forces," but with pragmatic and practical applications:

> The color of the orgone is *blue* or bluish grey. In our laboratory, the atmospheric orgone is accumulated by means of especially constructed apparatus. A special arrangement of materials ... make[s] it visible. The stoppage of the kinetic energy of the orgone expresses itself as temperature rise. The concentration of the orgone energy is reflected in the varying speed of the discharge in the static electroscope. The orgone contains three different kinds of radiation: bluish gray fog-like formations; deep blue-violet expanding and contracting dots; and whitish, rapidly moving dots and lines. The color of the atmospheric orgone is apparent in the blue sky and the bluish haze which one sees in the distance, particularly on hot summer days. Similarly, the blue-gray Northern lights, the so-called St. Elmo's Fire and the bluish formations which astronomers recently observed during a period of increased sunspot activity, are manifestations of the orgone energy. Cloud formation and thunderstorms – phenomena which to date have remained unexplained – depend on the changes in atmospheric orgone.... The living organism contains orgone energy in every one of its cells, and keeps charging itself orgonotically from the atmosphere by the process of breathing.[52]

Reich's orgone was a kind of nebulous universal. Moreover, his rich descriptive language is evocative of the careful, observation-based tradition of natural history and even Romantic *Naturphilosophie*. By the late 1940s, he believed "he was studying the putative origin of all matter and energy in living beings and galactic systems," and started discussing UFOs and weather control.[53] Before reaching this holistic bioenergetic apotheosis, however, he remained interested in the practical, constructing a simple device to collect his all-pervading energy source. He called it the orgone accumulator.

Fleeing Europe in 1939, Reich relocated to New York, where he taught courses at the New School for Social Research ("Character Formation: Biological and Sociological Aspects" and "Clinical Problems in Psychosomatic Medicine"). He began looking for practical ways to use orgone, and built the first orgone accumulators in 1940. Initially designed for small lab animals, Reich enlarged the devices and started experimenting on their use with people. The orgone accumulator was a simple device – a box made of interspersed layers of wood and steel. He hoped they would allow people to absorb "concentrated orgone energy," and be used in the treatment of certain diseases, particularly cancer. Reich even made smaller, portable accumulator-blankets to apply to specific body parts, and was convinced that there was a physical effect associated with the device.

---

[51] Reich (1973b, 8).

[52] Reich (1968, 360).

[53] Pietikainen (2007, 152).

Leaving Europe, Reich was increasingly isolated but motivated to prove the existence of orgone to the scientific community. He had suffered attacks in the mainstream media after he published his findings about bions and orgone in Norway in the 1930s, struggling to achieve legitimacy.[54] With these thoughts in mind he approached Albert Einstein at Princeton in 1940. He visited Einstein in January 1941. The two talked for an entire afternoon, and Einstein agreed to test the orgone accumulator, which Reich claimed raised the temperature of objects without any apparent heating source. As this defied the laws of thermodynamics, Einstein felt it significant enough to merit some investigation. Reich argued "the matter is much too important as to expose it to the dangers of complete destruction through the irrationalism of the scientific world."[55] Eventually, Einstein found the changes in temperature were the result of convection. Chagrined, Reich continued to write the brilliant physicist and insist on further experiments. After a few more tests, Einstein felt the matter was "completely solved."

Reich, deeply disappointed by the experience, felt further isolated from the general scientific community. His thoughts about Einstein's theories reveal both bitterness and a curious insight all at once:

> Einstein succeeded in fascinating the first half of the twentieth century just because he had emptied space. Emptying space, reducing the whole universe to a static nothing, was the only theory that could satisfy the desert-like character of man of this age. Empty, immobile space and a desert character structure fit well together. It was a last attempt on the part of armored man to withstand and withhold knowledge of a universe full of life energy, pulsating in many rythms (*sic.*), always in a state of development and change; in one word, functional and not mechanistic, mystical or relativistic. It was the last barrier, in scientific terms, to the final break-down of the human armoring.[56]

Around this time Reich moved away from New York and in 1942 bought a piece of property (150 acres) in the Rangeley Lakes region of northwestern Maine. Spending more and more time away from the city, particularly in the summer, he eventually set up a lab in Rangeley. Reich had always been an outsider, but the Einstein encounter pushed him further to the fringe. It was a fierce reminder of his earlier disappointments with politics.

3 Reich, Revolution and Politics

Born in 1897 in Galicia (now the Ukraine) into a well-to-do Jewish family that had largely integrated into the "rural *bourgeoisie*" of the Austro-Hungarian Empire, Reich "spent his childhood in the countryside,"[57] and little in his upbringing

[54]Ironically, in *The Bion Experiments*, Reich thanks the Norwegians for their hospitality and generosity. Reich (1979, 6–7).

[55]Conspiracy, box 1, The Einstein Affair, E-1a, WR Archives.

[56]Conspiracy, box 1, The Einstein Affair, E-36, WR Archives.

[57]Pietikainen (2007, 130).

suggested any overt political influences. Quite the contrary – he fondly recalls the simplicity of his early years and the close exposure to the idyllic pleasures of farm life. This instilled in him the qualities of a budding naturalist:

> My interest in biology and natural science was created early by the life on the farm, close to agriculture, cattle-farming and breeding, etc., in which I partook every summer, practically, during the harvest. At eight to twelve years old, I had my own collection and breeding laboratory of butterflies, insects of various kinds, plants, etc., under the guidance of a private teacher. The natural life function, including the sex function, was familiar to me as far back as I can remember. That may well have determined my later strong inclination as a psychiatrist [looking] for the biological foundation of the emotional life, and also my biophysical discoveries in the fields of medicine and biology, as well as education.[58]

Echoes of an Emerson or Thoreau in Reich's description are no mistake. Elements of a species of transcendental naturalism, and even *Naturphilosphie*, can be found in Reich's life and work. "More than anything," one commentator says, "[Reich's] ideas resemble German Romanticism and especially *Naturphilosophie*."[59] This romanticism also had a revolutionary quality, and looking at Reich's involvement in politics this comes into hard relief.

Reich's childhood was certainly not universally idyllic and simple. When he was 12, Reich caught his mother having an affair with his tutor and the aftereffects of this incident destroyed the family. Stricken with guilt, she committed suicide in 1910. This completely broke Reich's father, and after trying himself to slowly commit suicide by standing for hours fishing in a cold pond, he fell ill and died of tuberculosis in 1914. Reich had been sent off to military school since his mother's death, where he was obviously unhappy and developed a skin condition, possibly psoriasis, which afflicted him his whole life. Clearly, Reich had personal understanding of the idea of "character armor."

Reich served as an officer in the Austrian army during WW I and was remarkably politically disinterested in the early interwar years. His military service and status as a veteran allowed him to complete medical school at the University of Vienna in 4 years instead of 6; he received his M.D. in 1922. More significantly, while still in school in 1920, he attained membership in the Vienna Psychoanalytic Association, and became deeply drawn, as mentioned, to Freud's work.

There were glimmers of politicization, as when he began to reflect on the social role of psychiatry in the mid-1920s; his early work on the "impulsive character" was based on patients he saw at the Vienna Psychoanalytic Polyclinic, many of whom had conditions he felt were exacerbated by poverty and strained social circumstances.[60] What prompted Reich's involvement in more serious political action was a disastrous workers' demonstration in Vienna on 15 July 1927.[61] The police

---

[58] Raknes (1970, 13).

[59] Pietikainen (2007, 154–55). Cf. Sharaf (1983, 55).

[60] Sharaf (1983, 67).

[61] Reich started thinking about social factors – the connections between sexual suppression and "capitalist bourgeois morality" convalescing at Davos in the winter of 1927, but the events in Vienna in July crystallized this nascent notion. Cf. Sharaf (1983, 120).

crackdown on the event resulted in a massacre (89 dead, over 1,000 wounded) and pushed Reich to volunteer for a medical group affiliated with the Communist Party.[62] Thereafter Reich became deeply involved in the workers' struggle, and soon adopted a clear ambition to "unite sexual enlightenment with radical politics."[63] Reich's political sensibilities evolved rapidly and with a deep sense of sincerity: "Unlike Freud, whose politics were tinged with skepticism, Reich was nothing if not righteous and impassioned, and his political credentials were, on the face of them, impeccably radical."[64] His romantic tendencies affected both his ideas and beliefs. There was, however, always a kind of humorless, exalted seriousness in his utopian views.

Soon after becoming involved in socialist politics, Reich founded the Socialist Society for Sexual Counseling and Sexual Research in late 1928, an organization devoted to his new ideas of sex-politics. Rather than spout Marxist dogma, Reich focused on pragmatic initiatives, and "propagated concrete improvements, such as better housing, legal abortion and women's right to divorce their husbands." With his wife and colleagues, Reich "went out into various districts in and around Vienna with a van to give consultation to (mainly working-class) men, women and children."[65] In these early years, Reich was intensely aware of the sexual roots of social and political problems, and viewed his slowly developing ideas about combating repression as a panacea. Some critics eventually saw this approach as too simplistic; an all-consuming, single-minded solution in the totalistic relationship between politics and sexuality. New Left pioneer and sociologist C. Wright Mills once derisively dubbed Reich's ideas as "the gonad theory of revolution."[66] One commentator noted about Reich's later years in the US that:

> Sex, for Reich, *was* politics, and the contentious language of his manifestoes, with its military metaphors of blocks and breakthroughs, made his system sound less like a retreat from the blows of history than a regrouping for a war of liberation against the residual Puritanism and production-oriented austerities of American life.[67]

In short, a revolutionary conception of the vital and liberating potentialities of sexuality in a repressive and authoritarian society.

Of course, the archetype of authoritarianism was Nazi Germany. As Germany fell under National Socialist rule in 1933, Reich wrote *The Mass Psychology of Fascism*, one of the earliest critiques and analyses of fascism and Hitler's Germany.[68] Seeing deep problems with the "mass-psychological" vulnerability of the exploited classes, Reich's study analyzed the profound emotional content of fascism, "pointing out

[62] Pietikainen (2007, 141).

[63] Ibid.

[64] Shechner (1985, 99). In *The Freudian Left*, Paul A. Robinson sees Reich as the quintessential practitioner of "Freudian radicalism." Cf. Robinson (1969).

[65] Pietikainen (2007, 141).

[66] Shechner (1985, 104).

[67] Ibid., 101.

[68] Pietikainen (2007, 145).

[that] the German masses were attracted to the Nazi movement not so much by its political platform (which was purposely vague) as by the emotional appeal of *mystical notions* of 'blood,' 'racial purity,' 'fatherland,' 'Master race,' etc."[69] *Mass Psychology* was based on a close reading of Nazi propaganda pamphlets and other materials, through which Reich brought to light "the underlying sexual content of these mystical notions." He argued, in fact, "that religious mysticism – indeed all mysticism – was a symptom of unfulfilled, repressed or distorted sexuality." The "'mystical longing'" he maintained, was "an 'unconscious orgastic longing'."[70]

Reich's pathologization of the Nazi regime was not necessarily that far removed from the ideas of Freud, or Erich Fromm, who saw in fascism the human desire to escape from freedom by submitting to the crass assumptions of the mass and the dictates of authority:[71]

> My medical experience with individuals from all kinds of social strata, races, nationalities and religions showed me that "fascism" is only the politically organized expression of the average human character structure which has nothing to do with this or that race, nation or party but which is general and international. In this characterological sense, *"fascism" is the basic emotional attitude of man in authoritarian society, with its machine civilization and its mechanistic-mystical view of life*.[72]

It is hard to imagine a more vitalistic conception of politics, particularly in its vilification of the ordered mechanistic impulses in mankind. Also apparent is Reich's growing suspicion of mysticism, and as we have seen, the full flowering of this critique comes with *Ether*.

What really set Reich apart from his contemporaries was the explicitly sexual focus of his political reflections. As he urges in *Mass Psychology*: "*the mechanism which makes the masses of people incapable of freedom is the social suppression of genital love life in children, adolescents and adults*."[73] During this overtly politicized period in the mid-1930s Reich saw the true cause of human misery arising from the interplay between "sexuality, politics and psychology." He saw phenomena like neuroses, anxiety and alienation sprouting from structural sources, challenging Freud directly by suggesting that "'destructive drives' are not biologically but socially determined, and that the most abhorrent form of misery, sexual repression, is of a social origin." In his exploration of authoritarianism and its expression, for example, in the traditional patriarchal family, he saw repression as "an important ideological weapon, because it binds people to the Church and to the bourgeois social order, rendering them incapable of taking up critical attitudes."[74]

In contrast to Freud, Reich saw suppression as avoidable *and* curable: "social suppression is not part of the natural order of things. It developed as a part of

---

[69] MacBean (1972, 3).

[70] Ibid.

[71] Fromm (1994 [1941]).

[72] Shechner (1985, 100).

[73] Ibid.

[74] Pietikainen (2007, 147).

patriarchy and, therefore, is capable of being eliminated, fundamentally speaking."[75] Reich was inspired in this idea by the research of anthropologist Bronislaw Malinowski in the Trobriand Islands, who found that sexual repression did not exist in the local population. This was due, Malinowski argued, to the society's matriarchal structure.

Reich's therapeutic approach was innovative; a deviation from his contemporaries:

> In narrowing down the problem of alienation to the sexual sphere, Reich rescued political psychology from tragic biology and delivered it into the hands of medicine – not, to be sure, conventional medicine, as the AMA and the Federal Drug Administration understood it, but medicine as premodern naturalists imagined it, as a branch of moral philosophy.[76]

There is a species of vitalism in this characteristically holistic approach to health; an appreciation of the complex interconnected synthesis of the psychic and somatic.

Reich was an idealist in difficult times. Unlike most other members of the German psychiatric community, he spoke out against the rising tide of Nazism. Further, following a trip to Russia in 1933, Reich began to criticize the Communist Party, calling communism a kind of "red fascism". In 1936, he wrote in his diary that, "Stalin 'will undoubtedly become the Russian Hitler'."[77] Reich's outspoken approach differed from most of his contemporaries, and his so-called *Sex-Pol* movement made practical demands about the politicization of sexuality that have a familiar ring to the modern ear. His concerns included "... the free distribution of contraception to those who cannot afford it, a complete abolition of anti-abortion legislation, an abolition of prostitution and all compulsory regulations concerning marriage, and the eradication of venereal diseases." Additionally, "he demand[ed] changes in education so as to prevent neuroses and sexual disturbances, and he call[ed] for reforms in the training of doctors, pedagogues and nurses in matters relating to sexuality."[78]

Though Reich once corresponded with Trotsky, impressing on him the importance of sexuality in the progress of revolution, he was out of synchronicity with the rest of the Communist movement, and, for that matter, much of the psychiatric field. In 1933, the Communist Party denounced him and in 1934 he was expelled from the International Psychoanalytic Association.[79] He spent a couple of years moving around Europe before finally settling in Norway, turning away from politics and his professional associations, and moving towards more "fundamental" researches.

In the end, Reich migrated away from a politicized approach in the mid-1930s, disillusioned with Freud and Marxism. 1936 marks a major break for Reich, away from larger social concerns and towards his increasingly esoteric scientific research into the orgone. An interesting story, as recounted by his son Peter, regarding his

[75] Ibid.

[76] Shechner (1985, 100).

[77] Pietikainen (2007, 144).

[78] Ibid., 145–146.

[79] Lore Reich Rubin sees Anna Freud as largely responsible for Reich's expulsion from the International Psychoanalytic Association. Cf. Reich Rubin (2003).

time in northern Maine at the research center he established, illustrates the eventual tangent of his politics. Peter recalled a time when "his father grabbed a gun and went to confront a bunch of Maine citizens who had marched up to Reich's research center shouting 'Down with the Commies, down with the Orggies.'" This was their term for Reich and his colleagues. Reich recounts the event, "explaining that he simply told the angry mob he was no more a communist than they were; that he, too – 'like everybody else' – had just voted for Eisenhower, and that, in fact, if they wanted to fight the Commies, he was glad, adding 'I've been fighting the Commies longer than you have.'"[80]

4 Reich, the Counter-Culture and the Popular Consciousness

Reich's life-path after he immigrated to America is the most interesting and also perhaps the most tragic. It also gives us deeper insight into his role as an alternative healer and the censure he suffered because of it. After his 1941 encounter with Einstein – a deep disappointment – Reich suffered what he saw as another injustice. When the Japanese attacked Pearl Harbor in December 1941, Reich was arrested by the FBI and imprisoned on Ellis Island under suspicion that he was a "dangerous enemy alien." Authorities suspected he might be a German spy.[81] Though freed in January 1942, the experience soured Reich on the US government and its motives, providing context for the development of his later conflicts – hereafter, Reich tries to escape New York, moving to the greener pastures of rural Maine.

By the end of 1942 Reich was spending more time up in Maine at the property he purchased – dubbed Organon – envisioning it as a permanent home and research center. In 1945, a student laboratory was built on site, and a few years later construction began on an Orgone Energy Observatory. Reich's projects were becoming self-sustaining, and he funded all of these investments through his work and from donations and contributions from students and supporters. He had by then migrated fully out of the mainstream scientific current, establishing his own journals as venues for publishing. However, Reich was still attracting popular attention – unfortunately, most of it was negative.

In 1947, two articles about Reich written by Mildred Edie Brady appeared in the popular press. The first, "The New Cult of Sex and Anarchy," was featured in the April issue of *Harper's*. The second, "The Strange Case of Wilhelm Reich," appeared in *The New Republic* in May. Brady's approach, evidenced by the article's titles, was hardly balanced. The second title implies a poor parody of Robert Louis Stevenson (and others like Poe and Lovecraft); and by extension a kind of morbid witch-hunt. In *The New Republic* she wrote: "Orgone, named after the sexual orgasm, is, according to Reich, a cosmic energy. It is, in fact, *the* cosmic energy.

---

[80] MacBean (1972, 6–7).

[81] Pietikainen (2007, 136).

Reich has not only discovered it; he has seen it, demonstrated it and named a town – Organon, Maine – after it. Here he builds accumulators which are rented out to patients, who presumably derive 'orgastic potency' from it." Her approach was completely confrontational; she argued that the "growing Reich cult" had to be dealt with somehow.[82] In the second article in *The New Republic* she also called for more stringent regulation of psychiatric practice in America. Following these two articles, the US Food and Drug Administration (FDA) assigned an investigator to look into the Reich case, and he thereafter found himself on the government radar. One biographer suggests that the FDA thought they were dealing with "fraud of the first magnitude" and believed that Reich's work involved a "sexual racket" of some kind. Despite being under investigation, Reich continued his researches in earnest, writing that he would "like to plead for [the] right to investigate natural phenomena without having guns pointed at me."

This late 1940s and early 1950s period marks the final tangent in Reich's life. He becomes progressively more paranoid, and, arguably, had fairly good reason for this. He had some success – holding, for example, two international Orgonomic conferences in 1948 and 1950 at the research center in Organon – but grew increasingly isolated and demoralized. The Oranur (for the acronym Orgone Against Nuclear Radiation) experiment in 1951 was, by many accounts, the final straw. Reich, as mentioned, was becoming increasingly drawn to the idea of a universal "Cosmic Orgone Energy," feeling it had tremendous power. He speculated it could counteract the negative effects of nuclear radiation, and set about to test this theory, in what came to be known as the Oranur experiment. It was a disaster. Reich reported that orgone actually amplified the effects of nuclear radiation, but it seems fairly clear that it was just a bad idea to conduct experiments with radioactive materials and improper safety controls. And yet, even here, Reich would prove to be influential.

From 1951-on, one witnesses a darkening in Reich's worldview. He spoke, for example of "deadly orgone" (DOR) – which he thought caused desertification and was partly responsible for creating the current "emotional plague" he felt the world struggled with. During a trip to Tucson, Arizona, Reich was "struck by parallels between the physical desert there and the emotional desert in man."[83] The parameters of orgone had expanded considerably from their roots in the libido to become a universal energy Reich felt mankind had corrupted through misuse. Orgone energy that could not circulate freely transformed into DOR, and in terms of atmospheric orgone, was "darkened" by pollution. In October 1951, DOR possibly caught up with Reich himself when he suffered a heart attack. His health would steadily deteriorate from the early 1950s on. According to his second wife, Ilse Ollendorf Reich, during the period from 1953 to 1954 Reich was in a poor emotional state and "often drank himself into an absolute stupor."[84]

[82] Cf. Sharaf (1983, 360–61).

[83] Ibid., 443. He took this analogy even further, comparing bristly desert plants with the prickly behavior typical of heavily character armored individuals.

[84] Pietikainen (2007, 138). Cf. Reich (1969).

At this point in his life Reich also fully migrated to the fringes, developing an interest in weather control, UFOs, and extraterrestrials. This interest should come as little surprise, since the phenomena had become widespread in the popular consciousness of the Cold War. Reich's cosmology – in its affinities with Giordano Bruno, one of the first men to speculate about intelligent life on other planets – certainly melded with this kind of curiosity. His functionalism placed life at the center of the purpose of matter itself. Reich's basic philosophy, after all, contended that the universe is permeated with life energy, and indirectly implied an idea of other inhabited worlds. The draft copy of Reich's *Contact with Space*, for example, is a bizarre collage of material, including extensive records of sky and weather watching in the desert near Tucson, as well as notes on weather control, UFOs and other strange phenomena.[85] Reich thought that perhaps the disastrous Oranur experiment had drawn UFOs to his humble locale in rural Maine. From late 1953 on, inspired by reading about UFOs, Reich started to seriously think extraterrestrial visitors were behind the general increase in DOR. He thought these UFOs were powered by orgone, and even developed an acronym to describe this mechanism: CORE (cosmic orgone engineering).[86] His construction of the "cloudbuster" was partly designed to communicate with (and possibly even combat) UFO visitors. For a time he speculated he might himself be a spaceman, finding the 1951 sci-fi classic *The Day the Earth Stood Still* depicted something of his own life biography.[87]

Reich became "an intellectual and scientific outsider in every sense of the term."[88] When in 1954 a Federal district court in Maine, acting on evidence collected by the FDA, ordered an injunction preventing him from promoting the use of his orgone accumulator and recommended the destruction of his books that dealt with orgone energy, there seemed to be no turning back from a full-on confrontation with the authorities. Violating this injunction by shipping accumulators across state lines he was finally sentenced to 2 years in jail. He did not fight the case and genuinely saw himself as a martyr for science *à la* Bruno. Writing about "My Unlawful Imprisonment," Reich argued:

> I have 'done wrong' in having disclosed to mankind the primordial, massfree Cosmic Energy that fills the Universe ...I have 'done wrong' in having discovered and made practically accessible the basic force in nature that for millennia was called 'GOD' in many tongues.... I am neither a lunatic nor a faker. My discovery obeys simple natural laws. It was anticipated by many scientists, philosophers and writers; it is well known and at home in all true world religions. I have discovered the LIFE ENERGY.[89]

When he died in prison in 1957 of a heart attack the authorities regarded him as a mentally disturbed charlatan. A psychiatric evaluation preformed after he was sentenced made the suggestion that "Reich probably suffers from Paranoid Schizophrenia."[90] But, of course, this picture is too simple.

---

[85]Conspiracy, box 14, Typed copy of *Contact with Space*, WR Archives.

[86]Sharaf (1983, 413).

[87]Wilson (1981, 244).

[88]Pietikainen (2007, 138).

[89]Personal, box 23, Letter dated 7 August 1957, WR Archives.

[90]Personal, box 23, Report by Clifford G. Loew, MD, dated 18 March 1957, WR Archives.

Reich eventually became a martyr, in part because of the harsh treatment he received at the hands of the US government. To understand Reich's influence on popular culture and consciousness one need go no further than the events of May 1968 in Paris. Two of the most influential figures floating around the intellectual scene in those heady revolutionary days were Herbert Marcuse (1898–1979) and Wilhelm Reich. One can see clear parallels in the work of these two thinkers, and Marcuse's classic *Eros and Civilization* (1955) owes a debt to the links between sex and society long explored by Reich. Reich's advocacy of a revolutionary approach to sexuality, and his early politicization of sex, struck a chord with the developing counter-culture – his neo-romantic utopianism was just what their kind of doctor ordered. The New Age-inspired alternative health movement also eventually eagerly adopted many of his ideas about orgone and life energy.[91]

In the politics of sexual liberation so characteristic of the late 1960s and early 1970s, there is much to be said for the influence of Reich. In fact, there is a revival of interest in Reich in the early 1970s, marked by new publications of some of his major works, a flurry of secondary literature, and even a few special issues of journals featuring Reich's ideas like *Liberation* and *The Radical Therapist*.[92] Of course, his influence on radical new elements of psychiatry as seen in the Esalen Institute and Erhard Seminars Training (est) is clear. Also appearing in 1973 was *A Book of Dreams*, his son Peter's fascinating biography, wherein the elder Reich is featured prominently.[93] This memoir became the basis for alternative chanteuse Kate Bush's "Cloudbusting" (1985), ostensibly about the younger Reich's fond memories of his father, the cloudbuster, and the construction of said machine.

Beat and outsider writers like William S. Burroughs, Allen Ginsberg, Norman Mailer and J.D. Salinger all found resonance in Reich's work. In Jack Kerouac's classic *On the Road* (1951), "Old Bull Lee" (a character modeled on Burroughs) extols the benefits of an orgone accumulator that he owns. Burroughs himself mentions orgonomic principles in a couple of his works. In *Junky* (1953), Burroughs' description of a fundamentally unhealthy place echoes Reich's arguments about DOR and the environment:

> A premonition of doom hangs over the Valley. You have to make it now before something happens, before the black fly ruins the citrus, before support prices are taken off the cotton, before the flood, the hurricane, the freeze, the long dry spell when there is no water to irrigate, before the Border Patrol shuts off your wetbacks. The threat of disaster is always there, persistent and disquieting as the afternoon wind. The Valley was desert, and it will be desert again. Meanwhile you try and make yours while there is still time....
>
> But a new factor, something that nobody has seen before, is changing the familiar aspect of disaster like the slow beginnings of a disease, so that no one can say just when it began.
>
> Death is absence of life. Wherever life withdraws, death and rot move in. Whatever it is – orgones, life force – that we all have to score for all the time, there is not much of it in

[91] Sharaf says: "Other popular, body-oriented [therapeutic] approaches such as primal therapy and Gestalt therapy borrow considerably from Reich with little acknowledgment of his contribution." (Sharaf 1983, 481).

[92] MacBean (1972, 2). Cf. Raknes (1970), Mann (1973), and Greenfield (1974).

[93] Reich (1973c).

> the Valley. Your food rots before you get it home. Milk sours before you finish the meal. The Valley is a place where the new anti-life force is breaking through.
>
> Death hangs over the Valley like an invisible smog. The place exerts a curious magnetism on the moribund. The dying cell gravitates to the Valley.[94]

In another piece, *The Job* (1970), Burroughs explicitly mentions DOR, discussing it in relation to his ideas about viral information and as a kind of viral force of nature:

> I have suggested that virus can be created to order in the laboratory from very small units of sound and image. Such a preparation is not in itself biologically active but it could activate or even create virus in susceptible subjects. A carefully prepared jaundice tape could activate or create the jaundice liver cells, especially in cases where the liver is already damaged. The operator is in effect directing a virus revolution of the cells. Since DOR (Deadly Orgone Radiation) seems to attack those exposed to it at the weakest point, release of this force could coincide with virus attack. Reactive mind phases could serve the same purpose of rendering subjects more susceptible to virus attack.[95]

Robert Anton Wilson is another more contemporary and curious character having a fascination with Reich. A beat holdover, Wilson was influenced by figures like Timothy Leary and Burroughs. In addition to delving into the nooks and crannies of the weird and the occult, Wilson was something of a popular philosopher, a modern Pyrrhonist and proponent of skepticism as manifested in a "generalized agnosticism" towards all dogmatisms (including agnosticism), whether in religion or science. Wilson saw Reich as an inspirational outsider whose engagement with the repressive and authoritarian elements of Cold War American society (as embodied in the FDA, AMA and APA) was revelatory. He wrote a play about Reich's life, *Wilhelm Reich in Hell* (1987), which takes the form of a bizarre circus and trial evaluating Reich's theories. In *Wilhelm Reich*, Wilson appropriately explores Reich's ideas in light of the 1980s US-Soviet arms race and the ever-widening sway of the emotional plague. Reich emerges as a lone critical voice opposing this mechanistic madness.[96]

Reich conforms, I believe, to all the "classic" elements of vitalism. His marginalization by the mainstream shows how much vitalist themes, conventionally understood, had been pushed to the farthest fringes of science by the mid-1950s, and how dominant a more strictly materialistic, mechanistic paradigm had become.[97] Writing to Einstein in a letter in the early 1940s, Reich at times argued for the value of Hippocratic principles in regards to the healing power of nature.[98] Again, there is a

[94] Burroughs (2002 [1953], 106).

[95] Grauerholz and Silverberg (1998, 304).

[96] Wilson (2007 [1987]).

[97] In *Listen, Little Man!* Reich says: "Once a great man showed you that machines follow certain laws; then you build machines for killing, and you take the living to be a machine also. In this, you made a mistake not for three decades, but for three *centuries*; erroneous concepts became inextricably anchored in hundreds of thousands of scientific workers; more, life itself was severely damaged" (Reich 1965, 52).

[98] He says: "The old medical principle that it is best to leave the process of healing to nature, and to help it along by scientific means, is still valid, in spite of all the irrational activities of our times." Conspiracy, box 1, The Einstein Affair, E-10a, WR Archives.

deep relationship here between Hippocratic ideas and vitalism. This is a theme Canguilhem, for example, discusses in his understanding of vitalism.[99] In 1955 a journal was started, *Orgonomic Medicine*, based on his suggestions and prompting focused on medicine in light of his theories. In his thoughts on the history of science Reich talks about how brief the legacy of mechanistic-materialism was, and how much progress lay ahead:

> The history of science is a long chain of continuation and elaboration, shaping and reshaping, creation and criticism, renewed shaping and reshaping and new creation. It is a hard, long road, and we are only at the beginning of this history. Including long empty spaces, it stretches over only about 2000 years. It always goes ahead, and, fundamentally, never backwards. The pace of life becomes accelerated, and life becomes more complicated. Honest scientific pioneer work has always been its leader and always will be. Aside from this, everything is *hostile to life*. This places an obligation on us.[100]

Indeed Reich had a fascinatingly broad, holistic and idealistic conception of science and knowledge, one that echoes many earlier vitalist perspectives. For him, the search for knowledge was an imperative with a sustained, evolutionary, and one might even say emergent, thrust. There is optimism in this approach, something that distinguished him from his mentor Freud, which he nominally tied to the country whose government, sadly, turned on him:

> If Freud, then, was the social philosopher for intellectuals who saw in the agony of Europe a picture of man's fate, Reich supplied the program for those who saw in America, an eroticized, Whitmanized America to be sure, a picture of man's hope.[101]

Reich saw science not as tool for the domination of nature, a trope too common in modern techno-science, rather he saw in the quest to inquire into the natural world a potential to come into harmony with it.[102] "Reich … felt that nature was good. [He was a] Romantic vitalist who, like members of the early mystery cults, worshipped life."[103] There is even a sort of desire to "re-enchant" the natural world, restoring some of what Max Weber suggests was lost with the emergence of a modernist perspective.[104] The focus on sexuality was thus more than just a fetish for Reich, instead it was a celebration of the essence of living nature that had deep consequences for many areas, fundamentals like epistemology and politics. His was

---

[99] On vitalism and Hippocratism, Canguilhem says: "As defined by Paul-Joseph Barthez, a physician of the Montpellier School in the eighteenth century, vitalism explicitly claims to belong to the Hippocratic tradition; this filiation is undoubtedly more important than the Aristotelian filiation, for if vitalism often borrows terms from Aristotelianism, it always holds on to the spirit of Hippocratism" (Canguilhem 2008, 62).

[100] Reich (1968, 57).

[101] Shechner (1985, 107).

[102] Again here there are echoes in the thought of Canguilhem: "If vitalism translates a permanent exigency of life within the living, mechanism translates a permanent attitude of the living human toward life. Man is here a living being separated from life by science and attempting to rejoin life through science. If vitalism, being an exigency, is vague and unformulated, mechanism, being a method, is strict and imperious." (Canguilhem 2008, 62).

[103] Fuchs (2011, 49).

[104] Berman (1981).

an unusual perspective; an almost "pre-modern" (and anti-mechanistic) sense of science, of its holistic and vitalistic association with all aspects of being. In this respect, one could say there was a kind of "incommensurability" between Reich and the modern scientific paradigm.[105] Talking about life, Reich sees its proper appreciation as being capable of banishing even political terror:

> He who does not have the confidence in that which is alive, or has lost it, easily falls prey to the subterranean fear of life which begets dictatorship. *That which is alive is in itself reasonable.* It becomes a caricature when it is not allowed to live. If it is a caricature, life can only create terror. This is why knowledge of that which is alive can alone banish terror.

His functionalism, with its notion of life as possessed of either positive, open and attracting or negative, closed and repelling energies, was a dichotomy wherein the living itself, whether simple or complex, was the essential universal. In his unending pursuit of knowledge, Reich was, in fact, pursuing life, and by extension, love. He elegantly expresses this at the close of *The Function of the Orgasm*:

> The investigation of living matter went beyond the confines of depth psychology and physiology; it entered biological territory as yet unexplored. The subject of "sexuality" became one with that of "the living". It opened a new avenue of approach to the problem of biogenesis. Psychology came to be *biophysics* and genuine, experimental natural science. Its center remains always the same: the enigma of love, to which we owe our being.[106]

By "love" Reich meant, to a significant degree, an organism's capacity to be in touch with, tap into and use the universal life energy around it, and its essential harmony with all living things.[107]

Reich was, in this sense, the last truly committed vitalist of the contemporary era. His instinctual, almost visceral, appreciation for the fundamentally functional realities of the living was a version of vitalism as yet unappreciated. Reich was neither mad nor mentally ill; he essentially possessed an understanding of the world that was fully out of phase with the society he lived in. Only after his death did his real influence begin to grow.

Reich's vision of health and healing extended far beyond the modern medical paradigm. He thus became a threat to the medical establishment and was vilified as a result. Taken collectively, his ideas were a powerful force, an energetic undercurrent of the developing counter-culture, underappreciated but nonetheless influential in artistic, literary and even medical circles. His notion of the importance of individual desire as an essential function of health and well-being has also been warped into darker directions in the quest to create demand and commodify just about every human pleasure. In an ironic twist, this path, initially mapped out by Reich, has often been navigated using Freud's notion of unconscious desire.

---

[105] Kuhn (1982).

[106] Reich (1968, 362).

[107] The following captures the deep essence of Reich's thought: "He was interested in the protoplasm in a person, in what he was like when he did not talk, in the way he breathed, and in the way he touched, he was interested in what made him a part of nature, not so much in what made him able to distinguish himself from the rest of nature, not in what allowed him to answer to a name." Radista (1978, 102).

5 Conclusion

One hopes that this piece has contributed to a more nuanced appreciation of his impact, but more importantly, that it has argued persuasively that Reich's cosmology can be seen as part of the deep tradition of "outsiders"[108] in the West inspired by vitalism. Echoing elements of many of the *outré* cosmologies of the Western tradition, from Manicheanism[109] to Gnosticism[110] to vitalism, Reich follows the pattern of most outsiders – initially tremendously successful within the inner circle of psychiatry, but eventually increasingly isolated and alone, both in real terms and in the sense of his theoretical and ideological commitments.[111] What follows from this is a deepening sense of persecution and paranoia. In Reich's case, interestingly, it is not entirely unjustified. Perhaps this is even the case with many other outsiders.

There is a general trend here in the outsider to attempt to flee from man and society towards a kind of idealism, and that certainly matches the broad outlines of Reich's life-path. Again, in parallel with Bruno, there is the sense of martyrdom, which also grows and arguably ends up eventually being fatal. Some of his commentators and admirers have even picked up on this – Norman Mailer, for example, called Reich an "intellectual martyr."[112] In Reich's writing, the sense of persecution, alienation and isolation are particularly apparent in *The Murder of Christ* (1953) and *Listen, Little Man!* (1948). In *Listen, Little Man!* Reich's tone is noticeably preachy and somewhat aloof. This only becomes more prominent in *The Murder of Christ.*

Much of what is addressed in the name of vitalism in this volume concerns epistemology, method and other nominally "scientific" considerations – issues of irreducibility and complexity in biology, questions of teleology and the nature of evolution and the like. With Reich, we have a fascinating case of an ontological commitment to vitalism in a thinker and researcher who tried his utmost to engage

[108]For the concept of the "outsider" cf. Wilson (1982). Interestingly, one of Wilson's outsiders, William Blake, regarded figures like Locke and Newton as "devils who killed the spirit by cutting reality into some kind of mathematically symmetrical pieces, whereas reality is a living whole which can be appreciated only in some non-mathematical fashion" (Berlin 2001 [1965]). Indeed, it is not a stretch to suggest Reich saw the society around him as "irrational" (Radista 1978, 105).

[109]The Manichean sense of spirit and germination inborn in nature is captured in the quote by Reich about the essential fecundity of the earth. Cf. p. 188 (fn. 48).

[110]This Gnostic sensibility is perhaps best reflected in Reich's belief in the essential nature of love, something that is also witnessed in Burroughs' cosmology.

[111]From the view of one observer, Paul Goodman, this solitary nature was Reich's true character, which the experience of being a modern scientist drew him away from: "What strikes me, indeed, is not evidence of abnormal derangement but how he harassed himself, wasted himself, and suffered by falling victim to a characteristic of our society, and of scientific society, that is now judged eminently normal. This was his compulsion to organize Institutes, to be a Public Scientist, a political influence, to be allied with Higher Powers (including the United States Air Force), to be busy with the Cold War like countless other maniacs, although he was an explorer and a loner, and a physician of souls." (Reich 1969, xv).

[112]Turner (2011, 429).

with the world in a stripped-down, empirical way. Or at least this was the initial intent. Very quickly, however, Reich ran up against a scientific world, whether in psychiatry or biology, so immersed in mechanistic paradigms that I believe he felt forced to adopt a more problematized and arguably more polemical stance.

In this respect Reich's vitalism is of the essence of criticality and skepticism, a perspective that sees the scientist as social critic and outsider, a role that has become increasingly rare in the post-WW II "technoscientific" world. Reich here is thus the last of an essentially extinct breed. His commitment to his ideas made him a fringe figure, a martyr and an anachronism. To many, this is seen as a fundamental flaw. But, perhaps, from the perspective of "critical" vitalism, these traits can be seen as virtues.

References

Aristotle. 1956. *De Anima*. Trans. W.D. Ross. New York: Oxford.

Ash, Mitchell G. 1995. *Gestalt psychology in German culture, 1890–1967: Holism and the quest for objectivity*. Cambridge: Cambridge University Press.

Bergson, Henri. 1911. *Creative evolution*. Trans. Arthur Mitchell. London: MacMillan.

Berlin, Isaiah. 2001 [1965]. In *The roots of romanticism, the A.W. Mellon lectures in the fine arts, the national gallery of art, Washington, D.C. Bollingen series xxxv:45*, ed. Henry Hardy. Princeton: Princeton University Press.

Berman, Morris. 1981. *The reenchantment of the world*. Ithaca: Cornell University Press.

Burroughs, William S. 2002 [1953]. *Junky*. London: Penguin.

Canguilhem, Georges. 2008. *Knowledge of life*, eds. Paola Marrati and Todd Meyers. Trans. Stefanos Geroulanos and Daniela Ginsburg. New York: Fordham University Press.

Corrington, Robert S. 2003. *Wilhelm Reich: Psychoanalyst and radical naturalist*. New York: Farrar, Strauss and Giroux.

Driesch, Hans. 1908. *Science and the philosophy of organism, the Gifford lectures delivered before the University of Aberdeen in the year 1907*. London: Adam & Charles Black.

Driesch, Hans. 1914. *The history and theory of vitalism*. London: Macmillan.

Driesch, Hans. 1933. *Psychical research: The science of the super-normal*. London: G.Bell & Sons.

Farley, John. 1977. *Spontaneous generation controversy from Descartes to Oparin*. Baltimore: Johns Hopkins University Press.

Fromm, Erich. 1994 [1941]. *Escape from freedom*. New York: Henry Holt.

Fuchs, Daniel. 2011. *The limits of ferocity: Sexual aggression and modern literary rebellion*. Durham: Duke University Press.

Garrett, Brian. 2003. Vitalism and teleology in the natural philosophy of Nehemiah Grew (1641–1712). *British Journal for the History of Science* 36: 63–81.

Geison, Gerald. 1974. Science, politics and spontaneous generation in nineteenth-century France: The Pasteur-Pouchet debate. *Bulletin of the History of Medicine* 48: 161–198.

Grauerholz, James, and Ira Silverberg (eds.). 1998. *Word virus: The William Burroughs reader*. New York: Grove.

Greenfield, Jerome. 1974. *Wilhelm Reich Vs. The U.S.A.* New York: W.W. Norton.

Grogin, R.C. 1988. *The Bergsonian controversy in France, 1900–1914*. Calgary: University of Calgary Press.

Kuhn, Thomas S. 1982. Commensurability, comparability, communicability. In *PSA: Proceedings of the biennial meeting of the Philosophy of Science Association*, vol. 1982, vol. 2: Symposia and invited papers, 669–688.

MacBean, James Roy. 1972. Sex and politics: Wilhelm Reich, world revolution, and Makavejev's WR. *Film Quarterly* 25: 2–13.

Mann, Edward W. 1973. *Orgone, Reich and Eros: Wilhelm Reich's theory of life energy*. New York: Simon & Shuster.

Normandin, Sebastian. 2007. Claude Bernard and an introduction to the study of experimental medicine: 'Physical Vitalism', dialectic and epistemology. *Journal of the History of Medicine and Allied Sciences* 62: 495–528.

Pietikainen, Petteri. 2007. *Alchemists of human nature: Psychological utopianism in Gross, Jung, Reich and Fromm*. London: Pickering & Chatto.

Radista, Leo. 1978. *Some sense about Wilhelm Reich*. New York: Philosophical Library.

Raknes, Ola. 1970. *Wilhelm Reich and orgonomy*. New York: St. Martin's Press.

Redi, Francisco. 1964. The development of maggots is shown by experiment to result from the eggs of flies. In *The origins and growth of biology*, ed. Arthur Rook, 73–79. London: Penguin.

Reich, Ilse Ollendorf. 1969. *Wilhelm Reich: A personal biography by Ilse Ollendorf Reich*. New York: St. Martin's Press.

Reich, Peter. 1973c. *A book of dreams*. New York: Harper and Row.

Reich, Wilhelm. 1961 [1949]. *Character-analysis*. New York: Farrar, Strauss and Cudahy.

Reich, Wilhelm. 1965 [1948]. *Listen, little man!* Trans. Theodore P. Wolfe. New York: Farrar, Strauss and Giroux.

Reich, Wilhelm. 1968. *The function of the orgasm: Sex-economic problems of biological energy*. Reprint. London: Panther.

Reich, Wilhelm. 1973a [1949]. *Ether, god and devil: Cosmic superimposition*. New York: Farrar, Strauss and Giroux.

Reich, Wilhelm. 1973b. *The cancer biopathy*. New York: Farrar, Strauss and Giroux.

Reich, Wilhelm. 1979. *The Bion experiments on the origin of life*. New York: Farrar, Straus and Giroux.

Reich Rubin, Lore. 2003. Wilhelm Reich and Anna Freud: His expulsion from psychoanalysis. *International Forum of Psychoanalysis* 12: 109–117.

Robinson, Paul A. 1969. *The Freudian left: Wilhelm Reich, Geza Roheim, Herbert Marcuse*. New York: Harper.

Rowland, Ingrid D. 2009. *Giordano Bruno: Philosopher/heretic*. Chicago: University of Chicago Press.

Seelow, David. 2005. *Radical modernism and sexuality: Freud, Reich, D.H. Lawrence and beyond*. New York: Palgrave.

Sharaf, Myron R. 1983. *Fury on earth: A biography of Wilhelm Reich*. New York: St. Martin's Press.

Shechner, Mark. 1985. Reich and the Reichians. *Partisan Review* 52: 99–108.

Smuts, Jan Christian. 1926. *Holism and evolution*. London: Macmillan.

Strick, James Edgar. 2000. *Sparks of life: Darwinism and the Victorian debates over spontaneous generation*. Cambridge, MA: Harvard University Press.

Turner, Christopher. 2011. *Adventures in the orgasmatron: How the sexual revolution came to America*. New York: Farrar, Strauss and Giroux.

von Reichenbach, Karl Ludwig Friedrich. 1968. *The odic force: Letters on od and magnetism*. Reprint. New York: University Books.

Williams, Elizabeth A. 2003. *A cultural history of medical vitalism in enlightenment Montpellier*. Aldershot: Ashgate.

Wilson, Colin. 1981. *The quest for Wilhelm Reich*. New York: Anchor.

Wilson, Colin. 1982 [1956]. *The outsider*. Boston: Houghton Mifflin.

Wilson, Robert Anton. 2007 [1987]. *Wilhelm Reich in hell*. Reprint. Tempe: New Falcon.

Wolfe, Charles T., and Motoichi Terada. 2008. The animal economy as object and program in Montpellier vitalism. *Science in Context* 21(4): 537–579.

Wright, John P., and Paul Potter. 2000. *Psyche and soma: Physicians and metaphysicians on the mind-body problem from antiquity to enlightenment*. Oxford: Oxford University Press.

Chapter 9
Vitalism and Teleology in Kurt Goldstein's Organismic Approach

Chiara E. Ferrario and Luigi Corsi

Abstract In this chapter we focus on the relationship at the turn of the twentieth century between vitalistic theories and a special case of a holistic approach to biology, Kurt Goldstein's organicism. We consider Goldstein's biographical and historical background and present the research cases that lead him to formulate his 'holistic-organismic' approach to the study of the brain and the mind, which developed into an ambitious theory on the nature of biological knowledge. Goldstein's organicism emerges as an antagonist to the aseptic and unsatisfactory framework of mechanistic biology. However, Goldstein's organicism strives to keep apart from some members of its own 'family' of anti-mechanistic approaches. It is especially wary of the metaphysical commitments of vitalistic hypotheses, and seeks alternative conceptual routes to traditional problems in the biological sciences, like teleology and the organization of the living. Goldstein's organicism will result in a rigorously materialist but non-reductionist epistemology of biology, which represents in his view the only feasible route to effective therapy. Goldstein's efforts to solve the problem of teleology, while incomplete, ultimately reveals the richest dimension of his intellectual legacy, that of an ethical stance towards science and medical practice.

Keywords Anti-mechanism • Aphasia • Epistemology • Holism • Kurt Goldstein • Medical ethics • Neurology • Organicism • Philosophy of Biology • Self-Realization • Teleology • Vitalism

---

C.E. Ferrario (✉)
School of History, Philosophy, Political Sciences and International Relations (HPPI), Victoria University of Wellington (VUW), NZ, Wellington
e-mail: Chiara.ferrario@vuw.ac.nz

L. Corsi
Philosophy, General Psychology, University of Pisa, Italy

S. Normandin and C.T. Wolfe (eds.), *Vitalism and the Scientific Image in Post-Enlightenment Life Science, 1800–2010*, History, Philosophy and Theory of the Life Sciences 2, DOI 10.1007/978-94-007-2445-7_9,

1 Introduction

In this chapter we explore the relationship, both inclusive and confrontational, between the organismic view of life and biology held by neurologist and physician Kurt Goldstein and a wider family of theories generally defined as 'vitalistic'. The second section introduces what Goldstein's organicism is (and what it is *not*) and how it fits in the historical and cultural background of the time. Goldstein's organicism is in fact quite distinct from other approaches recognized as 'organicist' or 'organismic'. The influence of previous philosophical debates is also outlined, for Goldstein's speculation is deeply indebted to the German tradition.

The third section sketches some biographical details, which were extremely influential on Goldstein's career and theoretical production. The author's truly holistic approach to knowledge seems to curiously mirror his professional eclecticism. After beginning as a scientist (neurologist and physician) with a marked sensibility for biological questions, Goldstein turned towards psychological and philosophical themes, and to the practice of psychotherapy, stepping progressively further away from theoretical biology and, importantly, from cutting-edge research in the field.

The fourth section concentrates on the empirical cases that led Goldstein to develop a holistic conception of the organism's functions and, subsequently, an organismically informed 'theory of biology'. This section offers a concise overview of Goldstein's most prominent and theoretically relevant work, *The Organism*. The book in fact constitutes the single manifesto of his organismic philosophy of biology. Goldstein's holistic theory initially sprang from data regarding the functioning of the brain and nervous system. It extended thereafter to the organism's functioning, to finally become abstract speculation on the nature and conditions of biological knowledge.

The fifth section explores the question of the relationship between Goldstein's organicism and vitalism. In particular, we contrast the results of organismic and vitalistic explanatory efforts in respect to teleology, which is the most interesting ground on which Goldstein refutes vitalistic explanations. Further, we suggest that recent developments in the discussion of teleology show that both organicism and vitalism were doomed to fail in their efforts to offer a sound solution to the problem.

Goldstein's organismic theory is essentially an 'epistemologically-minded' theory of biology. It is primarily concerned with biological *knowledge* and its conditions of attainment, rather than with the definition of the object of biological studies (life) or the recognition of what should count, or not count, as the subject of biology. This epistemological character is the focus of the final section. We illustrate how Goldstein's theory diverges from the dubious metaphysical claims of vitalism, and is configured as a purely epistemological and methodological speculation. We eventually advance a hypothesis on the aetiology of such a position, and identify it with the pragmatist conception of knowledge held by Goldstein. In the light of this final reflection, Goldstein's organicism is interpreted as an axiological theory of biology, or as a form of 'metatheoretical' and specifically ethical commitment.

2 Goldstein's Organicism at the Turn of the Nineteenth Century

When it comes to defining terms like *organicism* (or *vitalism*, or *holism*) the task of arriving at a standard, generally accepted definition can be frustrating. Like many other 'isms', organicism is a word with indeterminate meanings, but at the same time it has a rich and appealing space of connotations. For this reason it is inescapable, once one has described a theory as organismic (or holistic, or vitalistic), to dig into historical and theoretical details to try and delimit the specific nature of the *investigandum* and dissipate confusion.

The holistic-organismic approach to biology formulated by Kurt Goldstein (1878–1965) at the beginning of the last century is only one of many quite distinct schools of thought that historically have been or have called themselves 'organicist' or 'organismic'. Other forms of organicism in biology were elaborated in the late nineteenth and early twentieth century, essentially as a reaction to the inadequacies of the dominant mechanistic paradigm – but most of them had surprisingly little connection to one another. Organicist views have had increased popularity in biology in recent years: a first revival opposed molecular biology in the early 1960s.[1] From then on, the organismic perspective has been questioning cell theory in debates on the fundamental elements of life and in other areas of biology.[2] All of these currents can be said to represent instances of the same organicist or organismic perspective, but differ quite clearly in details and critical concerns.

What most organicist thinkers would nonetheless agree upon is their sympathy with the conceptual category of holism,[3] and the adoption of a holistic perspective in organicist talk. Some attempted definitions of organicism actually take it to be a particular type of holistic approach that appeals to living organisms as models of interpretation of other real systems, although such a characterization has been criticized as simplistic and confusing.[4] Holism tends to be a purely epistemological thesis, and can avoid direct reference to the metaphor of the organism as model of interpretation of reality, which generally characterizes organicism. So, while all organicisms are holisms, the contrary is not true, and holism finds wider application well outside the biological and social sciences.[5] Most authors would describe a theory as holistic if it claims that:

(a) the whole is not merely equal to the sum of the parts;
(b) the parts are determined by their participation or even juxtaposition in the whole, i.e. the whole determines the nature or significance of the parts;

[1] Hein (1969).

[2] For example in botany cf. Kaplan (1992); or in neurophysiology, cf. Nishikawa (1989).

[3] Etymologically, the term 'holism' comes from the Greek ὅλος, meaning 'totality', 'unitary whole'. Its first relevant occurrence is in the 1926 work by the South African statesman Jan Christian Smuts, *Holism and Evolution* (Smuts 1926).

[4] Phillips (1970).

[5] Bertalanffy (1950).

(c) consequently the parts cannot be fully understood outside the context of the whole; and
(d) the whole responds to a principle of organization, or the parts are dynamically interrelated or interdependent.

Other significant features of holistic approaches are a critically polemic attitude towards the mechanistic and analytic approach seen as predominant in the natural sciences; and more rarely the orientation towards epistemological questions debating the nature of the relation between the subject and object of knowledge, especially in physics. Essentially, holism assumes the logical supremacy of the whole over the parts and, to speak bluntly, claims that the process of knowledge (to guarantee a proper understanding of phenomena) should proceed through a 'top-down' dynamic. Hence it finds application in practically every discipline: holistic perspectives are found in physics, biology, medicine, linguistic, cybernetics, anthropology and mathematics. [6]

Kurt Goldstein spoke of his approach as a 'holistic-organismic approach to biology'. We can expect then to observe most of the above-mentioned traits, and his organicism is in fact an epistemologically oriented thesis (see Sect. 6). But it is now time to look at the historical peculiarities of Goldstein's proposal, and to give a sense of the historical background that so deeply influenced his thought. Goldstein's masterpiece *Der Aufbau des Organismus* (1934), later published in English as *The Organism* (1939), is the heir to a tradition of thought that was deeply rooted in Weimar culture (1919–1934). The Weimar Republic was an era that suffered upheaval and bursts of anti-conservative spirit in postwar Germany, and the intellectual and academic climate of the period mirrors the social and cultural dynamics. The domain of biological studies particularly reflected such trends, with numerous challengers to the mainstream establishment of mechanistic biology. Rebellious strands of biological thought fed on a background of philosophical influences as rich and varied as the romantic Goethe and the orthodox Kant, to flow into the groundbreaking psychological school of *Gestaltpsychologie*, with its futuristic links to physics and the natural sciences. *Gestalt* is a German word with evident holistic potential: it indicates a structure or pattern such that a totality cannot be inferred from the parts, because the sum of the parts doesn't merely equal the whole. The father of the biological concept of *Gestalt*, understood as "self-actualizing wholeness of organic forms"[7] is Johann Wolfgang von Goethe (1749–1832), who was the first to openly contradict Newton's physics, which he considered to be mechanistic and atomistic, and to oppose the Newtonian picture with his organic and holistic vision of nature, extended from philosophy to poetry to science.[8] *Gestaltpsychologie* (to which Goldstein belongs, however marginally) inherited and assimilated

[6]Cf. Phillips (1970) and Ferrario (2008).

[7]Cf. Ash (1998).

[8]Cf. Ibid., 186. It should also be noted that after 1918, along with the concept of *Gestalt*, the organismic one became the dominant paradigm from which to borrow bio-political metaphors with a reactionary matrix – although other political sympathies were "possible and persuasive," cf. Harrington (1996).

Goethe's intellectual legacy, to revolutionize, in pre-Nazi Germany, theories concerning the foundations of the emerging psychological theory of mind, against the associationist claim that mental life is the mere sum of isolated operations.[9]

Various authors paint a gloomy picture of the German academic and cultural scene at the turn of the twentieth century.[10] The challenge to positivism, represented by the central metaphor of the 'machine', intensified in the Germany of Wilhelm II (1888–1918), from the end of the *Kulturkampf* on, and reached its peak after Germany's devastating defeat in the Great War. Isolated voices from multiple cultural areas, such as *Principles of Psychology* by William James (1890) and *L'Évolution créatrice* by Henri Bergson (1907), came together in a wide-ranging reaction against *atomistic-mechanistic science* and its sociological counterpart, *machine society*. The emergence of 'organizers' was investigated in embryology and the principles of *Gestalt* were formulated in psychology, while the bugbear of soulless mechanisms was thought to threaten the sacred historical German mission to infuse new blood into an arid, degenerate and materialistic western world, and assumed the appearance of Republican liberalism, democracy, communism, capitalism, international Jewish conspiracy, and so on.

At this point, in the aftermath of military defeat, the so-called *Krisis der Wissenschaft* occurred; this was a heated debate about the foundations of physics and mathematics, which flared up during the 1920s and came to seriously question the assumption of deterministic causality.[11] Of course, all this only served to nourish a widespread irrationality, as evidenced by the enormous success of Spengler's monumental work *Der Untergang des Abendlandes* (*The Decline of the West*)[12] and the proliferation of so-called *Lebensphilosophie*. The charge frequently brought against the physicists, after their manifest failure to provide Germany with a victorious conclusion to the war, consisted in the *Entseelung*, namely the "destruction of the soul" of the world. The 'crisis' soon became one of those concepts that "synchronically thematize factual circumstances and diachronically their transformation."[13] This concept, having become so meaningful as to organize and direct collective action in various fields of social life, "certainly referred to the ongoing economic and political crisis, but was not limited to it; its fundamental aspect was a moral and intellectual crisis, a crisis of science and knowledge."[14]

---

[9]The first to export the concept of *Gestalt* in psychology, with his essay *Über Gestaltqualitäten*, was Ehrenfels (1890). The historic meeting of the three leading exponents of *Gestaltpsychologie*, Max Wertheimer, Wolfgang Köhler and Kurt Koffka took place in Frankfurt in 1910. After a period of great productivity and greater impact on German culture during the 1920s, the representatives of *Gestalt* psychology, nearly all Jews, suffered the growing Nazi persecution, and fled to the States.

[10]Among the most important contributions on this topic, cf. Lukács (1955), Ringer (1969), Forman (1971), Harrington (1996), and Ash (1998).

[11]Cf. Forman (1984).

[12]Spengler (1918).

[13]Koselleck (1979, 102).

[14]Forman (1971, 26–27).

Was there a glimmer of light in this 'Twilight of the Gods'? We will follow the trail of biology. It is Kant (1724–1804), in his third *Critique* (*Kritik der Urteilskraft*), who defined the a priori basis of the life sciences in terms of a 'teleological causality': the 'purposiveness' of nature (*nexus finalis*) was divided into external (between living beings) and internal (within an individual organism). It was the latter, with the concept of 'natural purpose' (*Naturzweck*), that became the epistemological foundation of biology as a science of living beings; that is, a principle able to perform a function similar to that carried out in the Kantian system by the categories of the Intellect in mathematics and physics.[15] But things were to grow quite a distance from this foundational systematization of the biological field.

In an 1899 monograph (*Die Lokalisation morphogenetischer Vorgange, ein Beweis vitalistischen Geschehens*) that opened the way to his career, Hans Driesch (1867–1941) "introduced an anti-mechanistic concept of the embryo as a 'harmonious equipotential system' [… which] was supposed to be moulded during development …by an autonomous, nonmaterial teleological principle that Driesch would soon christen the *entelechy*."[16]

A decade later, his biologist friend and collaborator Jakob Johann von Uexküll (1864–1944) wrote that "Driesch succeeded in proving that the germ cell does not possess a trace of machine-like structure, but consists throughout of equivalent parts. With that fell the dogma that the organism is only a machine …, the organization of a structureless germ into a complicated structure is a power *sui generis*, which is found only in living things."[17]Actually, there had already been the so-called 'teleo-mechanicists': biologists such as Johann Frederick Blumenbach (1752–1840), Karl Ernst von Baer (1792–1876) and Johannes Müller (1801–1858), who during the nineteenth century had supported the existence of "special emergent vital principles" in living organisms in order to explain the holistic phenomena of development and differentiation; but the fortune of Driesch's theories went much further.[18] Driesch's vitalism was progressively extended to human action in general, until the publication in 1908 of the volume *Science and Philosophy of the Organism* (Driesch 1908), where it assumes the status of true philosophy. Gestalt psychologists and Goldstein himself had to face this passage from the world of biology to a psychovitalism that closely resembles Bergson's *élan vital*. The issue of entelechy

---

[15]Cf. Kant (1790, §§ IV, VIII, 63, 65); concerning reflections on 'external' purposiveness, cf. ibid. §§ 79–80, 82–85.

[16]Harrington (1996, 51), referring to Driesch (1899). *Entelécheia* [ἐντελέχεια]: a term used by Aristotle (*Metaphysics*, Book IX) to designate the state of perfection (from the Greek *entelés* [ἐντελές], completed, whole) of an entity that has reached its 'end' [τέλος] by fully implementing its potential being. The activity that transforms the possible into the real is called *enérgheia* [ἐνέργεια], and in this sense is distinct from the entelechy of the entity, which is instead the result of full implementation of the possible. This distinction is very important, because in the first case we are dealing with a *state*, a *process* or a *final cause*, while in the latter case with an *immanent force*.

[17]von Uexküll (1920), cit. in Harrington (1996, 51).

[18]Lenoir (1982).

implied unequivocally metaphysical aspects and, as philosophers were not so allergic as biologists to this type of problem, "just when Driesch's ideas had ceased to be interesting to most biologists, they became so for philosophers and psychologists."[19]

Not all psychological schools, however, were interested in this sort of metaphysical approach. In psychology, the response to the crisis took the form of a humanistic current, whose most important members were Karl Jaspers (1883–1969), Eduard Spranger (1882–1963) and Ludwig Klages (1872–1956),[20] and above all appeared in the "new world view" that was developed by the Gestalt theorists Wolfgang Köhler (1887–1967), Max Wertheimer (1880–1943) and Kurt Koffka (1886–1941), active in Berlin at the *Friedrich Wilhelm Universität*. This is not the place to revisit the theoretical path of the so-called Berlin School, which expanded the category of Gestalt from perception to action, from productive thought to behaviour and even to physical systems, creating a real philosophy of nature; suffice is to say that this holistic current of thought was a response, often accompanied by heated controversy, against both mechanism and vitalism. *Gestalt-theorie* was able to show that the scientific study of consciousness, no less than that of the physical world, could reconnect with the overall structured and dynamic processes that shape life experiences, without atomistically disrupting their patterns nor presupposing in them the action of some mysterious vitalistic principle.

In addition, the enemy of Gestalt was not the 'machine', so much as *Chaos*, i.e. the pure non-order of the universal 'thermal death', as envisaged and inexorably predicted by the second law of thermodynamics. Köhler's approach (1920) was not of an 'organicist' type, but looked to Maxwell's physics of fields for the development of a complete monist scientific paradigm. Considering the neurophysiological processes underlying Gestalt phenomena "in terms of the physics of field continua rather than that of particles or point-masses," Köhler stated that "the structural essence of physical Gestalten, expressed in their mathematical laws, is the same as that of psychological Gestalten, even though no such laws have yet been derived from them."[21] With this postulate of *isomorphism* between physical and mental processes, he avoided the epistemological dualism implicit in the psychovitalism of Driesch. Goldstein took a more specific position in this regard after 1934, asserting that "the 'whole', the 'Gestalt', has always meant to me the whole organism and not the phenomena in one field ...."[22] He put forth an *organismic holism* (based, like that of Driesch, on biology), which was also *rationalist* and *monist* (as the Gestalt theorists intended). But we will turn now specifically to our author's views.

---

[19] Ash (1998, 83).

[20] Thus, abandoning W. Wundt's cherished dream of a rigorous experimental foundation, the psychology proposed by these three authors (respectively psychology of 'worldviews', 'structural psychology' and 'characterology'), instead of being 'scientific', becomes 'humanist', and obtains a place within the *Geisteswissenschaften* (Dilthey 1883).

[21] Ash (1998, 171–185).

[22] Goldstein (1995, 285).

3 The Holistic Champion

When *Der Aufbau des Organismus* was first published, its author was 54 years old and waiting in Amsterdam for a visa to the US, forced into exile from Germany, away from the tragic developments of the Nazi regime.

Fleeing to the US surely meant to Goldstein, who was born in a Jewish family, a chance of pursuing his academic job (as well as, probably, of surviving *tout court*), but also dramatically disrupted his research path and dragged him away from the cultural environment he strongly felt he belonged to. Marianne Simmel reports that Goldstein never felt completely at home in the US, never again in his element. Regardless of the fact that he had become a US citizen in 1940,

> ...his comment on news of victories [of the Allies] was typically: *'Das haben die Amerikaner doch eben grossartig gemacht'* ['So the Americans have ultimately, magnificently made it']. Not once did I hear him say 'we' in this connection. He always felt a stranger among friendly natives. He was grateful to the country where he and so many others had found asylum first, and then a new home – but it was still a home in exile.[23]

In his later years, and especially after the suicide in 1960 of his second wife and collaborator Eva Rothmann, who had fallen into depression since the end of War World II, Goldstein "seemed to suffer an increasing and finally tragic isolation and unfulfillment."[24]

Goldstein's intellectual biography may be divided into three significant periods for the purposes of the present chapter:[25] a first stage (1899–1914) of pure medical research; a second period (1914–1934) of intense clinical activity and theoretical elaboration on the project of a holistic epistemology of biology; a third and final period (1935–1965) spent in America, where he primarily devoted himself to psychotherapy, and his reflections departed quite a long way away from biological themes.

Born in 1878 and graduating with a degree in Medicine in 1903 from the University of Breslau (now Poland) under Ludwig Edinger and Karl Wernicke's tutelage, Kurt Goldstein's first interests were directed towards philosophy. The choice of studying medicine came as a compromise between the paternal ambitions (which considered philosophy a financially unwise choice) and young Kurt's inclinations, which leaned strongly towards the humanities.[26] His early Kantian enthusiasms were thus abandoned, but only temporarily, as they were to re-emerge later on in his reflections. This early 'holistically oriented' scope of investigation represents the hallmark of Goldstein's intellectual production, extending well into his maturity. The designation

[23] Simmel (1968, 9).

[24] Sacks (1995, 10).

[25] Cf. Simmel (1968) and Ferrario (2008).

[26] "When I had to decide between natural science and philosophy before entering the university, I did not know which to choose. In deciding on natural science I was certain that I would use it only as a basis for becoming a physician. Medicine alone appeared suited to my inclination – to deal with human beings" (Goldstein 1959, 109).

that better suits this eclectic-minded neurologist is perhaps the one Hans Jonas (1903–1993) chose in his speech for Goldstein's eightieth birthday: a "philosophical scientist,"[27] a designation chosen to stress how strongly his philosophical and epistemological concerns were adherent to, and motivated by, concrete facts of scientific research. A somewhat atypical figure in an era of growing specialization, Goldstein maintained himself equidistant from the boundaries of well-delimited disciplines, instead articulating a personal synthesis of medical, philosophical and biological notions, meticulously moulded onto a coherent (though at times repetitive) conceptual network. It is also emblematic that one of the most important and eminent recipients of Goldstein's ideas, Georges Canguilhem (1904–1995),[28] practiced himself cross-disciplinarity as can be seen in his pursuing a medical degree while ultimately devoting himself to a philosophical career. But we will get back to the fates of Goldstein's intellectual legacy later on, after following the developments of his career.

The first period, which we christen 'the pure researcher', saw Goldstein obtaining his degree with a dissertation on the organization of dorsal traits of the spinal cord, and subsequently wander through several laboratories and institutes in Germany, to become, as Sacks reports, "an astute and anatomically minded clinician,"[29] who attentively followed the steps of his former masters, the illustrious anatomist Ludwig Edinger (1855–1918) and neuropathologist Karl Wernicke (1848–1925).[30] Goldstein started as Edinger's research assistant at the *Seckensbergische Neurologische Institut* of Frankfurt, after which he moved on to the psychiatry department of the university in Freiburg, then a neurological polyclinic in Berlin, and finally to the psychiatric clinic of the university of Königsberg. At this latter institution he was struck by the meagre consideration paid by Emil

[27]"Kurt Goldstein is a philosophical scientist because he's a true scientist.... It is not that he 'also' has a philosophical penchant and, as a kind of reprieve from the rigors of science, sometimes permits himself flights into philosophy. On the contrary, the very intimacy with concrete problems issues into philosophical dimensions, just as it was already a philosophical awareness which made the problem visible, as such, in the first place. 'Method as well as theory must originate from nothing but the most concrete evidence', says Goldstein himself. But, of course, they really originate from the viewer of the evidence, he who makes the evidence – mute in itself – tell its story by asking it the right kind of question, having gathered it first with questions in mind: and when he gives himself account of what kind of questions are right, and even reflects on such things as evidence and questioning in general, he has turned philosopher without turning from the matter in hand" (Jonas 1959, 161).

[28]Canguilhem, in his *Essai sur quelques problèmes concernant le normal et le pathologique* (1943), explicitly acknowledges the major intellectual debt he owed Goldstein, whom he considers his first inspirational and insightful source of reflection on the concept of health and disease. Cf. especially chapter XII of *The Organism,* entitled *On Norm, Health, and Disease. On Anomaly, Heredity and Breeding*. Cf. also Gayon (1998) and Delaporte (1994).

[29]Sacks (1995).

[30]Edinger (1855–1918) was an eminent German anatomist and neurologist and is considered one of the fathers of comparative neuroanatomy; Wernicke (1848–1925) was a pioneer neuropathologist (also physician, anatomist and psychiatrist) in the study of aphasia. He insisted on the necessity of evaluating the psychological consequences of neurological damage. In the classification of aphasia symptomatology, receptive or sensory aphasia is named after him (Wernicke's aphasia).

Kraepelin (1856–1926)[31] to the individuality of psychiatric patients, and he started developing a critical eye towards classificatory methods in respect to psychiatric disorders, and in general towards mainstream neuropsychiatry. During these years (until 1914), he published an impressive number of papers (more than 60), delivering the results of his research in comparative neurology, neuroanatomy and neuropathology, in particular on the relationship between localized cortical damage and sensori-motor impairment, and on visual agnosia and visual perceptual disturbances due to optical nerve impairment and their relationship to tactile recognition.[32]

In the second period of our chronology a major change in Goldstein's career and in his intellectual production was to occur. It was 1914, and large numbers of brain-damaged soldiers began to be sent back from the battlefields. Goldstein was asked to organize a centre for their care and rehabilitation in Frankfurt. From 1914 till 1929, he successfully ran the *Institut zur Erforschung der Folgeerscheinungen von Hirnverletzungen* (Institute for Research on the After-Effects of Brain Injuries), under government administration and together with his friend and collaborator Adhémar Gelb (1887–1934). The Institute is remembered as an extraordinarily productive, cooperative and relaxed working environment, where Goldstein was finally able to extend the study of neurological damage to its psychological aspects, as his early mentor Wernicke always encouraged him to do, and to implement his conception of individually focused and holistically oriented healthcare. During the Frankfurt years Goldstein gradually came to believe that in each clinical case, whatever particular neurological deficit might be treated (but especially in patients that presented impaired language capacities), there was always a general reaction or change in the individual as well.[33] The substantial clinical experience he gained here became the basis for the conception of his holistic theoretical system. In this second phase, his scholarly articles are mainly focused on the role of the cerebellum in aphasia and voice tuning, and it is evident how such topics provided ample material for his first insights into the holistic functioning of the brain.

In 1930 Goldstein was summoned to direct the *Moabit*, the municipal hospital in Berlin, which probably represented the most prestigious position in a neurological institution in Germany at that time. He accepted reluctantly, on the condition of maintaining participation in the research activities of his Frankfurt Institute. The last 4 years before his expulsion saw the trend he was developing since the mid-1920s gain full strength: his writings show a growing interest in psychotherapy, while undertaking a critical examination of psychoanalytic thought and the traditional approach in psychology, as well as attempting to reorganize into a broader biological context all that he learned about neurological and psychiatric patients.

---

[31] Emil Kraepelin (1856–1926) is now remembered for his detailed research on the symptomatology of psychiatric disorders, and is considered a precursor of the contemporary classifications of the psychiatric pathologies (DSM, Diagnostic and Statistical Manual of Mental Disorders; or the ICD, International Classification of Disease).

[32] Simmel (1968).

[33] For more detailed discussion of the clinical cases see Sect. 4. For Goldstein's discussion of aphasia, cf. Geschwind (1964).

His holistic-organismic theory of biology was starting to take shape. This fecund and productive period was not to last long: after only 3 years, in 1933, Goldstein had to take refuge in Holland in order to escape Nazi persecution. While awaiting his visa to the US, he thought deeply about his rich and significant neurological experience, figuring out the precise contours of the holistic perspective that had gradually been taking shape in his mind for more than 20 years. He finally dictated to his secretary *Der Aufbau des Organismus*, in a matter of literally 35 days; the work was to be translated into English and published in 1939 under the title *The Organism*. Today this book remains the manifesto of his holistic-organismic approach to biology.

The forced exile in Amsterdam ended in 1935 when Goldstein sailed for the US. This is the beginning of the last section of our chronology (1935–1965), the American period. Goldstein continued to work relentlessly and successfully as a researcher, teacher and clinician, surrounded by the respect and affection of many friends, students and colleagues, but his theoretical production went into stasis, at least in respect to theoretical biology. During these years the neat experimental style of investigation that characterizes the first chapters of *The Organism* receded into the background, finally disappearing. Works such as *Human Nature in the Light of Psychopathology* (1940) and "The concept of health, disease and therapy" (1954) undeniably extend their scope into the social more than the biological sciences, with a major attention to psychology and psychotherapy, but also saw incursions into peripheral fields like anthropology ("Concerning the concept of 'primitivity'," 1960). There is almost no trace of the empirical minuteness and the wealth of data collection of the early years. Goldstein's final reflections belong in the field of pure philosophical elaboration, with his early philosophical interests eventually re-emerging and dominating his writings: a revival of his Kantian-idealistic and especially Goethean background, which eventually assumed the tones and perspectives of European phenomenology and existentialism, focusing on the analysis of concrete human experiences.

In the American years our author would add almost nothing to the original conceptual apparatus of the philosophy of biology formulated in *The Organism*, while most of the nuanced concepts that were initially formulated in a biological context came to enrich and implement his new psychological terminology, which was deeply influenced by European existentialism and phenomenology (see for example the notions of *milieu* and "sphere of immediacy," of "individual essential nature" and "self-realization," "catastrophic reaction," and the improved concepts of "abstract and concrete attitudes").[34] At the same time, his preferences firmly tilted towards the practice of psychotherapy over laboratory research, to the point that he seemed to regret having spent so long as a neurology researcher.[35] Goldstein died on September 19, 1965, in New York City.

---

[34] Cf. Goldstein (1940, 1957, 1959).

[35] Cf. Brown, in Calamari and Pini (2007, 161–164).

It is this psychological development of his late thought that represents, somewhat ironically, the most prominent aspect of Goldstein's intellectual legacy. His influence is limited to psychotherapy circles, and has left practically no trace in biological studies, despite contemporary biology witnessing a thriving renaissance of the organicist paradigm.[36] Goldstein was one of the main inspirations in the movement of humanistic psychology, otherwise known as the *Third Force* of American psychology, which developed on the side of psychoanalysis and behaviourism, emphasizing the importance of conscious experience and the role of growth, actualization of self-potential and a more holistic fulfilment of the person. Abraham H. Maslow (1908–1970) for example, one of the leading personalities of the movement, borrowed from Goldstein the term *Self-realization* (from the German *Selbstverwirklichung*) which had been presented in *The Organism*, and put it at the apex of his famous motivational hierarchy, or "hierarchy of needs" (Maslow 1943). Other minor approaches, like the so-called "humanistic body psychology" (or organismic psychotherapy) of Malcolm Brown,[37] were to draw even more liberally from Goldstein's *magnum opus*, where concepts such as sphere of immediacy, communion, individual essential nature, and organismic awareness still retain an enduring flavour of the European existentialist tradition.

4 From Experimental Evidence to *The Organism*

The Organism constitutes the most important theoretical work of Goldstein's career, particularly if one is interested in his contribution to theoretical biology. The William James Lectures that he delivered at Harvard University in 1938–1939, and which were published in 1940 as *Human Nature in the Light of Psychopathology*, should be considered in fact his last substantial contribution to the field before he retired to the practice of psychotherapy, and indeed consist almost entirely in a more approachable paraphrase of *The Organism*'s contents.

In this long and complex book, Goldstein develops a peculiar and internally cohesive language to describe the processes and functions that are at work in the organism, drawing from an impressive wealth of data regarding primarily his object of observation as a neurologist: the nervous system, brain, mind and behaviour (esp. Chapters I, III).[38] Eventually he extends the validity of his conclusions to the entirety of organismal functions: growth, reproduction, development and behavioural strategies (esp. Ch. III, IV, VI, VII). The book not only constitutes a detailed *pars construens* of Goldstein's holistic theory, but also contains a large *pars*

[36] Cf. Gilbert and Sarkar (2000) and Nicholson (2010).

[37] Brown (1990).

[38] "I shall confine my discussion essentially to the nervous system, since only in this realm do I feel confident that my judgment of the material will be sufficiently reliable. I believe, however, that the conclusions drawn will convince the reader that it is permissible to make some generalizations regarding processes in the other systems of the organism" (Goldstein 1995, 26).

destruens, formulated as an attack on the inadequacies of mechanistic biology and medicine (like the 'theory of reflex', see Ch. II, V). The last part is definitely the most 'philosophical' (Ch. XI, XIII), and is dedicated to defining an epistemological rationale to biological knowledge (Ch. IX) and medical practice (Ch. X, XII), solidly anchored in the empirical evidence and clinical experience collected and presented so far. Overall, Goldstein's organicism adheres quite literally to the metaphor of the living organism as the archetype of the organized whole, and draws generously from Gestalt theory (Ch. VIII).

During his clinical work alongside Gestalt psychologist Adhémar Gelb on the rehabilitation of brain-injured soldiers returned from the war, he realized that *the symptoms are not the direct consequence of the brain damage,* but *are the responses of the whole organism to the changes that took place after the damage itself.* 'Symptoms are answers,' and "recovery is a newly achieved state of ordered functioning, that is, responsiveness, hinging on a specifically formed relation between preserved and impaired performances ... in the direction of a new individual norm, of new constancy and adequacy."[39]

The Organism presents a host of specific terms, each of them having a particular and precise meaning, though all sharing their relation with the concept of health. In disease, in fact, the organism has lost its 'individual norm'; shows 'defective responsiveness', 'disordered behaviour', anxiety and catastrophic reactions; it is unable to use the 'abstract attitude' and so to realize the 'coming to terms' with its world in an adequate manner ('adequacy'). This behavioural "coming to terms with the world," in turn, 'shapes an environment' and creates the organism's 'milieu', which is not really the surrounding world but "represents only a part of the world – that part that is adequate to it."[40] Perhaps the most central notion in this constellation of concepts is that of 'order' and 'ordered behaviour': "An organism that actualizes its essential peculiarities, or – what really means the same thing – meets its adequate milieu and the tasks arising from it, is 'normal'. Since this realization occurs in a specific milieu in an ordered behavioural way, one may denote behaviour under this condition as normal behaviour."[41] We will now try and give an overview of the features that intertwine in Goldstein's description of organismic behaviour.

The unit of investigation chosen by Goldstein for the understanding of life processes is the concept of "performance" (or "performance field"), namely:

> ...any kind of behaviour, activity, or operation as a whole or in part that expresses itself overtly and bears reference to the environment ...[A] performance is a coming to terms of the organism with environmental stimuli by a behavioural act, be this eyelid closure under stimulation or a total movement like running toward a goal, or hearing, seeing, and so on. [42]

In defining performance as a "coming to terms," it is obvious that the organism as a whole is always called into question, and that the organismic behaviour will

[39] Ibid., 35, 334.

[40] Ibid., 85, 106.

[41] Ibid., 325.

[42] Ibid., 42. Our italics.

always be holistically oriented in line with this kind of 'bio-ecological' finality. Sacks see the absolute centrality of the "coming to terms" metaphor in the description of organismal behaviour, as a hint of adaptive thinking.[43] Certainly Goldstein conceives of behaviour in a new, lively and fine-grained way, with a striking awareness of the role of the species-specific environment (the *milieu*) in co-shaping the destiny of organismic morphology and general phenotype; but we think that any parallelism with evolutionary theory should be avoided in Goldstein's case, given his debatable take on Darwin's idea.[44]

Goldstein's theory suggests a neurophysiology of performances, understood as holistic processes whose common denominator consists in the goal of the individual 'coming to terms' with the world. The author shows, through examples of laboratory experiments, *the relative independence of the performances from the functioning of a specific locality to which 'normally' they are related.* Thus, for example, in the *sequelae of amputation* after awakening from anaesthesia, the operated animals learned immediately to bring on the 'right' movements and so to accomplish the performance that the situation required. The nervous system is a network that never rests and always functions as a whole, an apparatus well differentiated from an anatomical point of view but extremely dynamic and flexible in functional and adaptive terms. Within this system,

> the total excitation pattern is not confined to a definite anatomical structure but represents a definite excitation Gestalt that can utilize, for its course, any available structure .... The performance is based not on the activity of certain mechanisms but on certain potentialities of the organism that realize themselves by utilization of all sorts of substitute means when the 'normal' means are out of order.[45]

As other organicist thinkers such as Bertalanffy seem not to have understood,[46] an organism is a very peculiar system. It always functions as a whole but, and this is the most striking fact, it is also able to function in the atomistic way, according with the principle of the best performance required in the respective 'field'. And yet, in such a case, each 'performance field' represents again the same Gestalt figure-background formation. After a 'change', due to brain damage or other circumstances, we have a 'shift' from one Gestalt to another (see Sect. 6). "The shift occurs suddenly. It is not the result of training, and it happens without the knowledge of the patient."[47] For instance, when a patient is affected by hemianopia, which implies the complete loss of a half of each visual field, then in the functionally preserved half of the retina "[a] new region of best vision, a new fovea, a so-called pseudo-fovea, has developed. But with this alteration" – and here comes the holistic principle – "the function of every point on the retina must likewise have undergone transformation. Centrally located areas are now hypofunctioning, or, to express it

---

[43] Sacks (1995).

[44] Cf. Ferrario (2008).

[45] Goldstein (1995, 187).

[46] Bertalanffy (1969, 174–179).

[47] Goldstein (1995, 60).

otherwise, they now function as peripheral zones normally do."[48] The 'adjustmental shift' may also manifests itself in behavioural patterns completely different from the original ones, as in the guinea pig whose legs were amputated in the experiment by Martin H. Fischer: "Soon after awakening from the anaesthesia, the animal began to roll around its longitudinal axis. Rolling was now the only possible means of locomotion."[49]

Hence, it seems that the organism reacts globally; it reorganizes the average levels of 'constants' with extraordinary plasticity, and does its best to restore its performance and behaviour. Constants are possibly, in Goldstein's theory, the most essential characteristic of the organism. "In contrast to the diversified and even contradictory character of the partitive data," Goldstein writes,

> ...the organism properly presents itself as a structural formation that, in spite of all the fluctuations of its behavioral pattern in the varying situations and in spite of the unfolding and decline in the course of the individual's life, retains a relative constancy. If this were not the case, it would never be possible to identify a given organism as such. It would not even be possible to talk about a definite organism at all.[50]

This criterion of the maintenance of a relative constancy is elaborated as the "tendency toward the preferred behavior" and above all as the "basic biological law": equalization toward an 'adequate' average level in an 'adequate' time. Constants are very generically defined as the respective 'average mean' in any organismic phenomena. In around twenty pages constants appear everywhere: physiological, behavioural, affective, constants regarding the sensory and motor threshold, intellectual characteristics, constants as expression of the essential nature of the species and of the individual organism under consideration, time constants in the temporal and rhythmic course of processes and so on.[51] "Ultimately," we read in *Human Nature*, "these constants are basic traits of the constitutional and character make-up of the individual."[52]

Sometimes, and paradoxically, a complete loss of a performance field leads to a better rehabilitation than a minor impairment with a lesser defect, insofar as it requires the 'adjustmental shift'. Goldstein describes this evidence in two cases of mind blindness and defines it as *the two types of adaptation to a defect*.[53] An emblematic case, which has sometimes been called the "Anna O. of holistic neurophysiology" is that of soldier Johann Schneider, 25 years old, who in February 1916 was transferred to the Institute of Frankfurt due to a brain injury suffered the previous year when metal fragments from a mine explosion had penetrated the left parietal and occipital areas of his brain. While treating this case, Goldstein and Gelb discovered that the young patient was seriously handicapped regarding his

---

[48] Ibid., 61.

[49] Ibid., 189.

[50] Cf. ibid., 265 (a section entitled "The constants. Preferred and ordered behaviour").

[51] Ibid., 282–283.

[52] Goldstein (1940, 184).

[53] Goldstein (1995, 198–199). Also cf. ibid., 334–337.

perceptual and reading skills. To him, everything was chaos in which he could recognize only light and dark spots. In spite of this, he very soon learned to read, without anybody instructing him, by tracing through head or hand movements, stepwise along the light-dark margins (i.e. along the visually perceived contours of each letter). Via a kinaesthetic process, the movement experienced constituted a letter in the same sense as the visually perceived letter did for normal readers. The authors argue:

> …It is no question that he had achieved this kind of reading all by himself, really knowing neither how he developed it nor what he was actually doing. Not before we disclosed the nature of this procedure and had explained this to the patient did he become aware of the fact that he read differently than normal individuals do and than he himself formerly did. It is doubtful whether he ever understood completely in what the difference consisted. But he learned one thing: namely, to use his new way of proceeding with great virtuosity …[and he finally] attained such perfection that his 'detour' behaviour was hardly noticeable ….[54]

Through practically automatic modalities that were completely unknown to the subject himself, the behaviour of Goldstein and Gelb's patient appeared to follow a peculiar holistic principle of organismic self-regulation (Corsi 2012). In her comments on this case, Harrington writes: "Strikingly, the patient was in no sense conscious of having modified his accustomed reading habits; *in some unknown way*, his injured brain had established global compensatory strategies of which 'he' himself was ignorant."[55]

Goldstein often speaks also of the astonishing lack of awareness of their deficits in severely brain damaged patients, pointing out that the more severe the neurological damage, the more marked phenomenon appears (anosognosia). But to account for all that, no action of intra-psychic forces is postulated, such as 'instincts', 'drives' or 'defence mechanisms' or still other 'agents', which would work separately within organismic life. Instead, he interprets this phenomenon as a result of a 'biological' process that comes to the aid of the patient, allowing him to avoid catastrophic anxiety deriving automatically from a full awareness of his real situation. "We observe all these ways of escaping catastrophic situations," Goldstein writes,

> …not only in cases of major brain defects but also in severe bodily disease. Most people have heard about the characteristic euphoria in patients in the last stages of tuberculosis…. In this we meet with a very general biological phenomenon: what seems to be a kindness on the part of nature saves the organism from an experience too poignant to be borne.[56]

---

[54]Ibid., 196.

[55]Harrington (1996), our italics. The study of this case (published for the first time in 1918: A. Gelb and K. Goldstein, *Zur Psychologie des optischen Wahrnehmungsund Erkennungsvorganges*, 1–142, then reissued two years later in the official journal of Gestalt Psychology, *Psychologische Analysen hirnpathologischer Fälle*, 1920, I, 561), is also quoted at length by a maternal cousin of Goldstein, the neo-Kantian philosopher Ernst Cassirer, who frequently came to Frankfurt to the Institute for Research on the After-effects of Brain Damage, where he could observe the 'concrete' behaviour of brain-injured patients. Cassirer devoted an entire chapter in the third volume of his *Philosophy of Symbolic Forms* to language pathology and review of clinical cases lacking symbolic capacities (Cassirer, *Philosophie der symbolischen Formen, III, Phänomenologie der Erkenntnis*, 1929). The case was revisited quite often also in recent times (Goldenberg 2003; Marotta and Behrman 2004; Jensen 2009).

[56]Goldstein (1940, 108), our italics. Cf. also the 'protective measure of nature' in (Goldstein 1995, 339).

Following Goldstein, in human beings the only basic drive, if any, is to attain *Self-realization* and to avoid the opposite event, namely the experience of 'loss of existence' in *anxiety* and *catastrophic reaction.* All living things share the same destiny: their potentialities are driven to actualize themselves. The tendency towards Self-realization might be considered the only true *basic drive* not only in human beings but also in all living organisms.

So what is this *vis medicatrix naturae* unconsciously restoring the (as much as possible) ordered behaviour? Where does it come from? Is it legitimate to postulate a principle like *Self-realization* that, out of consciousness, actualizes all that? As we will see in the following Section, Goldstein was always absolutely critical with respect to vitalism. The logic of his argument in this regard is in some respect the same as that of the Gestaltists: vitalism is just a new dualistic theory, and the principle of *entelechy* is not a scientific but a metaphysical one. The task of our author will then be that of explaining how such *vis medicatrix* (and more generally the autonomous, unconscious self-organizing principle that governs the ordered behaviour of the organism) could be at work, where it belongs, and what its ultimate nature is, without appealing to any metaphysical substance, as vitalists did with entelechies. The fifth Section will try and illustrate whether our author succeeded in the task, and it will discuss the merits and weaknesses of his solutions.

5 Teleology: A Red Herring to Share

Organicism has been at times coupled and even identified with a variety of vitalistic positions,[57] especially when described in opposition to mechanicism; at times, perhaps more recently, it has been recognized as an adversary to vitalism.[58] In this section, we examine the divergent relationship between Goldstein's holistic-organismic theory and vitalism.

We begin by elaborating a definition of vitalism that suits Goldstein's interpretation, given that we lack a definition proposed by the author himself. We then identify the passages of *The Organism* where Goldstein deals directly with vitalism, to finally unravel the details of his critical position, hopefully showing some of its merits and flaws, but above all its peculiarities.

For one, we need to pinpoint a definition of vitalism to understand Goldstein's criticism, and to be certain of *what* he is actually criticizing. As with organicism however, and perhaps even more markedly, the re-emergence of vitalistic themes in different areas and periods of biological debate (from embryology to life definitions, from morphology to neurology) has favoured a multiplicity of positions, perspectives, and conceptual nuances that are hard to describe as a uniform doctrine and encompass in a definition. The word *vitalism* describes a family of approaches, rather than a single theory, broadly sharing the belief in an ontological distinction between animate and inanimate objects. Beyond such vague characterization,

[57] Cf. Hein (1969) and Allen (2005).

[58] Gilbert and Sarkar (2000).

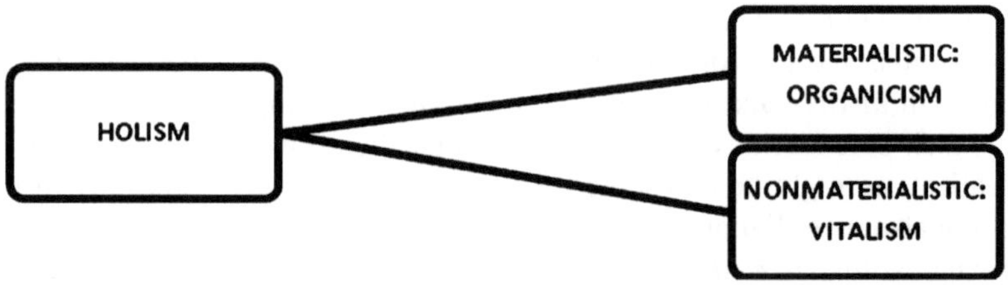

Fig. 9.1 How to think of the relationship between vitalism and organicism

however, there is little agreement on the details of a more precise conceptual analysis. The family is so diverse that some have asserted the need for a 'taxonomy'[59] – as opposed to a classical set of conditions – to provide an appropriate analysis of the concept (yet a taxonomy threatens to be potentially confusing to draw, for a greater distance sometimes divides two 'vitalist' thinkers than that between a mechanist and vitalist; and perhaps too ramified to be informative).[60] Goldstein himself, despite explicitly rejecting vitalism, never provides a clear definition of it, not even a tentative one. The reader is thus left with the task of figuring it out on his own. In several passages of *The Organism*, one could think that 'vitalism' is nothing but a synonym for 'entelechy', the latter being the only vitalistic notion exhaustively discussed in the book. However, when using the label 'vitalism', the author seems to be referring to a larger category of approaches, including those other than Driesch's.

We think of vitalism as a distinct, competing position with respect to organicism (or at least Goldstein's holistic-organismic theory, see Sect. 2) – in some aspects diametrically opposed to it. Organicism, although siding with vitalism in the quarrel against mechanicism,[61] is not to be identified or thoughtlessly paired with it, and it is even less a mere 'variety' of it: in fact, many organismic authors at the end of the nineteenth century[62] included *both* mechanicism and vitalism as polemical targets, and thought about organicism as an alternative to the latter. Even if our assumption is largely motivated by Goldstein's own alignment, which resolutely sets its holistic-organismic theory apart from vitalistic positions, it can also be seen as holding an overall validity in the field. Gilbert and Sarkar[63] have, for example, suggested a configuration of the relationship that we endorse for the purposes of our analysis as well (Fig. 9.1).

Once again, the above categorization is an arbitrary and perhaps incomplete sketch. How to trace flimsy boundaries as the ones occurring between vitalism, organicism,

---

[59] Wolfe (2008).

[60] Cf. Benton (1974).

[61] If we do not endorse an identity thesis, a strong affinity between vitalism and organicism is undeniable. Aside from being vivaciously allied against the dominant paradigm of mechanistic biology, the two approaches seem to us likely to share a character of 'meta-theoretical commitment', namely an attitude towards knowledge that is motivated from beliefs and concerns that do not necessarily belong to the field of investigation (cf. Greco 2005).

[62] Cf. Nicholson (2010).

[63] Gilbert and Sarkar (2000).

or holism is a debatable and, we believe, to a large extent subjective business. Other authors have chosen to depict the relationship between vitalism, organicism and holism in different ways.[64] Yet it is not within the scope of the present essay to evaluate merits and drawbacks of the different typologies; for the sake of clarity and for the purposes of this chapter, namely discussing Goldstein's aversion to vitalism, we assume a definition of vitalism as a form of *non-materialistic* holism.

In *The Organism* the discussion of vitalism is found essentially in three places: in Chapter II (*The Organism Viewed in the Light of Results Obtained Through Atomistic Method. The Theory of Reflex Structure of the Organism*), where Goldstein is delineating the flaws of the mechanistic analysis of the nervous system, and paving the way to present his own holistic solutions; in Chapter VIII (*On Gestalt Psychology and the Theory of the Physical Gestalten*), where he establishes in detail affinities and discrepancies between his own organismic theory and the 'sister' holistic theory of Gestalt, to which he only marginally feels he belongs to; and finally in Chapter IX (*The Nature of Biological Knowledge*), a theoretically crucial chapter, possibly the true core of Goldstein's philosophy of biology, in which he dedicates several paragraphs to the problem of alleged final causes in biology, the idea of entelechy, and "so-called purposiveness"[65] – or teleology.

Goldstein's critique of vitalism branches into two strongly interrelated but theoretically discernible conceptual lines, namely:

(a) the problem of teleology
(b) the problem of the whole and the parts.

The problem of teleology emerges particularly in Chapters VIII and IX. The present section will deal with (a), while the problem of (b) the part/whole relationship appears as the most effective argument to back up an important epistemological thesis, discussed in the next Section.

After a brief general-purpose exordium against vitalism, which according to the author does no better than mechanism in explaining the autonomous organization of life,[66] Chapter IX quickly gets into the 'meat' of Goldstein's discontent. In the sections

---

[64] Many authors have conceived of the relationship in different ways: Benton distinguishes between different degrees of 'epistemological scepticism' and 'metaphysical daring' involved in vitalistic theories, co-locating what we call here organicism on the sceptic side of the continuum (Benton 1974); Lenoir talks of 'vital materialism' (Lenoir 1982); Kaitaro prefers to call organicist thinkers 'materialist vitalists' (Kaitaro 2008); Allen talks of 'non-vitalistic holisms' (Allen 2005); Normandin (2007) of "physical vitalism"; and so on. Since in the end it seems to us that it is a matter of what aspect of theorization one finds more relevant, and this depends not on the intrinsic properties of a theory, as much as on the subjectivity of the observer, we want to insist on the adoption, for the purposes of this article, of the above-mentioned partition (i.e. organicism as non-materialistic holism).

[65] Goldstein (1995, 323).

[66] "We are not afraid of the term entelechy in so far as it is a metaphysical conception but primarily because it is much too general and undefined … [It] has too much the character of a correction, necessitated by errors made elsewhere … The cause of these errors rests in the conception of the organism as a mechanism … Since it was in no occasion necessary to assume mechanistic processes in order to understand life, we do not need to speculate on entelechy" (Goldstein 1995, 321).

on "Entelechy and 'Reason in Knowledge'" and "So-called 'Purposiveness'"[67] he discusses the problem of teleological explanations in biology. His main worry is the metaphysical suspicions that teleological explanations provoke in a respectable science, and the connection to vitalism for its deployment of teleological elements, such as Driesch's entelechy, resulting in an occluded and ambiguous ontology.[68] Goldstein is here making a move that is far from obvious. He is making an equation between teleology, or the concept of final cause, *and* a metaphysical commitment of some sort. As a consequence, he is treating teleology as an example of anti-rationalist ideology, and deeming vitalism anti-rationalist likewise, insofar as it contains teleological elements.

The whole argument may appear shaky, and in various senses it is. For one thing, the connection between teleology and vitalism is not as self-evident as Goldstein presents it. Secondarily, the equation between teleology and metaphysical commitments appears to be hashed out in a rushed and superficial way, and it is far from obvious on what elements of teleology it may hinge. As for the connection between vitalism and teleology, we think that it is largely justified by Goldstein's identification of vitalism with Driesch's theory of entelechies *tout court*. We will not try here to evaluate the rationale behind such identification, as it seems to us that it can be read as a straightforward personal bias (and frankly, an imbalance) of the author's views on the subject. We will concentrate instead on teleology, which is as a matter of fact the more fundamental source of the criticism towards vitalistic positions. Goldstein in fact shows a surprisingly limited understanding of the problem of teleology. Such limited appreciation becomes apparent when contrasting his approach with a more thorough analysis of the concept, such as the classic discussion of the evolutionary biologist Ernst Mayr. From similar works, it emerges clearly that teleology is a polysemic category, and should be treated as such; Goldstein has a partial understanding of the range of meanings connected to the category, and his reserves (the concern with the necessary metaphysical implications of teleology) actually apply only to some of these meanings – while others are completely ignored.

Teleology can be defined, by and large, as the view that legitimizes the use of finalistic explanations (i.e. involving final causes) in the natural sciences, in contrast with the dicta of physical reductionism – which admits nothing but straightforward causal explanations. As it calls into question the boundaries and statuses of sciences, the topic cannot exactly be quickly dismissed; however, as we have just seen, despite making a case against it, Goldstein avoids giving a neat definition,[69] or at

[67] Ibid., 320–324.

[68] By this argument, Goldstein is also trying to defend his own organismic position from charges of 'teleologism' or finalism, and to mark a clear boundary between organicism and vitalism: quite frequently organicism had been censured for superficial affinities with vitalistic positions, cf. Gilbert and Sarkar (2000).

[69] Once again, Goldstein does not provide it in *The Organism*. Actually, he thinks that the term *teleology* and all language involving finalistic gloss "would best be avoided altogether," or reduced to mere "descriptive use," and accordingly does not use it in the book. Unfortunately, this lexical accommodation leaves the problem substantially unaltered. As for vitalism, such vagueness represents a serious weakness of Goldstein's argument – see Sects. 5 and 6.

least a historical characterization of the problem. We believe that Goldstein's inappropriate vagueness depends on one hand on the misunderstanding of the "multiple meanings of teleological,"[70] which lead to a quick dismissal of the problem by means of a merely terminological distinction; on the other, on the inherent unfeasibility for organicism to offer a sound solution to the problems posed by teleology. Drawing from Ernst Mayr's historical analysis of teleology, we now try to dissipate terminological and conceptual confusion on the topic and hopefully show how the problems embedded in Goldstein's account can be accommodated.

Ernst Mayr provides a lucid and complete analysis of the notion of teleology and its history.[71] Teleology does not represent a genuine problem in biology any longer. Contemporary biology appears to serenely rely on teleological language and scientifically 'decent' forms of teleology (for example when discussing development, morphology, organic functions, genetic programs, intentional behaviour, and, at least in some sense, adaptationism), but it was not until the full absorption of Darwinian ideas, up to the late nineteenth century and beyond, with the modern synthesis, that mistrust surrounding the matter could be thoroughly dissolved.

Possibly the 'most influential' ideology of all times in biological thinking according to Mayr,[72] teleology is first discussed in ancient philosophy. It is from Platonic and Aristotelian philosophy that the distinction between intrinsic and extrinsic final cause first influences the philosophical debate. In the Aristotelian discussion of natural objects, the four causes (material, formal, efficient, final) overlap and the final cause represents the form or reason of being of the substance, configuring as a type of intrinsic teleology, while in Plato the final cause of the natural world is extrinsic, because it is governed by the Demiurge and oriented to the purpose of 'goodness'.

We do not wish to revisit a detailed history of the concept of teleology, which is at any rate best read in Mayr; suffice to say that the two perspectives re-emerged over time with differing fortunes. A large portion of the reflection on teleology in the natural world was for centuries a dominion of disciplines such as philosophy and theology. It is only in the sixteenth and seventeenth centuries, when the bases of modern biological science were being developed,[73] and even more in the nineteenth and twentieth centuries, when biology finally started to achieve a more definite disciplinary status, that teleology started to be considered as a problem specifically pertinent to the life sciences. Moreover, until that moment, the corpus of research that becomes the foundation of modern biological science is conducted in two virtually separate realms: on the one hand medical science, with physiological investigations of a mainly mechanistic sort, and on the other hand natural history, a discipline which was more easily infiltrated by the aforementioned philosophical or theological ideologies, and largely informed by religious cosmologies.

---

[70] Mayr (1988, 235).

[71] Especially in Mayr (1988) and (2004).

[72] Mayr (2004, 39).

[73] Cf. Mayr (2004).

In this heterogeneous pre-biological field of investigation, the concept of final cause targeted in particular two sets of problems that we would now label as 'biological': on one hand, scholars were trying to explain the harmonious and functionally coherent structures of the organisms, the purposively oriented direction of agency in animate objects, the mysteries of generation of new individuals (polemics on the origin of life among ovists, preformists, and epigenetists dominated the scene from classicism to Enlightenment); on the other hand, final cause was employed to explain the wonderful efficiency of species' adaptation, the concert of ecological interactions, the seemingly designed adaptability of organisms and morphological variation, the (hypothetical) hierarchy of living beings. In modern terms, and blatantly simplifying, one could say that in the first use, the idea of final cause was utilized as guiding principle for ontogenesis (intrinsic teleology), in the second for phylogenesis (extrinsic teleology). By and large, the first interpretation (ontogenesis) corresponds to what would later be revealed as 'good teleology', the second (phylogenesis) to what would be revealed as 'bad teleology'.

The history of teleology is one of 'bad intellectual company'.[74] Suspicious teleological concepts, such as cosmological teleology or the idea of design and intentionality inappropriately projected onto the natural world, ended up casting their shadow onto scientifically decent, although incomplete, explanations of ordered behaviour and other 'teleological' phenomena in the organism. Extrinsic, cosmological teleology had easy play in justifying with some *Deus ex machina* the finalistic organization of the natural world, while those who were trying to animate the world 'from inside' (intrinsic teleology), and had essentially materialistic sympathies, were constantly pilloried for the lack of convincing empirical evidence, and related conceptual tools to fill up gaps in their story and make it exhaustive and acceptable. It seemed inevitable that either they had to be contented with partial, elliptic accounts, or give up on their requirements of ontological monism. It is by combining the two sides of teleology into one polemical target that the dominant mechanism flourishing between the eighteenth and nineteenth centuries tried and succeeded in censuring the category of teleology altogether – which was inevitably, by that time, suspicious to anyone who wanted to line up in the fully materialist and rationalist side of the scientific endeavour, as mechanists aspired to do.[75]

Unfortunately for mechanism, the rigidly determinist causality governing the organism–machine was, in itself, a very poor conceptual tool to account for the problems of embryology and development, on which it eventually focused. It is not until Kant's charitable attempt to resurrect the legitimacy of final causes in his third *Critique* that the status of teleology was reintegrated in the scientific endeavour.

[74]Cf. Gilbert and Sarkar (2000) on the 'bad intellectual company' of organicism and vitalism.

[75]Mechanicism had dominated the scene in the natural sciences roughly since the Newtonian revolution in the seventeenth century. But it did not lack drawbacks: in the neat and entrenched model of the organism-machine, every finalistic consideration suddenly disappeared and left an explanatory gap that urged to be filled. This represented a problem especially in embryology. This persistent dissatisfaction with the mechanistic model in the biological sciences remained unchanged for a surprising amount of time. Cf. Micheli (1970) and Allen (2005).

Kant's concept of the 'inner purposiveness' of living entities (see Sect. 3), however, did not represent much advancement, in terms of scientific explanation, with respect to Aristotle's definition of organism and functions. A real solution was to come only very recently, especially from the developments of genetics, their full integration in the Darwinian paradigm, and from the introduction of the concept of *genetic program*. Regardless of this, even in contemporary times teleology has been surrounded by mistrust, a perpetual legacy of its tormented historical vicissitudes. Mayr identifies up to four critiques of teleology that have survived until recent times, and even after the above-mentioned discoveries.[76]

The real turning point in the conceptual discussion of teleology was the introduction of the notion of program, and in fact of genetic program. A program has been defined as "coded or prearranged information that controls a process (or behaviour) leading it toward a given end."[77] The change happened right after the discovery of nucleic acid in the chromosomes, and the understanding of its function as genetic material. Before such empirical findings, all the ideas on intrinsic finality and organizing principles of organisms, despite being somehow on the right track, could not be considered genuine scientific concepts.[78] In turn, the crucial breakthrough was made possible by the so-called 'molecular revolution', which gave an exceptional impulse to biological scientific advancement. In 1947 Oswald Avery (1877–1955) showed that genetic material was constituted by nucleic acids, with their coding structure, and not from proteins, as was earlier thought. Soon after, in 1953, James D. Watson (1928-) and Francis Crick (1919–2004) identified the double-helical structure of DNA. These discoveries opened a completely new era in genetics, which up to that time was growing exclusively as a mathematical and statistical treatment of the allelic frequencies expressed in a

[76]A first point was made about teleological language (language implying the recourse to final causes), which would entail the commitment to a metaphysical hypothesis. Mayr's reply is that no contemporary biologist needs to call for immaterial agencies to explain directed behaviour or development after the clarification of the concepts of genetic program, genetic information and evolution by natural selection. A second critique comes to teleology from physical reductionism. This has to do with the concern that accepting teleological explanations in biology would somehow subtract biology to universality of physical –chemical laws. The answer is here represented by the formulation of the concept of double causality in biology: biological objects would be subjected to a twofold source of causation, the proximate one constituted by their genetic code, and the ultimate one by chemical-physical laws. A third critique judges the assumptions of future ends and purposes to be logically contradictory with the principle of causality (it is essentially a critique of the very concept of final cause): the reply to this criticism is represented by nothing but the acceptance of teleological causality into logic, once the previous mentioned suspects are dispelled. A last criticism highlights how making use of concepts like plan, end, purpose, intention, could conceal the application of anthropomorphic qualities to completely unconscious processes. Mayr dispels this final doubt, emphasizing how contemporary biology can explain all physiological functions and all animal behaviour without having to appeal to consciousness or intentionality of any sort (cf. Mayr 1988; 2004).

[77]Mayr (1988, 49).

[78]Mayr (2004, 54).

population's phenotypes, and offered a solution to problems so far considered intractable. According to Mayr:

> ...What finally produced a breakthrough in our thinking about teleology was the introduction of new concepts from the field of cybernetics, and new terminologies from the language of the information theory. The result was the development of a new teleological language, which claims to be able to take advantage of the heuristic merits of teleological phraseology, without being vulnerable to the traditional objections.[79]

Mayr concludes by proposing a new terminology. We eventually arrive, by revising all those processes that are apparently directed to an end in nature, at the distinction of teleomatic processes (by assonance with 'automatic'), namely all these inorganic processes that proceed towards a final state in virtue of physical properties. Radioactive decay, for example, is a typical example of a teleomatic process, because it leads toward an end state of matter without being subjected to a program. Gravity and the second thermodynamic principle are the physical laws that most often govern teleomatic processes. Mayr coins a new definition for the legitimate category of the teleological in biology: 'teleonomic'. Teleonomic processes are those natural phenomena that are governed by a program. Specifically, a process or behaviour is teleonomic when it "*owes its goal-directedness to the influence of an evolved program*,"[80] where a program can be defined as "*coded or prearranged information that controls a process (or behaviour) leading it toward a goal. The program contains not only the blueprint of the goal* *but also the instructions for how to use the information of the blueprint.*"[81] Authentic teleonomic activities are those which depend on having an evolved program. Embryogenesis, development, regeneration of parts, the functioning of the nervous system and intentional behaviour are hence biological phenomena validly reintegrated into the boundaries of empirical science. The so-called cosmological purposiveness and apparently fine-tuned design of the universe (including literal adaptationism, see previous footnote) will instead be rejected. These will be explained by the completely chance-governed process of natural selection, and by the psychological propensity of the human lineage to anthropomorphise the surrounding world and find a meaning to even the most fortuitous event, which is otherwise so hard to accept.

---

[79] Mayr (1988, 39).

[80] Mayr (2004, 51). A different consideration holds for adaptiveness and adaptations, i.e. those features of an organism that are the product of natural selection and seem to perform a specific function. These are not genuinely teleonomic processes, for here the language employed is only apparently (so to say metaphorically) teleological. Adaptation is in fact an *a posteriori* phenomenon, due to the differential survival of phenotypic modifications produced by random processes of variation. Generally, adaptive features contribute to perform teleonomic activities, and are, so to say, executive organs of teleonomic programs. However, being themselves stable acquisitions and stationary systems, describing them in *genuinely* teleological terms is misleading.

[81] Mayr (2004, 53). In his latest work, Mayr goes as far as distinguishing *closed*, *open* and *somatic* programs (cf. ibid., 54–55) from the original insight into *genetic* programs. As the genetic one (which is a closed program), open and somatic programs are evolved to function as guiding information for biological processes that finally assume the appearance of goal-directed activities.

We hope that after this possibly tedious but in fact essential analysis of teleology the richness and depth of its spectrum of meanings will be clearer, as will the inadequacy of Goldstein's treatment. It should also be easier now to isolate the two particular meanings of 'teleological' that are unwelcome to Goldstein, which we could label 'teleology-as-purposiveness', and 'teleology-as-metaphysical ideology'.

Teleology-as-purposiveness provokes the concern of interpreting completely unconscious aspects of the organism and the natural world through the teleological category of intentional behaviour. The possibility of seeming like he is appealing to intentionality and purposiveness and adumbrating forms of anthropomorphism (or worse, of divine intervention) in his explanation is evident when Goldstein tries to reformulate his own teleological language in terms of 'ends', and to avoid the term 'purposes'. This lexical revision clearly does not do, but it is all we get from *The Organism*. The rest of Goldstein's solution consists in suggesting Kant's so-called 'inner purposiveness' as a possibly acceptable form of finality in natural explanations.[82] No further explanation is adduced to clarify why the Kantian notion would do better than other teleologies.

The other aspect of teleology that seems to worry Goldstein is the one more directly connected to the theory of entelechy, namely the preoccupation with introducing extra-materialistic ontologies in biological explanations. If entelechies are ontologically thick substances governing the coherent integration of core and peripheries in the organism during performances such as ordered or goal-directed behavior, then vitalism (or at least Driesch's vitalism) is relying on an anti-materialist and anti-rationalist ontological dualism. Goldstein has a strictly rationalist and materialist approach to science, and by no means intends to make concessions to matters reaching beyond the boundaries of physics. His solution to dispel metaphysical suspicion from organicism is to substitute entelechies with the explanatory category of *wholeness* or *totality* of the organism. *Totality* or *wholeness* is a purely epistemological category, hypothetically obeying the function of finalistic organization in the organism itself. As it is purely epistemological, or in Goldstein's terms a mere "reason in knowledge" (see Chapter IX and the next section of the present chapter), the principle of wholeness should avoid ontological and metaphysical entanglements. It is not clear, however, how this move could be effective, and the problem of explaining the finalistic organization of biological entities remains, we believe, open. For further discussion of this spectacular "leap into the epistemological," we point the reader to the next and final section.

Thus we can maintain that Goldstein was somehow on the right track in explaining the ordered behaviour of the organism, its constants and its flexible 'coping' with the *milieu*, but completely lacked the appropriate conceptual tools (of course connected to the delay in empirical discoveries) to keep the different meanings of teleology apart. However we argue that his rejection of vitalism on the grounds of teleology, despite being motivated and perhaps even necessary, falls short of its target, being a mere *pars destruens* and lacking most merits of a *pars construens*.

---

[82] "At most, the concept of the so-called inner purposiveness in the sense of Kant, could be taken into consideration" (Goldstein 1995, 323).

Organicism's inherent failure to provide a solution to the problem of teleology depends more on the lack of the relevant empirical discoveries, which were still to come, than on poor conceptualization. This may also explain Goldstein's reticence in digging out the details of the topic and suggesting instead a very superficial 'solution' to it. A clear recognition and in-depth analysis of teleology's significance would possibly have exposed the fragilities of Goldstein's own theory. Our conclusion is that, despite being justified in some respects, the whole critical argument against vitalism (i.e., against its teleological elements) is dubious: Goldstein is unable to provide a valid alternative explanation to the same problems that doctrines like vitalism or teleology were trying to explain, namely the finalistic organization of living beings. A final rehabilitation of teleology is indeed inescapable, for some biological phenomena, and specifically those Goldstein was interested in (i.e., ordered behaviour or the self-recovery of the nervous system) *truly* are teleological. Goldstein's 'solution' is desperately trying not to throw out the baby with the bath water (rejecting the 'good' teleological explanations with 'bad' ones), but there is little way out of there.

Goldstein died in 1965, a few years after the blossoming of molecular biology, the revolution of genetic programs, the spread of cybernetic language in biology, and the advancements in a more and more extensive and coherent formulation of evolutionary theory. How come the work of a man interested in 'biological knowledge' seems completely impermeable to these disciplinary revolutions? We can advance a hypothesis: we have already noted how Goldstein turned towards psychotherapy and suffered the forced exile from his Germany as an irreparable intellectual loss. He stepped away from the biology research community, and we find no mention in his last works of any of the new ideas that were revolutionizing the life sciences. Most strikingly, no mention is ever made of important debates in theoretical biology like that of the New Synthesis of Darwinism, which in many ways would have been extremely pertinent to Goldstein's theorization of biological knowledge. In this respect, it is extraordinary how Goldstein's reflection was never even marginally touched by ideas like the dynamism of organisms' evolution and the nature of phylogenetic relationships, and stayed exceptionally tied to old-fashioned conceptions of the hierarchy of organism and the 'centrament' of living beings.[83] In one of

[83]Goldstein's sympathies for Johann Wolfgang von Goethe (1749–1832) went well beyond always keeping a portrait of the German philosopher above his work desk (cf. Teuber 1966). Ferrario (2008) argues that it is exactly the conception of the organism as a prototype (Goethe's *Urbild*), to which all the organisms would tend in the process of realizing their 'essential natures' that inspires the bizarre view of the 'hierarchy of living beings' in Chapter XI of *The Organism*. The hierarchy of life ranks organisms (but also organs, and psychological types) on the basis of an hypothetical "degree of centering and richness" (Goldstein 1995, 370), i.e. the greater or lesser organization of the organism, which is manifest in a series of formal attributes (centering); and the capacity of an organism to "absorb the richness of content of the apprehended world," as well as the "richness of essential nature" (richness). In other words, centering and richness measure the approximation of the individual organism to the *Urbild*, aside from characterizing Goldstein's views on the topic as fairly anachronistic ones.

his last works, we read surprising statements on human evolution, which should leave no space for conceiving of Goldstein as an "evolutionary-minded neurologist," as Sacks[84] seems to do when comparing *The Organism* to Edelman's work on neural Darwinism.[85]

Goldstein never wrote again on philosophy of biology or produced a revision of his ideas. It is hard to tell whether his promising insights on biological questions would have led to further valuable developments had he not veered so decisively away from theoretical biology in the last years; despite appearing in many ways remarkably modern and ahead of its time, his conceptual plan was constrained by a number of old-fashioned and incongruous notions, and in some ways even by dubious cross-disciplinary contaminations that are not acceptable if his objective still was, as it seems fair to presume, the establishment of a comprehensive epistemology for the biological sciences.[86]

6 A Savvy Trick: Epistemological Acrobatics and the Ethics of Biological Knowledge

Vitalism is not only taxed by entertaining metaphysical views when trying to explain the teleological organization of the living organism. The critique of vitalism follows, as we anticipated in the previous Section, two main directions: on one hand, the teleology topic we dealt with above, on the other, the "problem of parts and wholes."[87]

What does Goldstein mean by this? Hardly any biologists would believe that a 'problem' phrased in such terms could be of interest to contemporary studies (which in fact it is not), but in Goldstein's conception of biology, and in a number of other historical debates, it happened to be indeed a crucial issue.[88] For one, it should be borne in mind that the 'parts' and 'wholes' we are talking about here are, in a physical sense, the parts versus the entirety of the organism, the living entity. In Goldstein's discourse, 'parts' of the organism can be morphological portions like the arm, finger, eye, root, wing or liver, or functionally isolated areas (i.e. the anatomical localizations of speech-dedicated, or motion-dedicated areas in the brain); or they can be functions or behaviours conceived as autonomous sub-units, such as the so-called 'reflexes' or instinctive triggers of nervous system, the 'drives' of psychoanalytic theory, or biological functions as 'digestion', 'locomotion', 'mating' or 'sleep'. The

---

[84] Sacks (1995).

[85] Cf. Edelman (1987) and Ferrario (2008) for an extensive refutation of the comparison.

[86] Cf. Ferrario (2008). The idea of hierarchy and 'centering' of living beings, the doubtful statements on evolutionary theory, and the position assigned to mankind with respect to the animal domain are among the most important elements of divergence from Darwinian theory.

[87] See the section on "The Problem of Parts and Wholes," Goldstein (1995, 302).

[88] Cf. for example the "certainly difficult and very serious question" of the being-alive of the parts, Ibid.

whole is on the contrary the entirety of the organism, be it animal or vegetal,[89] and of course, the totality of its functional dynamics.

However, 'parts' and 'wholes' correspond also to epistemological categories, the first embodying the correlate of the analytic, dissecting method of mechanistic science, the latter the holistic, top-down approach of Goldstein's own organicism and other holistic schools (like *Gestaltpsychologie*, vitalism and other organicisms). Therefore they represent not just alleged properties of the living being, but also conceptual kinds that we use to interpret reality. The confusion of such ontological and epistemological levels is at the core of Goldstein's discussion and critique of non-holistic epistemologies of biology, above all the mechanistic one. Goldstein refers to mechanistic, mainstream science as 'analytic', 'partitive', 'dissecting', to indicate the method of approaching the study of its object (analysis, decomposition, breakdown in simple subunits or 'components' to facilitate the analysis) and the correspondent frame of interpretation (talks of such subunits define genuine ontological categories). Analytic approaches in fact, as opposed to holistic ones, conceive of 'parts' of the organism (morphological parts, organs, functions, as said above) as if they were objective realities of living beings and could be considered having an independent existence. According to Goldstein, this is a gross ontological error, caused by the 'ontologization' of a category that is in fact only and solely epistemological (that of 'part').

In Chapter III and IV, Goldstein discusses the so-called 'reflex theory' and localizationism as examples of neurological theories of mechanistic inspiration. They are respectively accused of interpreting – and therefore inappropriately curing – the reflex as a phenomenon that happens locally and largely independent from the rest of the nervous system, and specific areas of the brain as connected to a specific function or behaviour by a deterministic, one-to-one causal relationship. To these, he contrasts holistic theories, like his own holistic organicism. The nervous system is thus seen as an integrated net: it always needs to be regarded and medically treated as a unitary system, even when impairment seemingly occurs in a well-determined area. Mechanistic and holistic approaches, however, do not seem to exhaust, in Goldstein's classification, the entire range of neurological theories under survey; between the two types, in fact:

> ...There are, however, other types of quasi-holistic approach. The relation between the parts and the whole is considered either as given in the organism itself – for example, the various biological organismic theories – or it is considered *adventitious*, as, for instance, in the form of an entelechy.[90]

---

[89]Problematizing the boundaries and individuality of the organisms, as in contemporary debates, was far from Goldstein's interests. His definition of the organism is a simple, intuitive, empirical one: the organism is the object of study of biology, the 'living being', which simply 'confronts us'. ("We stand in the presence of a multiformity of material that is scientifically undefined. This material is simply the world around us, in which certain phenomena immediately stand out as 'living,' without revealing to us the why and wherefore of this characteristic, or even challenging an inquiry concerning it. Life confronts us in the living being. These organisms, at least for the time being, provide our subject matter," Goldstein 1995, 26–27).

[90]Goldstein (1995, 87). Our italics.

The criticism of vitalism revolves around the fact that, in Goldstein's opinion, it conceives of the parts/whole relationship as *adventitious*. Though, the author does not offer any more extensive explanation of what he means by the term at this stage. It is from the discussion scattered through Chapters VIII, IX and XII that it becomes gradually clear that the relationship between parts and whole is 'adventitious' (accidental) exactly because it is obtained through postulating the further ingredient of entelechy. If the connection between two concepts requires a third hypothetical concept or entity, then the connection is *de necessitate* adventitious (i.e. accidental rather than necessary or constitutive). Finally, if this is the case, the connection inevitably implies that the elements involved are, firstly, ontologically separated. Thus, in vitalistic theories (or at least in Driesch's vitalism), parts and wholes are still given as autonomously existing objects, which become 'adventitiously' (coincidentally) joined through a third overlaid substance (entelechy).

The essence of this critique of vitalism is, in the end, based on the one directed at mechanism. In Goldstein's interpretation, by hypothesising entelechies, vitalism conceives of the natural world in terms of 'parts' and 'wholes', as if the properties of 'being a part' and 'being a whole' were objective features of it. Like mechanist thinkers, vitalists think of biological entities as if they were inherently 'divided in parts' or 'taken as wholes', which would then need to be combined and their interaction explained.

In Goldstein's holistic organicism, on the contrary, no biological entity is inherently divided in parts, and parts are just (pretty much illicit) abstractions, or illusions produced by our investigative efforts and attitude. Parts are nothing but projections created by our (necessarily) dissecting epistemological perspective; we do create parts (or the illusion of them) by focusing on different aspects of the object of knowledge – in fact, parts of living entities are the product of our 'abstract attitude'.[91] Parts, as opposed to biological wholes, have no ontological texture: in other words, the problem of the relationship between the whole and the parts is, at least ontologically, a false problem, for nothing like 'the parts' does actually exist but any biological entity is given as a totality and a functional whole.

It is interesting to note that Goldstein does not deny the benefits and even the necessity of the analytic method of data-gathering. He indeed claims that this is, in a first stage, the only method that guarantees the status of science to a corpus of knowledge, avoiding falling into 'fictitious generalities'.[92] What Goldstein is warning the reader to do is to *resist* the illusion of our innate epistemological attitude that

[91] "Abstract attitude" is an essential term in Goldstein's specific terminology (in Goldstein 1995) and indicates the propensity, in humans, to gain knowledge by analytic focus.

[92] Ibid., 28–29: "Apropos of methodology, one thing must be emphasized in advance. We will not be satisfied with any form of intuitive approach. Every natural science, indeed any science at all, must start with an analytic dissection. So, too, in biology we must first observe the 'parts' of the organism. We are forced to accept this point of departure because a naive approach to the phenomena is not feasible, unless one is to be content with fictitious generalities." Or: "Certainly, isolated data acquired by the dissecting method of natural science could not be neglected if we were to maintain a scientific basis" (ibid., 18).

instinctively highlights parts (because this is how it works and is effective), by avoiding considering them as real entities. The correct way to overcome the 'illusion' is, incidentally, provided by the Gestaltic perception of figure and background as part of the same single totality, only differentiated in a background (out of the focus of attention, but present) and foreground (under immediate attention). The natural modality of realization of our knowing processes (the so-called *abstract attitude*, over which we have no control) should not drag us into making claims about the reality of the partition in natural objects.[93]

Goldstein's position may sound extreme. The author is, however, extremely careful in stressing that, by negating the reality of parts, he does not intend to negate the specificity of biological functions, or the morphological differentiation of the perceived 'parts' of the organism. He makes this clear, for example, when he addresses the problem of localizationism. Goldstein acknowledges that there is anatomical differentiation in the brain tissue, and certainly this corresponds to specificity in function, but it does not amount to a *real* (both in the sense of complete and objective) distinction of areas. We cannot presume that mental activities occur in this or that specific part of the brain, because they always happen in the ultimately indiscernible totality of it. Goldstein is extremely clear about the fact that talk of specialization or differentiation is absolutely legitimate in biology,[94] but it is a completely different thing from treating the organism as the locus of reflexes or drives. Such theories act as if the object of their studies (reflexes, instincts, psychic drives, dedicated areas of the brain) were authentic ones, real-world entities with a definite and autonomous ontological status, and provide medical help targeted and confined to them – ignoring the rest of the organism, which is instead always affected. Pursuing erroneous targets cannot obviously result in a valuable therapeutic outcome: misconceptions occurring initially at a pure epistemological and theoretical level can ultimately affect important pragmatic achievements like neurological rehabilitation.

We are now, with the notion of pragmatically effective knowledge, close to the core of Goldstein's philosophy of biology. The whole critique of mechanism and vitalism has revolved around a subtle but categorically held discrimination: even if parts are legitimate epistemological categories, they are not nearly ontologically legitimate as much as wholes are. Parts do not correspond to any true fact about the biological world; and the belief in the objective existence of particular entities in the natural world is a false one. But why should we think that 'wholes' do any better than 'parts' as natural entities, in principle?

---

[93]The author often mentions Rubin's famous picture, which may appear as a white vase on a black background or as two black profiles on a bright background (Goldstein 1995, 125; 1940, 19–21). Goldstein uses the image quite freely; first of all he applies it to the 'figure' of excitement within the nervous system, in order to enact a behavior whatsoever, but he later extends its application to several levels of arguments: reflex alternations, performances, strategies of knowledge (atomistic-holistic), existential dimensions (sphere of adequacy-sphere of immediacy), cf. Corsi (2012).

[94]Goldstein (1995, 198) (a section entitled "Localization and specificity").

After all, Goldstein himself speaks in favour of the analytic procedure of data gathering; the so-called 'holistic' perspective is inapplicable at early stages of biological research. Thus it is unclear why and how at a certain point we should detach from the analytic method; but above all, it is unclear why we should feel tricked and dismiss the useful and innate representation of parts as objective entities. Goldstein has a *tout simple* (in his view) and direct answer: we should prefer the holistic perspective, as a comprehensive interpretative glance at biology, *because that is the one that delivers the best therapeutic results.* Likewise, the analytic procedure proves flawed exactly because it reports unsatisfactory clinical results, and in most cases leads to erroneous diagnoses.

The attainment of a more or less correct grasp of 'biological ontologies' is evaluated through a completely pragmatic perspective: clinical and therapeutic effectiveness.[95] It seems therefore legitimate to speak of a pragmatist conception of knowledge and truth in Goldstein: the criterion that should guide our beliefs about the world is what we get in terms of positive feedback from it. Treating patients by the dicta of mechanistic neurology leads to poor rehabilitation? Then we should discard the dicta and if necessary, after careful consideration, the method altogether. Goldstein is not scared of admitting that knowledge must in a first place be useful: can we get knowledge in an imprecise way, but for a useful purpose? Then we should get it with no hesitation.[96]

Following the same logic Goldstein solves the questions that internally arise in his holistic system. For example, a common objection questions why, if wholes are logically dominant over parts, should we stop at the level of the individual organism? Could not the organism itself be part of a 'greater whole', and appear illusorily to us as a self-contained totality?[97] Goldstein's answer is that we should not bother with the problem of larger, cosmological wholes, just because by considering the individual organism we arrive at sufficiently satisfactory epistemological results.

This is not the place to attempt an appraisal of the cohesiveness and logical feasibility of Goldstein's pragmatic conception of knowledge. Suffice it to say that the author's project is an extremely ambitious one, and it certainly leaves lots of questions unanswered under the profile of a biological epistemology – above all, for example, what are the criteria by which we might judge a medical intervention to be successful, and by what theoretical framework we can rightfully employ them. On the other hand, Goldstein's holistic approach with its concepts of the self-organizing totality of the organism, organism and milieu responsiveness, and enriched causal explanations, has merit in striking early and interesting analogies with many modern fruitful ideas in biology, such as *emergent properties*[98], developmental or

[95]"Whether or not both [subject matter and methodology, which are interrelated] are adequate instruments of science can be verified by only one criterion: fruitfulness in their respective fields. We must attempt to understand living organisms in the most fruitful way" (Ibid., 28–29).

[96]Cf. Ibid., 316 on the incompleteness of biological knowledge.

[97]Ibid., 302.

[98]Cf. Laughlin 2003.

complex systems and their related properties, or the growing role of the environment in shaping developmental and evolutionary dynamics. Even more, it is an admirable example of integrity and ethical stance in medical practice, one that should be seriously taken into account.

It is interesting to highlight here a hypothesis of aetiology for Goldstein's unusually distinct 'epistemologically flavoured' position. We believe that this can be sought in his predominant professional orientation, which is undoubtedly, as emerges among his diverse talents in the late years, that of the physician. It is success in therapy that constitutes for Goldstein the criterion for establishing the truth (or falsity) of beliefs concerning the organism. In other words, Goldstein's epistemology of biology is ultimately founded on a series of practical reasons, rather than theoretical ones, reasonably extrinsic to the discipline of biology itself. It is in this light that Goldstein's organicism might be interpreted, like vitalism has previously been, as a form of metatheoretical commitment (following Hein 1968: "an intellectual mind frame deeply based on motivations that are not necessarily embedded in the context of the topic itself") of a specific kind, namely the medical or ethical kind.[99]

After having grasped the centrality of the link between pragmatic and epistemological claims in Goldstein's reflection, it is worth getting back finally to the dispute with vitalism. We said before that Goldstein was going to offer a personal solution based on a "leap into the epistemological": it is now time to clarify the meaning of this expression. The critique of vitalism branches, as we saw, into two closely connected but at a closer look discernible conceptual lines, namely the problem of parts and wholes and teleology. The problem of conceiving of parts and wholes as separates entities, which then need to be connected by a third coincidental link, has been denounced as a false problem. Parts are in fact nothing but an epistemological illusion (an illusion relating to our knowing procedure), and do not possess any ontological quality in themselves. Therefore, the necessity of postulating entelechies is a false one, because parts themselves, in contrast with totality, are, so to say, nonexistent. With respect to the problem of parts and wholes Goldstein recurs thus to an epistemological adjustment to dissipate a problem that initially, but erroneously, appeared ontological (i.e. related to the essence of the subject of investigation). Such a move is a first signal of the centrality of the epistemological dimension that will emerge when dealing with the question of teleology.

This centrality is legitimized by the pragmatic principle of truth determination. If we can gain therapeutic success through a specific method of investigation, then we should have confidence that such method of investigation (epistemology) reliably delivers true beliefs about the organism. Anything further than a mutual validation of utility and correctness of knowledge seems to be beyond Goldstein's scope, but nonetheless the latter inspires a pervasive confidence in epistemological claims. In almost no part of *The Organism*, contrary to other organicist authors,[100] does Goldstein make explicit ontological assertions about the essential nature of the

---

[99] Cf. Williams (2003): Wright and Potter (2000).

[100] Cf. Nicholson (2010).

biological world, on the contrary he always stresses that his discussion of biology is a purely methodological one.[101]

It is worth emphasizing that Goldstein's critique of vitalism, despite being ultimately based on his more fundamental critique of mechanism, is not identical to it. As opposed to mechanists, vitalists are partially aware of the problem engendered by the analytic approach to knowledge. The theory of entelechy constitutes in fact an attempt to respond to the mechanistic-bred problem of the relationship between the parts and the whole (specifically in the nervous system), but in Goldstein's opinion fails to do that. Vitalism addresses the wrong aspect of the problem; it fails to realize that it is not an ontological addition that is required, but actually an epistemological subtraction. Besides, postulating new substances like the entelechies that 'holistically' connect parts to the entirety of the organism causes a further metaphysical complication, which is the one discussed in the previous Section, namely the problem of teleology. Thus Goldstein got rid of parts, and got rid of entelechies; but did he truly get rid also of the original problem of mechanism, namely explaining the ordered behaviour of organism, and all of those biological phenomena that appear to proceed according to a predetermined plan, like embryogenesis, development, or simply the unconscious strategies of recovery that he observed in so many of his patients? Goldstein's answer is yes. Unfortunately, it is not quite a convincing answer. In spite of the entelechy or any other form of vital force, he postulates the 'totality', the wholeness of the organism as an intrinsic organizing principle. In the organismic tendency of "coming to terms with the world," Goldstein does not see the action of some mysterious vitalistic principle or psychic instance, but only a general biological law of the organism as a whole, that always tends to *self-realization* or the *actualization of its own essence*, which is determined by its normal constants. However, it is not any clearer why this organizing principle would be less suspicious than vital forces. The author insists the fact of totality as an organizing principle is just a "reason in knowledge" (Chapter IX), and does not have the slightest ontological and metaphysical connotation:

> ...The idea of a definite end ... must also be taken only as a guiding notion for the procedure of knowledge rather than in a metaphysical sense...In this sense, one can describe the concept of wholeness as a category, as the category that substantiates and encompasses the subject matter of biology.[102]

Thereby "definite end," "actualization of essence," and "wholeness" are synonymous and, in our author's viewpoint, should replace or overcome the embarrassing frame of reference of teleology. But, again do these concepts suffice to *explain* the purposive holistic behavior, in disease as in health, of the organism? It does not seem to be enough. Goldstein's *wholeness* appears very much like a substitute, not really

[101] "*The Organism* consists mainly of a detailed description of the new method, the so-called holistic, organismic approach.... We were confronted then with a difficult problem of epistemology. The primary aim of my book is to describe this methodological procedure in detail, by means of numerous observations" (Goldstein 1995, 18).

[102] Ibid., 324.

more explanatory, to entelechies or vital forces. It might be that organicism, by the way of its 'leap into the epistemological', appears 'cleaner' than vitalism under the metaphysical aspect; but it certainly does not provide a real answer to the problem of finalistic organization of biological entities. The black box remains locked; and a further atomic unapproachable nucleus (the category of wholeness) is accepted at the core of the theory, and is not further investigable. The transfiguration of the problem of teleology and organization of the living into a matter of purely epistemological significance is a final signal of the peculiarly biased epistemological character of Goldstein's approach to biological studies. We believe that this character constitutes the originality and a valuable aspect of his thought, but, as we pointed out in the previous Section, the problem of teleology will actually have to wait for the concept of genetic program, which introduces that special but perfectly materialistic, scientifically acceptable 'double causality' that characterizes living beings.

A last word on the epistemological developments that followed after *The Organism*, which will give the measure of how far from the original biological themes Goldstein's reflections ended up. The concept of 'biological law' ended up being so inflated that it seems to account also for those psychodynamic processes attributed, in psychoanalytic theory, to the conflict between the *Id* and the conscious *Ego*. We must remember that, in Goldstein's theory, Self-realization (i.e., the actualization of a Being's essential nature) is synonymous with Existence, which at the same time is equated to an epistemological principle.[103] The operation to trace existence back to a mere epistemological issue might appear reductive. To Goldstein, however, knowledge represents that specific "coming to terms between the organism and its world', which is intrinsic not only to human nature, but to all living beings. The way the individual apprehends is isomorphic to his essential nature, to his *in-der-Welt-sein*, to his *existence*. The original German term *Selbstverwirklichung* has been translated in English as *Self-realization*. However, we must bear in mind that the term "means nothing but *the realization of all capacities of the organism in a harmonious way so that the 'existence' of the organism is guaranteed*.... To avoid misunderstanding I shall use the term *realization of the particular nature*, instead of self-realization."[104] Therefore, the meaning of this '*Self-*', which might lead us into thinking of the 'self' of academic psychology, can actually be traced to the concept of 'nature' according to Goldstein – a concept which is somewhat unscientific, mystifying and which can ultimately be defined as 'romantic'. It rather resembles the *project* of *Dasein*, the existential structure 'in-view-of', as stated in Heidegger's philosophy.[105]

[103]In one of his last papers, outlining the differences between his point of view and that of existential psychiatry, the old Goldstein was to state: "I agree with the existentialistic concept in so far as I also deny that biological phenomena, particularly human existence, can be understood by application of the method of natural science. But I differ in the meaning of the term existence. It means for me an *epistemological concept* based on phenomenological observations, which enables us to describe normal and pathological behavior and to give a definite orientation for therapy. It is a kind of philosophical anthropology" (Goldstein 1959, 13, our italics).

[104]Goldstein (1957, 179–180).

[105]Heidegger (1927, §§ 31, 48, 53).

Ultimately, Goldstein's holistic biology seems to be almost magically transformed into a philosophy of existence.[106] The organism is a 'Being' in a temporal succession of definite forms. The functional significance or value of performances has nothing to do with mere survival, nor with an instinct of self-preservation and the like. All this is "itself a symptom of abnormality,"[107] because "... under adequate conditions, the tendency of normal life is toward activity and progress."[108] The *value* of self-realization consists in actualizing all individual potentialities, which in our species also include psychological dynamics and spiritual experiences.

References

Allen, Garland E. 2005. Mechanism, vitalism and organicism in late nineteenth and twentieth-century biology: The importance of historical context. *Studies in History and Philosophy of Biological and Biomedical Sciences* 36: 261–283.

Ash, Mitchell G. 1998. *Gestalt psychology in German culture, 1890–1967*. Cambridge: Cambridge University Press.

Benton, E. 1974. Vitalism in nineteenth-century thought: A typology and reassessment. *Studies in History and Philosophy of Science* 5: 17–48.

Bergson, Henri. 1907. *L'Evolution Créatrice*. Paris: Alcan.

Brown, Malcolm. 1990. *The healing touch – An introduction to organismic psychotherapy*. Mendocino: Lifetime Rhythm Press.

Calamari, Elena, and Mauro Pini. 2007. *Kurt Goldstein – Il Concetto Di Salute E Altri Scritti*. Pisa: ETS.

Canguilhem, Georges. 1943. *Essai sur quelques problèmes concernant le normal et le pathologique*. Clermont-Ferrand: La Montagne.

Cassirer, Ernst. 1929. *Philosophie Der Symbolischen Formen, III, Phänomenologie Der Erkenntnis*. Berlin: Bruno Cassirer.

Corsi, Luigi. 2012. Organismic self-regulation in Kurt Goldstein's holistic approach. *International Body Psychotherapy Journal* 11(1): 57–65.

Delaporte, François (ed.). 1994. *A vital rationalist: Selected writings from Georges Canguilhem*. Trans. Arthur Goldhammer. New York: Zone Books.

Dilthey, Wilhelm. 1883. *Einleitung in die Geisteswissenschaften*. Leipzig: Duncker & Humblot.

[106]For example, readers may wonder whether the writer is a scientist or a mystic when they read: "...the organism is a Being enduring in time, or if we may say so, in eternal time; for it does not commence with procreation, certainly not with birth, and does not end with death. What we mean by the terms 'birth' and 'death' are merely certain landmarks like others, for example, like puberty and menopause. Their real *nature* is yet to be determined. But they belong essentially to existence ..." (Goldstein 1995, 387, our italics). It remains an open question regarding the possible substantialistic or metaphysical implications of this term Being, written capitalized (in German *Sein*): implications that the author would vehemently deny. But in the light of the last quotation, the logical step comes to mind, with which Freud envisioned a possible hypostatization of the libido. So we understand the remark of Walter Riese, when he wrote that in Goldstein's theory, "The organism thus appears as a kind of a higher truth ..." (Riese 1968, 28).

[107]Goldstein (1995, 47).

[108]Goldstein (1940, 142).

Driesch, Hans. 1899. *Die Lokalisation Morphogenetischer Vorgänge, Ein Beweis Vitalistisch Geschehens*. Leipzig: Engelmann.

Driesch, Hans. 1908. *Science and philosophy of the organism*. London: A. & C. Black.

Edelman, Gerald M. 1987. *Neural Darwinism, the theory of neuronal group selection*. New York: Basic Books.

Ferrario, Chiara Elettra. 2008. *Kurt Goldstein: Towards a biologically based epistemology*. MA Dissertation, University of Milan.

Forman, Paul. 1971. Weimar culture, causality and quantum theory, 1918–1927: Adaptation by German physicists and mathematicians to a hostile intellectual environment. In *Historical studies in the physical sciences*, vol. 3, ed. R. McCormmach, 1–115. Philadelphia: University of Pennsylvania Press.

Forman, Paul. 1984. *Kausalität*, *Anschaulichkeit* and *Individualität*; or how cultural values prescribed the character and the lessons ascribed to quantum mechanics. In *The sociology of knowledge*, ed. N. Stehr and V. Meja, 333–348. New Brunswick: Transaction Books.

Gayon, Jean. 1998. The concept of individuality in Canguilhem's philosophy of biology. *Journal of History of Biology* 31: 305–325.

Geschwind, Norman. 1964. The paradoxical position of Kurt Goldstein in the history of aphasia. *Cortex* 1: 214–224.

Gilbert, Scott G., and Sahotra Sarkar. 2000. Embracing complexity: Organicism for the 21st century. *Developmental Dynamics* 219(1): 1–9.

Goldenberg, Georg. 2003. Goldstein and Gelb's case Schn.: A classic case in neuropsychology? In *Classic cases in neuropsychology*, vol. 2, ed. C. Code, C.W. Wallesch, Y. Joanette, and R. Lecours, 281–299. New York: Psychology Press.

Goldstein, Kurt. 1934. *Der Aufbau Des Organismus. Einführung In Die Biologie Unter Besonderer Berücksichtigung Der Erfahrungen Am Kranken Menschen*. Den Haag: Martinus Nijhoff.

Goldstein, Kurt. 1939. *The organism. A holistic approach to biology derived from pathological data in man*; with a foreword by K. S. Lashley. New York: American Book Co.

Goldstein, Kurt. 1940. *Human nature in the light of psychopathology*. Cambridge, MA: Harvard University Press.

Goldstein, Kurt. 1957. The smiling of the infant and the problem of understanding the "Other". *Journal of Psychology* 4: 175–191.

Goldstein, Kurt. 1959. Notes on the development of my concepts. *Journal of Individual Psychology* 15: 5–14.

Goldstein, Kurt. 1995. *The organism. A holistic approach to biology derived from pathological data in man*; with a new foreword by Oliver Sacks. New York: Zone Books.

Greco, Monica. 2005. On the vitality of vitalism. *Theory, Culture & Society* 22: 15–27.

Harrington, Anne. 1996. *Reenchanted science – Holism in German culture from Wilhelm II to Hitler*. Princeton: Princeton University Press.

Heidegger, Martin. 1927. *Sein Und Zeit*. Tübingen: Max Niemeyer Verlag.

Hein, Hilde. 1968. Mechanism and vitalism as meta-theoretical commitments. *Philosophical Forum* 1: 185–205.

Hein, Hilde. 1969. Molecular biology vs. organicism: The enduring dispute between mechanism and vitalism. *Synthese* 20: 238–253.

James, William. 1890. *Principles of psychology*. New York: H. Holt.

Jensen, R.T. 2009. Motor intentionality and the case of Schneider. *Phenomenology and the Cognitive Sciences* 8(3): 371–388.

Jonas, Hans. 1959. Kurt Goldstein and philosophy. *The American Journal of Psychoanalysis* 19: 161–164.

Kaitaro, Timo. 2008. Can matter mark the hours? Eighteenth-century vitalist materialism and functional properties. *Science in Context* 21(4): 581–592.

Kant, Immanuel. 1790. *Kritik Der Urteilskraft*. English Trans: 2001. *Critique of the power of judgment* (trans: Paul Guyer). Cambridge: Cambridge University Press.

Kaplan, Donald R. 1992. The relationship of cells to organisms in plants: Problem and implications of an organismal perspective. *International Journal of Plant Sciences* 153(3): S28–S37.

Köhler, Wolfgang. 1920. *Die Physischen Gestalten In Ruhe Und In Stationären Zustand. Eine Naturphilosophische Untersuchung*. Brunswick: Vieweg.

Koselleck, Reinhart. 1979. *Vergangene Zukunft. Zur Semantik Geschichtlicher Zeiten*. Frankfurt: Suhrkamp Verlag. English Trans. 2004. *Futures past: On the semantics of historical time*. New York: Columbia University Press.

Lenoir, Timothy. 1982. *The strategy of life: Teleology and mechanics in nineteenth century German biology*. Dordrecht: D. Reidel Publishing Co.

Lukács, György. 1955. *Die Zerstörung Der Vernunft*. Berlin: Aufbau-Verlag.

Marotta, Jonathan, and Marlene Behrman. 2004. Patient Schn: Has Goldstein and Gelb's case withstood the test of time? *Neuropsychologia* 42: 633–638.

Maslow, Abraham. 1943. A theory of human motivation. *Psychological Review* 50(4): 370–396.

Mayr, Ernst. 1988. *Towards a new philosophy of biology*. Cambridge, MA: Harvard University Press.

Mayr, Ernst. 2004. *What makes biology unique?* Cambridge: Cambridge University Press.

McLaughlin, Brian P. 2003. Vitalism and emergence. In *The Cambridge history of philosophy 1870–1945*, ed. Thomas Baldwin, 631–639. Cambridge: Cambridge University Press.

Micheli, Gianni. 1970–1972. Caratteri E Prospettive Del Meccanicismo Nel Seicento. In *Storia Del Pensiero Filosofico E Scientifico II*, ed. Ludovico Geymonat, 409–431. Milano: Garzant.

Nicholson, Dan J. 2010. *Organism and mechanism: A critique of mechanistic thinking in biology*. Ph.D. dissertation, University Of Exeter.

Nishikawa, Kiisa C. 1989. "Organismal" vs "Mechanistic" biology. *Herpetologica* 45(4): 473–479.

Normandin, Sebastian. 2007. Claude Bernard and an introduction to the study of experimental medicine: 'Physical Vitalism', dialectic and epistemology. *Journal of the History of Medicine and Allied Sciences* 62(4): 495–528.

Phillips, D.C. 1970. Organicism in the late nineteenth and early twentieth centuries. *Journal of the History of Ideas* 31(3): 413–432.

Riese, Walter. 1968. Kurt Goldstein – The man and his work. In *The reach of mind: Essays in memory of Kurt Goldstein*, ed. M. Simmel. New York: Springer.

Ringer, Fritz K. 1969. *The decline of the Germans mandarins – The German academic community, 1890–1933*. Cambridge, MA: Harvard University Press.

Sacks, Oliver. 1995. Foreword. In *The organism. A holistic approach to biology derived from pathological data in man*. New York: Zone Books.

Simmel, Marianne L. 1968. *The reach of mind: Essays in memory of Kurt Goldstein*. New York: Springer.

Smuts, Jan Christian. 1926. *Holism and evolution*. London: Macmillan.

Spengler, Oswald. 1918. *Der Untergang Des Abendlandes. Umrisse Einer Morphologie Der Weltgeschichte*. München: Beck.

Teuber, Hans L. 1966. Kurt Goldstein's role in the development of neuropsychology. *Neuropsychologia* 4: 299–310.

von Bertalanffy, Ludwig. 1950. An outline of general system theory. *The British Journal for the Philosophy of Science* 2: 134–165.

von Bertalanffy, Ludwig. 1969. *General system theory*. New York: G. Braziller.

von Ehrenfels, Christian. 1890. Über Gestaltqualitäten. In *Vierteljahrsschrift Für Wissenschaftliche Philosophie* 14: 249–292.

von Uexküll, Jacob. 1920. *Staatsbiologie (Anatomie-Physiologie-Pathologie des Staates)*. Berlin: Verlag von Gebrüder Paete.

Williams, Elizabeth. 2003. *A cultural history of medical vitalism in enlightenment Montpellier*. Aldershot: Ashgate.

Wolfe, Charles T. 2008. Vitalism without metaphysics? Medical vitalism in the enlightenment. Introduction to special issue of *Science in Context* 21(4): 461–463.

Wright, John P., and Paul Potter (eds.). 2000. *Psyche and soma: Physicians and metaphysicians on the mind-body problem from antiquity to enlightenment*. Oxford: Oxford University Press.

[illegible]

Press.

[illegible]

S[illegible], Kib[illegible] C. 1986. "Circumstantial" [illegible]

[illegible]

Sacks, Oliver. [illegible] New York: [illegible] Books.

Simmel, Marianne L. 1968. *Reach of mind: Essays in memory of Kurt Goldstein.* New York: Springer.

[illegible]

Spengler, Oswald. 1918. *Der Untergang des Abendlandes.* [illegible] München: Beck.

[illegible]

[illegible]

[illegible]

Chapter 10
The Origins of Georges Canguilhem's 'Vitalism': Against the Anthropology of Irritation

Giuseppe Bianco

Abstract Canguilhem's work in epistemology and in the history of the life sciences rests on a double deontological dogma: a 'vitalism of norms' and, as a consequence of this, a 'normative vitalism'. According to Canguilhem, life consists in the plastic power, proper to all organisms, of creating qualitatively new norms; if life is essentially a potentiality, then this means that the living being is not simply a machine, an assembly of pieces reacting to the environment, but is what modifies and creates it. If the organism is not a mechanism, then it also means that pathology cannot be described as a deficit or a disorder of a supposed normal state, but it is just a qualitatively different norm proper to the living being confronted with an obstacle.

This simple stance also entails an anthropology: both Canguilhem's theory of knowledge and social theory are vitalist insofar as they are deeply rooted in this minimal definition of life. Both technology and society are conceived as external organs (prostheses) created by the human animal and science and morals, value judgments and judgments of fact are a reflection on the reason for the failure of those organs. On a deontological level (methodological and ethico-political), it follows, finally that, from the perspective of life, 'vitalism' as a doctrine is the most 'vital' stance one can adopt both epistemologically and politically.

While not systematically formulated, these ideas are sketched for the first time in *The Normal and the Pathological* and in a series of essays written at the beginning of the 1940s (and later published in *The Knowledge of Life* in 1955). Therefore, if we look at Canguilhem's intellectual trajectory before World War II, and, even more, before 1935, the moment at which Canguilhem begins his medical studies, it seems that Canguilhem was far from being a 'vitalist', even in this peculiar sense, and far from presenting his work as a historical epistemology of the life sciences. On the contrary, he was a harsh critic of vitalism and finalism and

G. Bianco (✉)
Department of Philosophy, University of Warwick, Coventry CV4 7AL, UK
e-mail: g.bianco@warwick.ac.uk

S. Normandin and C.T. Wolfe (eds.), *Vitalism and the Scientific Image in Post-Enlightenment Life Science, 1800–2010*, History, Philosophy and Theory of the Life Sciences 2, DOI 10.1007/978-94-007-2445-7_10,

a strong defender of "transformism" in its strictly mechanistic, Lamarckian version. Faithful to Alain's theory of knowledge and to Auguste Comte's sociology, he strictly distinguished human mind (*l'esprit*) – conceived as the only source of judgments and volition – from the human body, conceived as a machine.

Why such a change? This essay aims to describe Canguilhem's first 10 years of activity, the implicit theoretical framework of his intellectual work, the relation between his anthropology and the doctrines of Alain, Comte and especially Broussais's theory of irritation. Finally, it aims at explaining the social and political reasons at the base of his later vitalist philosophy of life and his implicit rejection of Kantianism and Comtism.

Keywords Alain • Anthropology • Canguilhem • Comte • Epistemology • French philosophy • Irritation • Sociology • Vitalism

1 Introduction – Vital Rationalist? Rational Vitalist?

Mentor of Michel Foucault, Jacques Derrida, Alain Badiou, Gilles Deleuze and Michel Serres, comrade and friend of Jean Cavaillès during the Resistance, belonging to the same generation as Jean Hyppolite, Jean-Paul Sartre and Maurice Merleau-Ponty, heir of Gaston Bachelard at Sorbonne's Institute for the History and Philosophy of Sciences; Georges Canguilhem occupies a key place in twentieth century French philosophy and, furthermore, has deeply influenced the perception we have of the totality of French thought.

Canguilhem is, first of all, known for his studies in philosophy and the history of the life sciences and, more generally, for having constituted a link in the 'chain' of the tradition of 'French epistemology', between, on the one hand, Bachelard and, on the other hand, Michel Foucault, Louis Althusser and the *Cercle d'Épistémologie* (Jacques-Alain Miller, Alain Badiou, etc.…). This image of Canguilhem's position in contemporary French philosophy has been canonized by Foucault in the famous essay "Life: Experience and Science" first published as an introduction to the English translation of Canguilhem's *The Normal and the Pathological.* Here Foucault opposed a Bachelardian "tradition" – a philosophy "of the concept," devoted to the study of science's structure and mutations – to a phenomenological-Bergsonian one – devoted to the subjective analysis of experience – to which Sartre and Merleau-Ponty belonged.[1]

But, before Foucault, this idea of a clear-cut cleavage between two supposed 'traditions' had already been sketched out by Canguilhem himself in a strategic intervention into the notorious polemics between 'structuralists' and 'existentialists.' Consequently, it is not incorrect to say that Canguilhem is at the very origin of this image. In 1967 he had defended Foucault against Sartre's attacks[2] with the famous article, "The *death* of *man*, or, the exhaustion of the *cogito*?"[3]; in this

[1]Foucault (1998).

[2]Sartre (1966, 96).

[3]Canguilhem (2005, 87).

review of *The Order of Things,* Canguilhem tried to protect Foucault from Sartre's accusations – Sartre, "pretending to represent humanism," reproached Foucault with ignoring human praxis and constituting the bourgeoisie's "last barricade" against Marx.[4] In the attempt to prove that the idea of a supposed immorality of 'structuralism' was nonsense, Canguilhem invoked the example of Jean Cavaillès, a mathematician and member of the Resistance who had defended both the primacy of freedom over Nazi barbarism and "the primacy of the concept, the system and the structure" over "the primacy of living consciousness."[5]

Much later, this duality would be revived by Alain Badiou in his *Deleuze,* in certain conferences and, finally, in *Logics of Worlds.* Badiou contrasts the tradition "of the concept," that he calls the "thread of mathematical Platonism" – Brunschvicg, Bachelard, Lautman, Cavaillès, Lévi-Strauss, Lacan and himself – to the "vitalist mystic" one. The latter would be constituted by a Bergson-Deleuze axis that would include figures as different as those of Sartre, Simondon, Foucault and...Canguilhem!...

But there is another surprise that has to be taken into account, which is the object of this essay: Canguilhem's shift from a rigid, rational and mechanistic Cartesianism to a particular kind of vitalism, actually inspired by Bergson and Nietzsche.

On the one hand, contrary to what Foucault argued in his essay, Canguilhem's reflection belongs to the same paradigmatic space as Marxist phenomenologists like Sartre and Merleau-Ponty. He shared with them not only a constellation of problems – determined by a common space of practices and discourses – but also a 'conceptual *Gestalt*' that he was trying to articulate and define in his own way. This constellation was shared by thinkers as different as Sartre, Jankélévitch and Politzer and was opposed to that of the philosophers belonging to the sequence of the "Belle Epoque" (Bergson, Brunschvicg, Alain and the other founders of the *Revue de métaphysique et de morale*).

On the other hand Canguilhem's 'vitalism' made his position completely original. Since the 1940s he had been criticizing the use of biological philosophy as a simple instrument for "a reactionary or counterrevolutionary politics," and as a "suspect excrescence growing over positive science and suitable to serve the most shameful political plans,"[6] but, at the same time, he was one of the few philosophers

---

[4] Sartre (1966, 87).

[5] Canguilhem (2005, 88).

[6] Canguilhem (1947, 324). In "Aspects of vitalism," an essay written during the 1940s and later published in *The Knowledge of Life* (Canguilhem 2008) Canguilhem tries to defend vitalist models in the life sciences, but denounces their use by Nazi ideology (e.g. Driesch, who viewed entelechy as the "organism's Führer"). According to Canguilhem, it is not a question of connivance between ideology and scientific models, but rather of a simple "parasitism of biology": the ideological use of models does not diminish their scientific importance. If "cellular theory" had been inspired by images deriving from the "imaginary of the discontinuous," it does not mean that those images were mere myths. Starting from the second half of the 1930s Marxist philosophers were on the contrary inspired by a strong Cartesianism aimed at opposing all philosophical currents that were supposed to prepare the terrain for nationalist ideology. Politzer denounced Rosenberg's propagandist book *Der Mythus des Zwanzigsten Jahrhunderts* where he pretended to create a "new kind of man starting from a new Life myth."

of his generation who supported a reflection on life based on a peculiar vitalism partially inspired by Bergson and Nietzsche. Canguilhem's vitalism was implicitly opposed to both existential phenomenology – which considered a book like *Creative Evolution* as naively metaphysical – and Marxism – as found in Politzer's[7] and Lukács' work – which had denounced the politically dangerous 'mystifications' embedded in all biological philosophy. In fact, according to the Marxists, Bergsonism and, more generally, all vitalism, had ideologically prepared French philosophy to accept the 'Blut und Boden' Nazi myths, thus participating in a movement directed towards the "destruction of reason."[8]

But even if Canguilhem was very prudent concerning the risks of any possible appropriation and distortion of biological philosophy by reactionary ideologies, he also affirmed its importance. In a 1947 review of Ruyer's book *Éléments de psychobiologie*, bearing the title "Note sur la situation faite en France à la philosophie biologique" ("Notes pertaining to the situation of biological philosophy in France"), Canguilhem praises Bergson and complains about the absence of a tradition of biological philosophy in France, since and because of Descartes's dualism and mechanistic theory of life.[9] If he treated *Creative Evolution* as the "most clairvoyant essay ... to complete the [biological] explication of mechanisms,"[10] what was crucial for him was not simply to "adhere to Bergsonism, but to express a perplexity toward what ... [was] the philosophical task of the moment."[11] In that precise moment the importance of Ruyer's book and of Bergsonian biological philosophy lay in their "interpretation of the fundamental biological phenomena ... starting from psychological models."[12]

But, before 1947, even before the publication of *The Normal and the Pathological*,[13] Canguilhem had already been a very active and prolific writer, publishing several articles in political journals such as Emile Chartier's *Les libres propos* and Roman Rolland's *Europe*. What is extremely striking is that Canguilhem's philosophical ideas were completely different from the ones that he expressed 10 years later. Except for two short articles, he did not write on biology or on problems related to life.[14] And, above all, *his refusal of vitalism was absolute and radical.* In a short review of a book by the biologist Louis Vialleton,[15] ironically titled "La

[7]Cf. Politzer (1967).

[8]Cf. Lukács (1980).

[9]Canguilhem (1947, 323).

[10]Ibid., 332.

[11]Ibid.

[12]Ibid.

[13]Mike Gane points (correctly, but too quickly) to the naivety of the majority of Canguilhem's readers, who consider this author – who has stressed intellectual discontinuities – to be a "relatively non-problematic" figure (Gane 2003, 136). Cf. Gane (1998).

[14]Canguilhem (1929a, b).

[15]Vialleton (1929).

renaissance du vitalisme?"[16] ("The rebirth of vitalism?"), he expressed arguments which were the exact opposite of the ones found in the 1947 article: by stressing the rigorous Cartesian separation between thought and matter,[17] Canguilhem warned against all the dangerous "connections that had been established between life and thought" and against the "metaphors deriving from an imaginative representation of spirit."[18] When he comes to Bergson, Canguilhem accuses him of being "responsible for *Creative Evolution*," a book which, according to him, had permitted "all sorts of confusion"[19] between what belongs to the subject and what belongs to the object.[20] It was precisely because of Bergson that the "new vitalists," such as Louis Vialleton, dared to compare living beings, which are simple machines governed by the law of causality, to mind (*esprit*), which is spontaneity, will, creation and the condition of all experience.

What happened between 1930, when Canguilhem wrote his essay against vitalism, and 1947, when he wrote in favor of it, to provoke such a change? Answering this question not only sheds light on the genesis of Canguilhem's 'biosophy', but also on the apparently unjustified "passage through medicine" that he undertook between 1936 and 1943 and that would later transform him into one of the tutelary figures of the "French epistemological tradition."[21]

The main problem that one has to face in this quest is Canguilhem's absolute silence: he never talked explicitly about his medical training, and he only very rarely mentioned his intellectual itinerary before the publication of his medical doctoral dissertation, *The Normal and the Pathological*. In a letter sent to Michel Alexandre's wife at the end of the 1960s, he simply admits that he did not regret the "writings or acts expressed in the '*Libres Propos*' collection," the journal in which he wrote between 1927 and 1934. But he added that they were just the expression of a "starting position" (*position de départ*) with respect to where he later took his mark. Canguilhem's position had changed – he wrote – "on a few fundamental points."[22] Now, what are those "fundamental points"?

---

[16] Canguilhem (1929a).

[17] "Concerning thought one can read Descartes, who will explain the separation of thought and extension," Ibid., 139.

[18] Ibid.

[19] Ibid.

[20] "According to Vialleton it is easy to admit that thought is in things and creation in the object.... Biologists do not know very well what is thinking, and there's no psychologist that could teach them, insofar as psychologists are learning from biologists. For a long time, in psychology thinking is considered as ... 'a thing like other things [*une chose parmi les choses*].' This is why a biologist cannot take into account thinking as a factor without making great mistakes. And calling it *hormé* doesn't change the situation, as do Monakov and Morgue, who are quoted by Vialleton," Ibid., 138.

[21] Canguilhem wrote to the Clermont-Ferrand faculty of medicine in 1936, and defended his thesis in 1943.

[22] Canguilhem (1968, 48).

2 Education and Action

As usual, to understand a theoretical change on some "fundamental points," one has to first look at the "practical" change. The discrepancy of Canguilhem's positions during the 1930s and 1940s is also tied to a profound discrepancy between two different "intellectual styles" or two different models of philosophical activity that he practiced. As a matter of fact, the Canguilhem who, beginning in 1950, becomes known as a philosopher of the life sciences and one of the most important representatives of the French epistemological tradition, was, from 1927 to 1934, the most faithful pupil of a famous neo-Kantian and pacifist philosopher, Emile Chartier, better-known under the pseudonym of Alain or "the man" (*l'homme*). Canguilhem was considered as nothing less than the "official heir of the master's thought."[23] It is very difficult to briefly summarize the importance that, in the post-war period between 1920 and 1930, Alain had for the circle of students gathered around him in his preparatory class at the Henri IV lycée in Paris. It is equally difficult to describe the complex network of pacifist politicians and intellectuals that surrounded Alain such as George Demartial, Romain Rolland and Félicien Challaye, and critics from the *Nouvelle Revue Française* like Paul Valéry. We can restrict ourselves to the idea that Alain's importance inside the intellectual field was enormous and he embodied a very precise idea of philosophy, of the educator's and the philosopher's role, as well as of citizens' rights and obligations in regard to the institutions of the Third Republic. Canguilhem is at the center of this complex network formed by other networks: he is Alain's pupil at Henri IV, a pacifist agitator at the Ecole Normale Supérieure; he collaborates frequently with Alain's journal, *Libres propos*, that he edits starting from 1931; he periodically writes in Romain Rolland's review, *Europe*; he follows the debates promoted by the League of Human Rights and by the Committee of Anti-Fascist Intellectuals promoted by Alain together with communist intellectuals such as Paul Langevin.

But, starting in 1936, Canguilhem disappears from this context: one cannot find a single article by him in the *Libres propos* or *Europe*. That year, a 31 year-old Canguilhem begins a new course of study at the faculty of medicine that will lead him, in the spring of 1943, to present his doctoral dissertation, titled *Sur quelques problèmes concernant le normal et le pathologique* (*On Some Problems Pertaining to the Normal and the Pathological*). In the meantime, he fights the "phony war" and he engages in the partisan war, becoming the paradigm of the *Résistance* philosopher, the fighting philosopher. Canguilhem's disappearance from Alain's circles and, more generally, from directly politically engaged circles,

[23]The description given by J.-F. Braunstein is very clear: "Canguilhem writes several essays against war and militarism. In 1935, he is at the origin of a normaliens' petition against the Paul-Boncour military law and, in 1927, on the occasion of the Ecole normale's annual review, he mocks the military preparation. In 1931, he reacts against the 'enregimentation of intellectuals'" (Braunstein 2000, 11).

has often been explained in a simplistic manner: Alain's pacifist politics would have been powerless in the face of the ascent to power of the fascists: "It's impossible to negotiate with Hitler" Canguilhem once said, referring to his master's pacifist politics.

This 'Alainist' past has always been treated as a kind of 'pre-history', flatly separated from the theoretical content of Canguilhem's supposed 'real work' that follows *The Normal and the Pathological*. The leading thesis of this book is that the pathological state of an organism *differs qualitatively from the state that we call 'normal.'* In his Nietzschean theory of vital values – which is not the object of this essay – Canguilhem opposes the thesis concerning the nature of disease proposed by the positivist doctor, Victor Broussais.[24] Broussais, and especially his famous *De l'irritation et de la folie* (*On Irritation and Madness*) had a huge influence on the whole of nineteenth century sciences of man: quoted or not, he inspired the works of Claude Bernard, psychologist Théodule Ribot, but most of all the sociologists Auguste Comte and Émile Durkheim. Broussais's thesis – that Comte was the first to call the 'Broussais principle' – affirms the identity of the normal and the pathological, their purely quantitative difference, that is to say that one can be more or less sane, according to a range. What I would like to do in the following pages is link Canguilhem's political rupture with pacifism to the theoretical rupture in his philosophy and the discussion of this 'Broussais principle', and the consequent emergence of a 'vitalist' theory of norms.

Let us take as our point of departure Canguilhem's political activism during the 1920s and his pacifism. His journalistic or direct political interventions[25] aimed at supporting the right to conscientious objection, stressed the real causes of the First World War, fought against the demonization of Germany, denounced the political and social consequences of an overrated victory, and studied the political developments determined by the treaty of Versailles. Most of the reviews and short articles he wrote in *Europe* and in the *Libres Propos* concerned books that dealt with World War I. Canguilhem's explanation of war paralleled that of Alain: social inequalities and the unequal distribution of powers provoke passions, which, in turn, divert man from his rational nature. If, according to Canguilhem's formula, borrowed from Alain, "all power corrupts all leaders"[26] ["tout pouvoir corrompt tout dirigeant"], pushing them to violence, and if violence cannot but provoke more violence, then power, which may be the cause of passions, has to be controlled through the use of judgment, which is the expression of rationality.

When she uses her judgment with the aim of reducing passions and protecting peace, the 'radical'[27] philosopher has to take into account two aspects of reality: (a) the structure and the functioning of the body, which determines the passions

---

[24]For Broussais, cf. Braunstein (1986).

[25]For an accurate (yet not exhaustive) list of Canguilhem's published writings, cf. Camille Limoges's bibliography in Canguilhem (1994, 385–454).

[26]Canguilhem (1996, 90).

[27]I am taking the adjective in the sense Alain gave it.

(physiology), (b) the sociological laws regulating life in common (sociology). Beside this, the radical philosopher has to communicate this wisdom to the citizens through pedagogical exercises, which are to be practiced inside the educational institutions (high school and university), but also within public space.

Canguilhem's education and activity followed these two lines.[28] On the one hand he read the most common manuals of physiology (it's no accident that Canguilhem signs a lot of articles with the pseudonym of "G.C. Bernard"[29]); on the other hand (like two of his alainist friends, Georges Friedmann and Raymond Aron) he collaborated with the École Normale Superieure's Centre de documentation en sciences sociales.[30] This centre was directed by Célestin Bouglé, a sociologist and a philosopher, an "ambivalent Durkheimian"[31] and one of Alain's closest friends.[32] Canguilhem's DES dissertation,[33] supervised by Bouglé, dealt with Comte, who was one of Alain's most admired and quoted authors.[34] The subject Canguilhem chose, *Théorie de l'ordre et du progrès chez Auguste Comte: Étude sur la méthode statique et dynamique*,[35] dealt with one of the problems – that of the nature of historical time – at the center of intellectual preoccupations during the 1920s in the aftermath of World War I.

Canguilhem's cartography of the sciences and his philosophical *habitus* was entirely alainist. On the one hand philosophy, conceived as a reflexive and critical knowledge, which is grounded on the regulative ideas of truth and justice. On the other hand physiology and sociology, which are crucial if one wants to understand the mechanism of *affection*, the manner in which passion diverts rational judgment. The denunciation of passions originated by power and the importance given to mind (*esprit*) as a point of dissemination for value and factual judgments are the two

---

[28] See the lecture notes written between 1922 and 1927 and preserved in the *Fonds Georges Canguilhem* of the *Centre d'archives de philosophie, d'histoire et d'édition des sciences* in Paris, pressmark GC. 5. 4.

[29] The only essay concerning medical problems is a review of the book by the psychoanalyst and doctor René Allendy, *Orientation des idées médicales*. Canguilhem (1929b).

[30] As a rider to this, cf. Mazon (1983).

[31] Cf. Vogt (1979). François Souillé has called Bouglé a "mediator" between the idealist intellectualism of the *Revue de métaphysique et de morale* and Durkheimian sociology (Cf. Souillé 2009, 232). Cf. also Logue (1979, 159); A. Policar, "Sociologie et morale: la philosophie de la solidarité et Célestin Bouglé," in Ferréol (1997, 287–320), Policar (1999, 2000), Spitz (2005).

[32] Bouglé's influence on the students formed at the École Normale (Aron, Canguilhem, Friedmann and Weil) has been stressed by Sirinelli, who, however, did not emphasize his friendship with Alain (Sirinelli 1984).

[33] The DES, acronym for 'diplome d'études supérieures' is the dissertation that French students were supposed to write at the end of their third year at the university.

[34] The nature of the relations between Alain and Comte have never been seriously studied. Cf., e.g., the quite disappointing analysis by Jacques Muglioni (1987) and Michel Bourdeau (2004).

[35] Canguilhem (1926). Starting from 1923 and until the completion of the dissertation, Canguilhem wrote almost 300 pages on Comte (Cf. Canguilhem 1923–1926).

invisible axes around which all of Canguilhem's productions gravitated between 1927 and 1934.

But something changes in Alain's circles at the beginning of the 1930s. Facing the sudden appearance on the international scene of the national-socialist party, which displayed a very aggressive foreign policy, Félicien Challaye, one of Alain's closest friends, published a pamphlet titled *Pour une paix sans aucune réserve* (*Towards a Wholehearted Peace without Reserve*).[36] Challaye expressed a categorical refusal of any kind of war, even of a war of defense (as World War I had been declared by French propaganda), but, at the same time, he supported the idea of a legitimacy of oppressed people's revolts against their governments. Challaye argued that it is possible "to be a pacifist and to participate in an insurrection against a criminal ... government."[37] Later, the German political situation changed even more radically. At the end of 1933, at the moment of the uncontrollable success of the Nazi party – which constituted both a threat to German democracy and a potential threat to France – Challaye used, in a second pamphlet entitled *Pour la paix désarmée même en face de Hitler* (*Towards an Unarmed Peace even when faced with Hitler*), the slogan "better foreign occupation than war." Two months later, in February 1934, in the unstable political situation that followed the economic crisis of 1929, the 'fascist' Leagues, composed of war veterans and far right militants gathered in the Place de la Concorde denouncing the 'Republic of scandals' and threatening the government. For the first time since the beginning of the Third Republic, France seemed at risk of a *coup d'état*. Alain, with other socialist and communist intellectuals, formed the Committee of Antifascist Vigilance. Canguilhem not only joined it, but also wrote, anonymously, a short essay in a series of brochures promoted by the group. Canguilhem's *Le fascisme et les paysans* [*Fascism and Peasants*] focused on the way in which fascism tries to take advantage of peasants' interests in order to acquire power.[38]

During the period that precedes 1935, Canguilhem's theoretical and political positions seem slowly to be changing. In 1933, in an article entitled "Pacifisme et revolution," despite his pacifism concerning foreign policy, he defends the right to revolt affirmed by Challaye. But, already in *Le fascisme et les paysans* he implicitly and politely disagrees with the theoretical root of Alain's and Challye's political positions. At that moment Canguilhem is partially following some Marxist analyses and those of the *Annales* School: he affirms the epistemological imprecision of the concept of peace and affirms the necessity to act, even violently, against the oppressing governments.

The following year, Canguilhem's short review of Alain's *Les Dieux*[39] represents the last homage paid, before the war, to the man that he described as "mon maître

[36] Challaye (1932, 8).

[37] Ibid, 3.

[38] Canguilhem (1935a).

[39] Canguilhem (1935b).

Alain."[40] After having become a professor in the *khâgne*[41] of Toulouse's Fermat High School, Canguilhem continued his antifascism militancy, but he totally abandoned the Comité and the *Libres Propos'* editorial board. More generally, he abandoned Alain's group of pacifist intellectuals. The new socio-political mutations were hardly explicable in terms if one followed Alain's politics, grounded in a wholehearted pacifism. According to Alain one has to refuse all kinds of violence when confronted with fascism. Following a motto first formulated in Alain's *Le Citoyen contre les pouvoirs*, composed during the 1920s, "one must not change the powers" but "moderate them (*les assagir*)." On the contrary, according to Canguilhem, the pacifist thesis was impracticable because the political events had shown that what appeared to be a state of peace was nothing but the concealment of a social war. Like many of his contemporaries (Robert Aron and Arnaud Dadieu of the *Ordre nouveau* journal, Georges Lefebvre of *Avant-poste*, Georges Bataille of *Contre-attaque*, and Simone Weil), Canguilhem tried, starting with *Le fascisme et les paysans*, to elaborate new conceptual schemas likely to configure a new front.[42] In 1936, he began his medical studies, which led to his doctoral dissertation, *The Normal and the Pathological*.

Canguilhem does not say much about his medical studies and about their original motivation. At the beginning of the introduction to the *Normal and the Pathological* he simply wrote that he was expecting from medicine "an introduction to some concrete human problems"[43] and, during an interview with his former students Jean-François Braunstein and Jean-François Braunstein, he admits that he was searching, in medicine, for "practical knowledge." In 1935, Canguilhem's problem was without any doubt that of an effective political action, which would have had to follow the Comtian motto: "know in order to foresee and foresee in order to act" ("*savoir pour prévoir et prévoir pour pouvoir*"). If, as Canguilhem wrote in the first lines of the first chapter of *The Normal and the Pathological*, "to act, it is necessary at least to localize,"[44] then medicine appears as a "technique": it consists in an operative knowledge that aims at action. Without any doubt, Canguilhem was hoping to find in medicine some elements useful to ground his philosophy (and politics) of action. Actually, in the *Normal and the Pathological*, he admits that his objective was that of "integrating into philosophical speculation some of the methods and acquisitions taken from medicine."[45]

---

[40] Canguilhem (1995, 18).

[41] Hypokhâgne and khâgne are, respectively, the first and second year of preparation (*classes préparatoires*) to the entrance examination for the École normale supérieure.

[42] Canguilhem is interested in the German situation starting from the early years of the 1930s. In a review of a book by Pierre Viénot, *Incertitudes allemandes*, he stresses "the German hesitation is relative to an entirely relative conception of Civilization" (Canguilhem 1931, 514). Quoting Valéry's words, "We, civilizations, we now know that we are mortal," he stresses how they can "resume the dramatic state of German consciousness," which explains Hitler's success.

[43] Braunstein et al. (1998, 8).

[44] Canguilhem (1991, 39).

[45] Braunstein et al. (1998, 8).

But why did Canguilhem choose *precisely* medicine?[46] To understand this passage we have to go back to Alain.

3 Alain and Broussais's Theory of Irritation

I tried to briefly explain that, according to Alain, philosophy is an ethics, which aims at the realization of a wisdom by purifying the mind of the passions that affect it and divert rational judgment. According to the famous formula that opens one of Alain's first books, the *Quatre-vingt et un chapitres sur l'Esprit et les passions*, philosophy consists in a "solid judgment opposed to death and disease, to dreams and disillusions." If *false* is the result of a malfunction of mind, caused by passions, *disease* is a malfunction of the body, caused by an external aggression. What could the relation be between those two malfunctions? And how could man conquer disease with his judgment or his reason alone? To answer these questions, crucial for philosophy conceived as a form of wisdom, means to answer the key question of the relations between the mind and the living body, but also between the two disciplines that study them: philosophy and medicine.

A careful examination of Alain's "propos,"[47] and especially the ones composed during the 1910s and 1920s, reveals that he paid a great deal of attention to medicine, both to physiology (blood circulation, the nervous system, respiration, etc.) and to pathology, and most of all to mental pathology. What is commonly named 'madness' is a disease concerning both the body and the mind, thus it constitutes the privileged theater in which to study their relations. Alain affirms, for example:

> all knowledge in anatomy or physiology is good .... The documents about those topics abound and one can choose whatever one wants, provided that now and again he reads with

---

[46]The answers that had been given to this question are not entirely convincing or they simply do not constitute a real answer. In his reconstruction of Canguilhem's itinerary Jean-François Braunstein rightly argues that "the specialization, the diversity of Canguilhem researches seems to be the exact opposite of Alain's indifference for the particular" (Braunstein 2000, 16). Dominique Lecourt advocates the view that Canguilhem's career is not that of a 'canonic' epistemologist, but he does not explain why medicine had constituted the best "unknown material" (*matière étrangère*) (Lecourt 2008, 33). The idea that philosophy is a discipline that has to deal with an "unknown material" ("Philosophy is a reflection for which all unknown material is good, and we would gladly say, for which all good material must be unknown," Canguilhem 1991, 33) came from Brunschvicg. In an essay from the early 1920s he argues that "philosophy does not have its own matter; because its own matter is mind [*esprit*], how it appears when one studies history, science and esthetics; it is on this mind [*esprit*] that it uses its reflection in order to make its unity appear" (Brunschvicg 1921, 123).

[47]The 'propos' (that we can translate as remarks) is the chapter-form in which Alain structured almost all his books (the reader *Éléments de philosophie* is one of the exceptions). Those short chapters (two or three pages-long) are supposed to be read independently from each other and take as points of departure observations concerning man and society. They use a common language insofar as they are supposed to be addressed to every man, every citizen who is in search of wisdom. At the beginning the 'propos' were published in popular journals (such as *La Dépêche de Rouan*) and, nearly always, they bear, in addition to their title, the date they were written.

careful attention the best physiology treaty, so that the details will take their order following the indivisible human form.[48]

What Alain calls "the best physiology treaty," is the one written by the nineteenth-century positivist François Broussais: *De l'irritation et de la folie*. What Alain appreciates most in this treatise is its theory of disease, which is tied to his theory "of feelings, passions and temperaments."[49] In a draft of his intellectual biography he sent to Aline Textier in 1937, this book is included in the list of the ten most important books for the philosopher.[50] Moreover, Alain very often makes allusions to it, and calls its author the "illustrious Broussais" or "the great Broussais." Alain's idea of pathology is faithful to Broussais': if we consider man as an animal, namely as a living being subjected to the same biological laws to which each organism is subjected, disease does not consist in a configuration of the organism *qualitatively* different from the one proper to the normal state. Disease is a state of the organism, which is just *quantitatively* different from the normal one. According to Alain, disease consists in an *irritation*. This word appears in almost all of Alain's books, sometimes dozens of times. He even devotes a chapter to it ("Irritation," composed on the 5th of December 1912) in his *Propos sur le Bonheur*.[51] Naturally, at the same time, the concept of irritation is directly inherited from Broussais. This concept was already widespread in medical milieus during the second half of the eighteenth century,[52] but its meaning had been ratified in *De l'irritation et de la folie*. Irritation consists in a disproportionate reaction to an exterior excitation; this reaction causes a disorder in the organism and a consequent trouble. Irritation appears when the intensity of the stimulation of the organic tissues by what Broussais calls the "modifier" (*modificateur*) provokes a reaction "above the normal level." Excitability or irritability is the "tissues' faculty of moving following contact with a stranger body,"[53] then called the modifier. In *The Normal and the Pathological*, Canguilhem later argued "Broussais saw the vital primordial fact in excitation. Man exists only through the excitation exercised on his organs by the environment in which he is compelled to live."[54]

In *Les passions et la sagesse*, Alain writes that "what Broussais called irritation emerges around a lesion and propagates inch by inch; all our actions irritate us this way." [55] In the *Esquisses de l'homme* he adds that, even if it's not true that "there is no difference" between the normal state and the pathological one, it's always a question of a "difference of degree" and Broussais is responsible for "this brilliant

[48] Alain (1991, 327).

[49] 'Propos' originally included in *Sentiments, passions et signes* (1926), republished in Alain (1970, 520).

[50] Cf. Alain (1960, 2). This interest in Broussais and, more generally, for the medical and physiological treaties, had been inherited from Alain's "maître," Lagneau. Lagneau was fond of quoting excerpts from Broussais and Bernard during his lectures (see Ragghianti 1993, 71).

[51] Alain (1956, 139–40).

[52] Cf. Starobinski (2003).

[53] Broussais (1828, 22).

[54] Canguilhem (1991, 54).

[55] Alain (2002, 399).

idea which concerns health and disease: there's always a small difference of degree between them."[56] A sick man, he adds in *Eléments de philosophie*, 'is a man who's not able to adjust himself to the physical milieu, who does not govern his machine,' he is a man who is *diminished* because of his exterior actions.[57]

It is clear that Alain's idea of disease, conceived as a lower degree of health, is attached to a Cartesian idea of the body,[58] conceived as a "composite machine,"[59] as a "living mechanics"[60] which is strictly separated from mind. This mechanist conception, ratified in physiology during the nineteenth century, is directly opposed to the vitalist one, which conceives the body as the theater of fights between opposed forces. Canguilhem explained, in *The Normal and the Pathological*, that this latter conception, prior to the Pasteurian age, was attached "to an ontological conception of disease."[61] In a passage from *Vigiles de l'esprit*, Alain actually writes:

> Man is made of a brain, heart, two lungs, and other organs. Good health and disease apparently depend on quarrels and accords between those tragic protagonists…. But the medical geniuus … dissolves those tragic protagonists.[62]

In the *Système de Beaux-Arts*, the body is described as a "mechanism which *irritates* … itself following its own peculiar laws without caring about our own judgments." The body is a "little kingdom which is too close to us; everyone relies too much on it."[63] Thus, to efficaciously accomplish his critical and pedagogical task, the philosopher has to know "following a summary physiology, those bizarre regimes of movement and rest which are totally independent of our will and he has to transform them with compensatory actions."[64]

One of the main laws governing the body is that of "spontaneous movement" or "reflex:" "excitation radiates through the centers and quickly conquers the whole body."[65] "If I get pricked, if I get burned, if I make a false step – writes Alain – my

[56] Alain (1964, 34).

[57] Alain (1991, 323). Italics added by the author.

[58] Olivier Raboul rightly remarks that Alain had "never abandoned Cartesian dualism": "To tie the involuntary to the animal-machine and to reduce the ego to the thinking subject, this is exactly Descartes's legacy …: our screams of fear, anger, hate, and, in general, of sufferance, do not have more sense than those of a dead chicken" (Raboul 1968, 155). Alain will always consider "Cartesian mechanism" as "the universal type of explanation" (ibid., 125) and, through the theory of the man-machine, he will always consider the human body, as Descartes did as a "simple mechanism, refusing any idea of a vital force …. Thus physiology takes into account and will always take into account nothing but mechanisms without thought" (Ibid., 162).

[59] 'Propos' originally published in *Propos sur le bonheur* (propos du 23 mars 1922, entitled 'Médicine') republished in Alain (1956, 379).

[60] 'Propos' originally published in *Mars ou la guerre jugée*, republished in Alain (2002, 609).

[61] Canguilhem (1991, 39).

[62] 'Propos' originally published in *Vigiles de l'esprit*, republished in Alain (1956, 786).

[63] 'Propos' originally published in *Système des beaux-arts* (1922), republished in Alain (1958, 229). Italics added by the author.

[64] Ibid.

[65] Ibid.

entire body immediately trembles."[66] Thus, if it is not controlled, a traumatic event can led to an upheaval that propagates throughout the entire organism. Alain compares this situation to a crowd's spontaneous movements, which reacts with "an alert, a tumult, an alarm, and an effervescence."[67] Alain provides multiple examples of an irritation, but the one that is the most frequent is the one of the coughing caused by a bad deglutition:

> When one swallows the wrong way, a great tumult affects the body, just as an imminent danger was announced to all its parts; each one of the muscles pulls in its way, the heart participates in the tumult; it's a kind of convulsion.... Most people cough as they scratch; they are victims of a kind of fury. This generates crises which make people tired and irritated.[68]

Alain uses also others examples, inspired by Descartes' *Traité de l'homme* and *Traité des passions*: a stitch that causes someone to cry out, a muscular traumatism provoking a cramp, the introduction of an insect into the eye inducing a wild scratching. In all of these cases, a being endowed with intellect and free will, a being which, consequently, is able to intervene rationally in his "machine," is able to redress it and to stop the propagation of the irritation. According to Alain, the principal remedy consists in diverting the attention that we automatically direct towards the concerned part of the body. In fact this attention is the cause of the excessive influx of blood to the irritated part and of the consequent functional disorder and, finally, of the pathology. If, after having introduced some saliva into the airways, instead of coughing, one relaxes the muscles and expulses the saliva,[69] the "tumult" stops; if, after an intense muscular tension of the leg, instead of tensing the leg more even, concentrating the attention on the irritated muscle, one places the foot on the floor, one is "immediately healed;" if, instead of scratching the eye, one directs one's gaze towards the extremity of the nose, one is freed from tears.

Thus, man, who, contrary to the animals, is a rational being, is able to intervene in the process of irritation and to find a remedy to a pathological or potentially pathological situation. In *Vigiles de l'esprit*, Alain describes the birth of medicine as the moment man realized he was able to *act* upon the source of the pathology and developed knowledge of the body-machine (physiology) that excluded all metaphysical characteristics, all vitalist interpretations and all organic teleology. In modern medicine, conceived as *technique*, man-mind (*homme-esprit*) acts on the man-machine and manages to cure it.

[66] 'Propos' originally published in *Vingt leçons sur les beaux-arts* (1931), republished in Alain (1958, 515).

[67] 'Propos' originally published in *Système des beaux-arts*, republished in Alain (1958, 221).

[68] This excerpt is taken from the 'propos' "Irritation," originally published in Alain's 1928's *Propos sur le bonheur*. Now in Alain (1958, 139).

[69] "What can we do [when we cannot stop coughing]? Can we avoid following and undergoing all those reactions? .... The main cause of those tensions and seditions ... is precisely the fact that we absolutely don't know what to do. In our example, what one has to do, is precisely to relax the whole body, and, especially, instead of inhaling vigorously, which would cause the disorder to deteriorate, expel the little parcel of liquid which was introduced into the wrong pipe. This means, also, chasing away fear which, in this case as in other cases, is entirely harmful" (Ibid., 140).

4 Passions and the Body Politic

But then, it is the very 'spiritual' nature of man, which is likely to provoke the opposite effect and cause an auto-irritation. This auto-irritation, the passions, is the cause of mental diseases. Passions are also irritations provoked by a modifier: faced with an excitation coming from the exterior, instead of thinking calmly, the passionate man (also called *irritable*) reacts instinctually, mechanically, directing all his attention towards the interested part, which gets irritated. This disorder is not located in man's mind, but in the body, which – Alain writes – "is a cause, not an effect." "We're made in such a way that all of our emotions are misfortunes, and this happens because of this law of irritation ... which governs them."[70] Passions provoke, in turn, errors of judgment, "errors of interpretation": "The one who doesn't understand – writes Alain in *Le citoyen contre les pouvoirs*[71] – gets irritated. The one who gets irritated is not able to hit the nail on the head." These "errors of interpretation" are linked to a series of bad behaviors, which are likely to cause a series of habits, which can become mental diseases. For instance, according to Alain in *Propos sur le bonheur*, when one is depressed and 'sees everything as black', his opinions are not rational, but they are "gut reactions" [*opinions d'estomac*],[72] that is to say that they are caused by passions. This is the reason why 'bad temper' has to be treated as 'a disease', which is comparable to the pathological result of a painful calf cramp or to excessive lacrimation.

Thus, the word 'irritation' has a double meaning that Alain never ceases to emphasize, it indicates both a modification in the organism and the strongest of all the passions.

> A man who scratches his scars ruins his doctor's work; but there's more than just one way to scratch himself. When our attention is directed towards a certain part of the body, the blood goes in that direction; this is the reason why the liar blushes. We can thus understand that the best way to prevent someone to cough is not that of thinking about the throat. *Thinking about our own illness can do nothing but irritate it more. This word, irritation, has a double sense, which is admirable.*[73]
>
> Everyone would have to think about the meaning of the word irritation. Because of the revelatory nature of language, it designates also the most violent passion. I don't see any difference between a man who loses himself in a fit of anger and a man who has a coughing fit. Fear is a kind of anxiety proper to the body; something we cannot fight. The worst error, in this case, is to use thought to serve the passions, to abandon ourselves to fear or to anger with a wild enthusiasm.[74]
>
> Anger is one of the most common passions, a paroxysm .... We can see how man is able to forget his interests .... It is very common that anger, even when it is without good reason, pushes us to extravagant actions – such as beating, breaking and even insulting things. I would dare to say that the most profound anger is the anger caused by the fact of being

[70] 'Propos' originally included in *Vingt leçons sur les Beaux-Arts*, republished in Alain (1958, 515).

[71] Alain (1970, 401).

[72] Alain (1958, 376).

[73] 'Propos' originally included in *Esquisses de l'homme*, republished in Alain (1956, 575).

[74] 'Propos' originally included in *Propos sur le bonheur*, republished in Alain (1956, 144).

> angry and to know that we are going to abandon ourselves to it, that we are going to feel it rise as a physical storm. *The word 'irritation', in its double meaning explains it* .... The child screams more and more loudly because he gets irritated at the fact of crying, as others are irritated by the fact of coughing.[75]
>
> Irritation is the most fundamental fact, as Broussais was the first to notice; the double meaning of this wonderful word teaches us much more than a treatise on the passions constructed in a Machiavellian way.[76]

If we take into account man's capacity to irritate himself, the task of the doctor – a profession that Alain, like Comte, admired – is particularly difficult.[77] The doctor has to try to become a *philosopher* to understand the logic of irritation and its relation to the mind, to understand the relations between the mind and the body. As a matter of fact, announcing a disease, thus provoking the concentration of the patient's attention on the part of his body concerned, the doctor can provoke others diseases, what Alain calls "imaginary diseases." These diseases are not really imaginary because they're *concretely* caused by passions, and so by an irritation: "by fear ..., anxiety, or anger, [thus they] are absolutely not imaginary according to the proper meaning of the word." Imaginary diseases consist in "a series of concrete and real movements of one's body, sometimes very violent ones; these movements disturb circulation, digestion, secretions, as we can notice in lacrimation. Each one of us knows that a man can harm himself and even get destroyed by his inconsiderate movements."[78] This is why the doctor's profession is difficult, "always confusing and ambiguous," because, concludes Alain, "medicine [is] similar to politics, it can make strides only through the works of those who do not practice it."[79]

---

75 'Propos' originally included in *Mars ou la guerre jugée*, republished in Alain (2002, 548).

76 'Propos' originally included in *Sentiments, passions et signes* (1926), republished in Alain (1970, 531).

77 Cf. e.g. the 'propos' originally included in *Propos sur le bonheur* and entitled, "Sur la médecine," Alain (1956, 375).

78 "Thaumaturgie et médecine," 'propos' originally included in *Esquisses de l'homme*. Alain (1956, 575): "If one studies fear and uneasiness from a physiological perspective and very carefully, one would discover that they are diseases that add to others and accelerate their development, so that the sick man who knows that in advance, thanks to the doctor, will be twice sick. I admit that fear will help us to fight the disease through cure and remedies; but what kind of remedies will cure us of the fear? This vertigo that suffocates us when we're high up is a real disease which derives from the fact that we're mimicking the fall and the desperate movement of a falling man. This trouble is an imaginative one. The diarrhea of the student too; thus the fear of giving the wrong answers is as powerful as castor oil. Try to imagine the effects of a continuous fear. To be prudent towards prudence, one has to take into account the fact that the movements of fear will naturally worsen the trouble. The one who fears not falling asleep will not be well-disposed towards sleeping, the one who fears about stomach-ache will not be well-disposed toward digestion. Therefore it would be better to mimic health than disease. This gymnastics is not well known in its details, but one can bet that kindness and benevolence are tied to health, following this theorem according to which the signs of health are nothing but the movements conforming to health; conversely the good doctors are the ones who ask you, as usual: 'How are you?', and then do not even listen to the answer," "Médecine," 'propos' originally included in *Propos sur le Bonheur*, Alain (1956, 375).

79 "Thaumaturgie et médecine," 'propos' originally included in *Esquisses de l'homme*. Republished in Alain (1956, 575), italics added by the author.

This reflection concerning the nature of disease has a direct consequence on a political level: just as philosophy – which aims to liberate the mind from the passions and to teach man how to judge better – has to play on this ambiguous terrain in which the laws of the body influence the mind, and the other way around, in the same way politics has to take into account the double feature of man. Politics is a *technique* enabling man to intervene in society following the laws formulated by a precise science, *sociology*. Society and humanity, life in common and passions are tightly linked – they are inseparable. Alain sticks to Comte's idea according to which "there's no society but the human and there's no thought but in society." Both governors and governed are at the same time rational beings and biological bodies likely to be affected by passions provoked by envy, thirst for revenge or for power.

In his writings – especially starting from 1912, when he begins reading Comte intensively, Alain multiplies the analogies between the human body ("le corps humain") and the body of the state ("le corps de l'État"). The double use of the verb "to govern" (*gouverner*) – indicating the action of man's intellect on his and other men's bodies – is the main signal of that analogy. The basic idea is that society, composed of human animals, obeys a series of specific laws, but which are nevertheless subjected to biological laws which regulate the mechanism of the passions; consequently society, as with the human body, is likely to be modified through intellect.

This parallel between society and the human body concerns both the "normal" and the "pathological" state. In his book *Eléments d'une doctrine radicale*, Alain remarks that human groups, "nations, corporations, congregations, churches, cooperatives, all kind of bodies are subjected to all kind of diseases, anemia, convulsions."[80] Society, writes Alain, is most of all affected by "mental" diseases, especially neurasthenia, feminine and "Bergsonian" diseases, consisting in an excess of sensibility. This excessive sensibility prevents man from acting. In the eponymous chapter of *Mars ou la guerre jugée*, Alain argues "the State is very often neurasthenic.... This particular disease seems to me ... peculiar to all States; this is the reason why this great body is always sad and often dangerous."[81] This is why, according to Alain, the wise politician has to learn a lot from doctors and from medicine: "the State would have to show at the same time wisdom and a good knowledge of medicine."[82]

How is it possible to apply medicine to the State's governance? I have shown that a violent reaction of the body caused by a 'modifier' could cause an irritation that ends in a disease. This disease does not change anything in the structure of the organism, but alters its functioning, and lowers the level beyond the "normal" one; in the same manner, a violent movement of one of society's parts – as happens during popular uprising or wars – cannot change its structure, but only generate other passions. This wave of passions is followed by the tightening of society's

[80] 'Propos' originally included in *Eléments d'une doctrine radicale*. Republished in Alain (1993, 431).

[81] Alain (2002, 664).

[82] 'Propos' originally included in *Propos sur le bonheur*. Republished in Alain (1956, 109).

body, which very often coincides with the instauration of less liberal powers. It is just after a long period of 'disease' that the situation of 'normality' is restored.

If both wars and revolutions are nothing but irritations of the body of the State, then the role of the good citizen is not that of fighting against powers to depose them – as a man who contracts his leg following a cramp would do – but that of opposing all abuse of power, trying first to understand its logic. The Alainian motto is a result of these considerations: "do not change the powers, but rather render them rational [*les assagir*]." All powers are corrupted and corruptor and they constitute a potential source of irritation.

Thus medicine and sociology are two disciplines that are very close and have to be known by the philosopher, who possesses a knowledge precious for all citizens. Once again, the inspirer of this doctrine is Auguste Comte, as Alain admits in *Le citoyen contre les pouvoirs* and, later, in *Eléments d'une doctrine radicale*:

> A contested power becomes immediately tyrannical; ... it establishes itself, it reigns, it defends itself. During these fights, right perishes; the rebels are always right and, later, are tyrannically governed; ... with the hundredth part of the energy that we employ to fight a bad master, we would be able to render him good ... Comte, an avant-garde man, had understood that the discussions concerning the origin and the legitimacy of powers are metaphysical and that the citizen's positive function is rather that of overseeing and limiting the action of powers.[83]
>
> *Auguste Comte used to call metaphysics all discussion concerning the origin of powers.* It's a badly directed effort.... The big problem, for me, as a citizen, is not that of choosing my own peaceful friend, ... but rather that of preventing the leader, whoever he will be, from preparing for war. The more pacific among men will prepare for war if they do not at all times exercise a strong resistance to passion.[84]

It is worth quoting one more example, inspired by World War I, which Alain had fought in. In *Mars*, he argues that the war did not change any of his positions because big events cannot leave a long-lasting trace on a mind that has accumulated habits. At the end of the war, even the most traumatized men, like the mutilated soldiers, have preserved the "same ways of thinking, approving, blaming, despising anterior to the terrible event."[85] At this crucial point Alain names the inspirer of his position: Auguste Comte.

> [following this theory] proposed by Comte, who was following the famous Broussais, ... the most profound modifications, compatible with life, are reduced to variations of intensity.... A man who was once irritable, will come back from war more or less irritable, but always following his structure and his familiar gestures, without any profound modification of this law of equilibrium in movement that defines the individual.... This is what war is able to do thanks to its immense means: it can destroy the individual, but it cannot change it. And, following the law of life, the one who has not been destroyed by the excess of movement, finds himself and restores himself as he was before.[86]

---

[83] Alain (1979, 150).

[84] 'Propos' originally included in *Eléments d'une doctrine radicale*. Republished in Alain (1970, 407).

[85] Ibid.

[86] 'Propos' originally included in *Mars ou la guerre jugée*. Republished in Alain (2002, 629), italics added by the author.

Comte had published an "Examination of Broussais's *Treatise*," republished as an appendix to his famous *System of Positive Polity*,[87] a book which inspired a great deal of Alain's analysis. Comte and Broussais argued that the living beings, when confronted with an obstacle, react in one and only one way, which varies exclusively in intensity. This means that biological bodies are machines without any intrinsic plasticity or teleology. This plasticity is tied to mind (*esprit*) which is present only in man.

> We always fall back on a mechanism: because it is clear that it is the animal-machine, which is answering, insulting, acting monotonously and convulsively. But, very easily, we forgive what belongs to man in this animal tumult, that is to say a genre of misfortune or even a type of humiliation which isn't conscious, but which, nevertheless, is marked by thought.[88]

The Durkheimian Lévy-Bruhl – in his book on Comte,[89] published in 1900 and republished in 1925, namely some months before the writing of Canguilhem's DES dissertation – had been the first scholar to stress the importance of Broussais' principle (according to which "morbid phenomena follow the same laws as the normal ones"[90]), and to emphasise its application to sociology. Much later, in *The Normal and the Pathological*, Canguilhem wrote that Comte had been influenced by Broussais' idea of a "real identity of the pathological phenomenon and the corresponding physical phenomenon" and he gave to this principle, which "he called Broussais's principle, universal significance in the order of biological, psychological and sociological phenomena."[91] In his review of Broussais' *De l'Irritation et de la folie*, Comte had underlined the importance of Broussais' idea that disease consists in "the excess or lack of excitation in the various tissues above or below the degree established as the norm."[92]

This principle has an heuristic importance that Lévy-Bruhl stressed: given the fact that in sociology one cannot experiment as in natural science, sociology can try formulating the laws regulating society's changes by observing spontaneous 'experiments.' This is the case with the social body's 'pathologies' such as wars and revolutions.[93] Therefore it is not a *static* method – which aims at analyzing society's "normal" state – but rather a *dynamic* one – the one that analyzes its development through crises – which takes advantage of Broussais' principle. However, order and progress, statics and dynamics are hierarchically ordered. Lévy-Bruhl argues that

[87] Comte (1877).

[88] 'Propos' originally included in *Les Idées et les âges*. Republished in Alain (2002, 18).

[89] Lévy-Bruhl (1900).

[90] Ibid., 239.

[91] Canguilhem (1991, 47).

[92] Ibid., 47–8.

[93] "It is the case of pathologies which, unluckily, are too frequent in the life of societies, perturbations more or less severe, caused by accidental or temporary causes. Those are the revolutionary times, which correspond to living bodies' diseases. If one applies, as it is advisable to do, Broussais's principle, that is to say if one admits that morbid phenomena are produced by the effect of the same laws of normal phenomena, the study of social pathology will replace, in a way, experimentation" (Lévy-Bruhl 1900, 278–279).

"social statics" grounds "social dynamics:" the variation of an existing order is subordinated to another order which is *coherent* with the former.

In fact, the place of Broussais' principle in Comte's system is tied to another principle that the latter borrowed from physics and according to which progress is nothing but the simple development of order. Given a certain order, this order can only change in intensity, but not in structure. *Social progress* is obtained only through little, imperceptible, variations. Those variations have to be commanded by the intellect.

This analogy of physiology and sociology, medicine and sociology was used by Durkheim, whom Lévy-Bruhl considered to be Comte's "real inheritor."[94] Durkheim uses – without mentioning Broussais – the 'principle' in *The Rules of Sociological Method*. Its central chapter bears the title, "Laws concerning the distinction between the normal and the pathological." Even if Durkheim opposes the 'biologization' of the social fact, he follows Comte and uses the medical paradigm to give an account of the order or disorder of the social body and to evaluate the possibility (or otherwise) of societal change. Like Alain, Durkheim argues that the governor has not to force changes but, as a doctor "he forestalls the outbreak of sickness by maintaining good hygiene, or when it does break out, seeks to cure it[95]" In *The Division of Labor in Society* Durkheim classifies the main "abnormal forms" of the division of labor. This classification is inspired by the natural sciences and presupposes the possibility of deducing the normal state starting from the pathological, following the principle of spontaneous experimentation: "as in biology, the pathological will enable us to understand better the physiological."[96]

Lévy-Bruhl, Alain and then, finally, Canguilhem, had also been following this principle. In a 'propos', written on June 10th 1912, Alain stresses that "Auguste Comte had correctly argued that.... one has not to create a new order, but to modify the one that already exists" and that he perfectly applied this principle to sociology and then to politics.[97] After him, Canguilhem, writes, in 1926, in his dissertation:

> The base of progress has been found by Comte, who expressed it in a biological theory which had a great influence on him. Broussais's doctrine, as it is presented in the *Traité de l'irritation et de la folie*, presents living beings' pathological manifestations, even those which apparently are the most abnormal, as they are entirely reducible to the conditions of the normal functioning. Since 1828, in the examination of Broussais's *Treatise*, Comte signals with enthusiasm the new conception concerning the relations between the normal and the pathological, reducing diseases to a simple change of intensity of the natural phenomena and not as a radical perturbation of the biological order. It is the extension of this principle to all human knowledge which is for Comte the real ground for a theory of modifiability (*modifiabilité*) that he simply sketches, leaving to his successor the task of establishing the doctrine.[98]

[94]Lévy-Bruhl (1900, 413).

[95]Durkheim (1982, 104).

[96]Durkheim (1994, 7).

[97]Raboul (1968, 268).

[98]Canguilhem (1927, 139).

The negative conclusion of those two principles is that progress cannot proceed through jumps, through the violent intervention of a group of men.[99] Sudden changes are, on the contrary, the origin of dysfunctions, which can cause an irritation, a disease, and a consequent retarding of society's progress. Alain explains that very well in his *Mars*:

> A conversion obtained through the intervention of an external violence or of a brutal experience renders us dependent on the events. Despotism, which pretends to forge new men through constraint, in doing that submits them and submits himself to forces' indefinite action. All revolution is at the same time despotic and fatalist because of this pretension to change abruptly everyone's vital equilibrium. On the contrary, true ideas concerning freedom and progress are contained in Comte's remark that *individual natures are modifiable only by little causes, and are never modifiable by the big ones.* And I firmly believe that, against injustice, and even against war little modifications are enough.[100]

After him, Canguilhem adds in his 1926 dissertation, quoting Comte's *System of Positive Polity*: "The first symptom of revolutionary blindness consists in desiring reforms which would be, at the same time, immediate and radical."[101]

The well-being of the body and of the society's body (their state of "good health") has to be obtained not through sudden movements which aim to destabilize powers, but through rational control of them, just as the will aims to control the human body and its mechanical reactions. Revolutionary movements seem to have a rational motivation, but, in reality they obey much more animal dynamisms, because they are dependent on passions. In *Le citoyen contre les pouvoirs*, in a 'propos' paradigmatically entitled "The revolutionary spirit," Alain underlines the complexity of the revolutionary phenomenon. If during a revolution "the idea and the violence are tightly tied together," nevertheless, revolution has always to be condemned, because "it ultimately depends upon the most instinctual reasons, common to human nature." Revolution is dependent upon passions, most of all the "thirst for revenge." Alain writes in *Eléments d'une doctrine radicale* that revolution is useless because it is based on an abrupt passionate movement: "anger, which is peculiar to revolution … is not revolutionary at all; it is rather the mind's sleep and the sleeper's anger against those who woke him up."[102] Consequently Alain's

[99] Guillaume Le Blanc correctly argues that "social life is regulated by a limited number of fundamental norms, starting from which all differences of intensity and speed are ordered, respectively by static and dynamic sociology. Comte criticizes the idea that several heterogeneous qualities would exist in social life. In social life every qualitative variation is just a difference of quantity. Thus all modifications find their justification starting from a series of norms which are ordered starting from a standard norm which is, according to Comte, the most normal state of society" (Le Blanc 2003, 138).

[100] Alain (2002, 629). Italics added by the author.

[101] Canguilhem (1927, 39).

[102] Republished in Alain (1997, 136). Again, Guillaume Le Blanc is right, when he writes that, according to Comte, "revolution is the pathology *par excellence* … but which has the advantage of making visible, to the eye of the good social physician, the normal type of society. In this respect positivism aims at ending the revolution, ending the state of crises that have engendered it." He is right, once again, when he writes that Broussais's principle offers an answer to the question of the rupture provoked by the events that can be understood only starting from a "normal type that gives them sense." Revolution, according to Comte is considered as a "transition" (Le Blanc 2004, 141–143).

radicalism is not a revolutionary and violent one. In the *Eléments d'une théorie radicale,* radicalism is described as a "revolt against the power" ["révolte contre le pouvoir"] but, at the same time, as rational and peaceful. This radicalism needs a rational "doctrine able to reconcile negations and affirmations, and to organize freedom inside Society. Without this doctrine we would only have an anarchical instability where all citizens and power itself fight against power, and where government itself is the opposition."[103]

5 Conclusions

Thus we can conclude that Broussais' principle applied to the analysis of society possesses a value which is at the same time theoretical and normative: given the fact that it is not possible to change a living being's behavior through a sudden modification, in the same way society, composed of living human beings, cannot be changed by a large, violent event. All forced and violent modifications are traumatizing, they cannot be long lasting, because they address the passionate, irritable, and thus animal part of men. *Pacifism* is the only rational politics that one can apply to the body of the State: it calms 'social irritations', which are the potential sources of pathology and restore the state of society's 'normality', helping its progress.

It is finally clear that Alain's philosophical practice and pacifist politics depend on three elements, which are tied together in his theory of passions, inherited from Descartes, Comte and Broussais:

1. Cartesian mechanism, which is the ground for the theory of the man-machine. This theory implies a complete confidence in technology, conceived as a simple extension of science. Medicine is just an extension of physiology, the science concerning the laws that govern the 'animal' part of man. Politics is just an extension of sociology, the science of the collective behavior of man, who is, at the same time, 'animal' and 'spiritual.'
2. Broussais's theory of irritation, which affirms the identity of the normal and the pathological, differing only in degree.
3. The analogy between the living body and the body of the State.

In 1926, Canguilhem was still convinced that the "theory of order and progress" underlying Alain's pacifism was "without any doubt the social conception that allows us to grasp today's problems better."[104] This conception was suitable for a society in a state of homeostatic equilibrium, a society similar to France's Third Republic. Ten years later, this theory was unable to give an account of events, discontinuities, ruptures, and qualitative changes in human societies. Moreover, at the beginning of 1936, the French radical leftist party suffered a terrible defeat in the elections: that was the last confirmation of the failure of Alain's pacifism, unable to understand the new 'pathologies', the new social upheavals.

[103]Republished in Alain (1997, 146).

[104]Canguilhem (1927, 139).

During a long medical training and a long reflection – which led him to *The Normal and the Pathological* and to the essays contained in *The Knowledge of Life* and, later, to his Ph.D. dissertation in philosophy on the concept of reflex[105]–, Canguilhem puts into question exactly the three elements of Alain's theory that I have briefly described. At the end of this long reflection, Canguilhem refuses Alain's philosophy and replaces it with his own 'vitalist' anthropology that Alain obviously despised.[106] This anthropology was structured following three main assumptions or postulates:

1. Life is a normative power. Each organism differs from matter, because it is capable of producing norms according to its proper *milieu*. The difference between men and animals, is that, in man, this normative power is 'reflexive'.
2. Pathology is a state qualitatively different from the 'normal.' A sick organism is a coherent totality organized in a completely different way than the healthy one.
3. There is no possible parallelism between the organism and society. Canguilhem conceives the latter as an instrument – or as a prosthesis – devoid of an intrinsic teleology, but with a teleology given to it by man.

But the concrete elaboration of this 'vitalist' theory is another story, which took place elsewhere, far from Alain's group, between Toulouse, Clermont-Ferrand, and in the scrubland.

References

Alain. 1956. *Propos I.* Paris: Gallimard. Selection of 'propos' by Maurice Savin.

Alain. 1958. *Les Arts et les dieux*, Paris: Gallimard. Containing: *Histoire de mes pensées* (1936), *Système des Beaux-Arts* (1920), *Vingt leçons sur les Beaux-Arts* (1931), *Entretiens chez le sculpteur* (1937), *La Visite au musicien* (1927), *Lettres au docteur Mondor* (1924), *Stendhal* (1935), *En lisant Dickens* (1945), *Avec Balzac* (1935), *Définitions* (1953), *Préliminaires à la mythologie* (1943), *Les Dieux* (1934).

Alain. 1960. Les livres d'Alain. *Bulletin de l'association des amis d'Alain* (65): 2–3, June.

Alain. 1964. *Esquisses d'Alain.* Paris: Gallimard.

Alain. 1970. *Propos I.* Paris: Gallimard. Selection of 'propos' by Samuel Sylvestre de Sacy.

Alain. 1979. *Le Citoyen contre les pouvoirs [1926].* Geneva: Slatkine.

Alain. 1991. *Éléments de philosophie (1942).* Paris: Gallimard.

Alain. 1993. *Les propos d'un Normand de 1908.* Paris: Institut Alain.

Alain. 1997. *Les propos d'un Normand de 1911.* Paris: Institut Alain.

Alain. 2002. *Les Passions et la sagesse.* Paris: Gallimard. It contains: *Les Idées et les âges* (1927), *Les Sentiments familiaux* (1927), *Les Aventures du cœur* (1945), *Souvenirs de guerre* (1937), *Mars ou la guerre jugée* (1921), *Souvenirs concernant Jules Lagneau* (1925), *Abrégés pour les aveugles* (1942), *Platon* (1928), *Descartes* (1928), *Hegel* (1932), *81 chapitres sur l'esprit et les passions* (1917), *Entretiens au bord de la mer* (1931).

---

[105] Canguilhem (1955).

[106] Cf. Murat et al. (2012, 16). In a 1937 annotation from his diary, Alain is very ironic concerning Canguilhem who, he writes, "has sworn to invent a philosophy and who has just to read Kant [*n'a qu'à ouvrir son Kant*] to outclass his rivals."

Bourdeau, Michel. 2004. Alain, Comte et le pouvoir spirituel. In *Alain dans ses œuvres et son journalisme politique*, 221–233. Paris: Institut Alain.

Braunstein, Jean-François. 1986. *Broussais et le matérialisme. Médecine et philosophie au XIX siècle*. Paris: Méridiens-Klincksieck.

Braunstein, Jean-François. 2000. Canguilhem avant Canguilhem. *Revue d'histoire des sciences* 5: 9–26.

Braunstein, Jean-François, et al. 1998. *Actualité de Georges Canguilhem. Le Normal et le pathologique*. Le Plessis-Robinson: Institut Synthélabo pour le progrès de la connaissance.

Broussais, François-Joseph. 1828. *De l'irritation et de la folie*. Paris: Baillière.

Brunschvicg, Léon. 1921. *Nature et liberté*. Paris: Alcan.

Canguilhem, Georges. 1923–1926. *Histoire de la philosophie*. Auguste Comte, Paris: Centre d'archives de philosophie, d'histoire et d'édition des sciences, Fonds Georges Canguilhem, Box 4, Pressmark: GC. 4. 3, 275 handwritten pages.

Canguilhem, Georges. 1927. *La théorie de l'ordre et du progrès chez A. Comte*. 155 handwritten pages. Paris: Fonds Centre d'archives de philosophie, d'histoire et d'édition des sciences. Pressmark: GC. 6.1.

Canguilhem, Georges. 1929a. La renaissance du vitalisme? *Europe* 23: 134–139.

Canguilhem, Georges. 1929b. A la Gloire d'Hippocrate, père du tempérament. *Libres Propos*. 297–298, 20 August.

Canguilhem, Georges. 1931. Pierre Viénot, *Incertitudes allemandes. Libres Propos*. 514, November 1931.

Canguilhem, Georges. 1935a. *Le Fascisme et les paysans*. Paris: La Laborieuse.

Canguilhem, Georges. 1935b. Alain, *Les Dieux*. *Europe* 37: 445–448.

Canguilhem, Georges. 1947. Note sur la situation faite en France à la philosophie biologique. *Revue de métaphysique et de morale* 52(2): 322–332.

Canguilhem, Georges. 1955. *La formation du concept de réflexe aux XVIIe et XVIIIe siècles*. Paris: PUF.

Canguilhem, Georges. 1968. Lettre du 20 janvier 1968 à Jeanne Alexandre. *Bulletin de l'Association des amis d'Alain* 48–49, 26 May.

Canguilhem, Georges. 1991. *The normal and the pathological*. Trans. Carolyn R. Fawcett and Robert S. Cohen. New York: Zone Books.

Canguilhem, Georges. 1994. *A vital rationalist. Selected writings from Georges Canguilhem*, ed. Delaporte François. New York: Zone Books.

Canguilhem, Georges. 1995. Raymond *Aron* et la *philosophie* critique de *l'histoire*. In *Raymond Aron, la philosophie de l'histoire et les sciences sociales*. Paris: Rue d'Ulm Editions.

Canguilhem, Georges. 1996. Discours de Charleville. *Cahiers philosophiques* 69: 87–92.

Canguilhem, Georges. 2005. The death of man, or exhaustion of the *cogito*? Trans. Catherine Porter. In *The Cambridge companion to Foucault*, ed. Gutting, Gary. New York: Cambridge University Press.

Canguilhem, Georges. 2008. *Knowledge of life*. Trans. Stefanos Geroulanos and Daniela Ginsburg. New York: Fordham University Press.

Challaye, Félicien. 1932. *La paix sans aucune réserve: thèse de Félicien Challaye*. Paris: La Laborieuse.

Comte, Auguste. 1877. *System of positive polity: Or, treatise on sociology, instituting the religion of humanity*. Trans. Robert Congreve. London: Longmans.

Durkheim, Emile 1982. *The rules of sociological method*. Trans. William D. Halls. New York: Free Press.

Durkheim, Emile. 1994. *The division of labor in society*. Trans. William D. Halls. New York: Free Press.

Foucault, Michel 1998. Life: Experience and science. In *Aesthetics, method, and epistemology*, 297–333. Trans. Robert Hurley. New York: The New York Press.

Gane, Mike. 1998. Canguilhem and the problem of pathology. *Economy and Society* 27(2–3): 298–312.

Gane, Mike. 2003. *French social theory*. London: Sage.

Le Blanc, Guillaume. 2004. *L'esprit des sciences humaines*. Paris: Vrin.
Lecourt, Dominique. 2008. *Georges Canguilhem*. Paris: PUF.
Lévy-Bruhl, Lucien. 1900. *La Philosophie d'Auguste Comte*. Paris: Alcan.
Logue, William. 1979. Sociologie et politique: le libéralisme de Célestin Bouglé. *Revue Française de Sociologie* 20: 141–161.
Lukács, György. 1980. *Destruction of reason*. Trans. Peter R. Palmer. New York: Merlin Press.
Mazon, Brigitte. 1983. La création du Centre de documentation sociale. *Études durkheimiennes* 9: 15–20.
Muglioni, Jacques. 1987. *Alain lecteur de Comte, Alain lecteur des philosophes, de Platon à Marx*. Paris: Bordas.
Murat, Michel, et al. 2012. *Alain philosophie et littérature mêlées*. Paris: Éditions de la Rue d'Ulm.
Policar, André. 1997. Sociologie et morale: la philosophie de la solidarité de Célestin Bouglé. *Recherches sociologiques* 28(2): 85–110.
Policar, André. 1999. Science et démocratie: Célestin Bouglé et la métaphysique de l'hérédité. *Vingtième siècle* 61: 86–101.
Policar, André. 2000. De la critique de la sociologie biologique à l'autonomie de la morale: itinéraire de Célestin Bouglé. *Mil Neuf Cent* 18: 137–166.
Politzer, Georges. 1967. *Ecrits*. Paris: Éditions Sociales.
Raboul, Olivier. 1968. *L'homme et ses passions d'après Alain*. Paris: Gallimard.
Ragghianti, Renzo. 1993. *Dalla fisiologia all'etica dell'effort*. Firenze: Le Lettere.
Sartre, Jean-Paul. 1966. Jean-Paul Sartre répond. *L'Arc* 30: 87–96.
Souillé, Charles. 2009. *Les philosophes en république*. Reims: PUR.
Spitz, Jean-François. 2005. *Le moment républicain en France*. Paris: Gallimard.
Starobinski, Jean. 2003. *Action and reaction: The life and adventures of a couple*. New York: Zone Books.
Vialleton, Louis. 1929. *L'origine des êtres vivants. L'illusion transformiste*. Paris: Plon.
Vogt, Paul. 1979. Un durkheimien ambivalent: Célestin Bouglé 1870–1940. *Revue Française de Sociologie* 20: 123–139.

Part III
Vitalism and Contemporary Biological Developments

Chapter 11
Homeostasis and the Forgotten Vitalist Roots of Adaptation

J. Scott Turner

Abstract Through most of the twentieth century, biology's image as a valid science has been gauged by how closely it adheres to the norms of "objective" sciences like physics, chemistry and mathematics. Strains of biological thought that depart from this norm are deemed non-scientific. This presumes that life is fundamentally a physical, chemical and thermodynamic phenomenon. While this approach has been very fruitful, it is questionable that it can lead to a coherent theory of biology. This is particularly the case for certain obvious (self-evident?) properties of living systems, including purposefulness, design, and intentionality. The tendency has been to treat these phenomena as illusions, as in numerous invocations of "apparent" design, "apparent" purposefulness and "apparent" intentionality. I argue in this essay that these phenomena are far from illusory, but are in fact quite real. I further argue that a coherent theory of biology must account for purpose, design and intentionality, and I offer one possible way to do so through the fundamental phenomenon of homeostasis.

Keywords Vitalism • Claude Bernard • Homeostasis • François Magendie • Genetics • Physiology

1 Introduction

Is vitalism dead? And if it is, should its demise be lamented?

One of the tropes of modern biology is the triumph of "materialism" over "vitalism." Usually, this is depicted as a triumph of reason over superstition, but it is

J.S. Turner (✉)
Department of Environmental and Forest Biology, SUNY-ESF, 1 Forestry Drive, Syracuse, NY 13210, USA
e-mail: jsturner@syr.edu

S. Normandin and C.T. Wolfe (eds.), *Vitalism and the Scientific Image in Post-Enlightenment Life Science, 1800–2010*, History, Philosophy and Theory of the Life Sciences 2, DOI 10.1007/978-94-007-2445-7_11,

actually shorthand for the twentieth-century conversion of biology into a sub-discipline of chemistry and physics. Molecular biology is the prime example of this, but placing a close second is the neo-atomist synthesis of Mendelian and population genetics that we now know under the broad label of Neo-Darwinism. In the one, life is the mindless churning of molecules, complicated and wonderful churning to be sure, but churning nonetheless. In the other, life is a phenomenon on the rack, slung between the determinacy of its thoughtless genes on the one hand and a merciless and indifferent environment on the other. From this – *mirabile dictu!* – comes the evolution of life's marvelous existence and diversity. Lest we are tempted to see purpose or design at work in this, never mind, it is only an apparition, a reflection of our brain's perverse tendency to see pattern and purpose where there is, in fact, none.

The triumphs of the materialist approach to the phenomenon of life are many and undeniable, not only in the practical benefits conferred, but also for the intellectual discipline it has imposed on the science of life. In a materialist 'culture of science', when there is a hard problem to crack, there is no easy refuge to be found in vital essences, *élan vital*, entelechy, Omega points, or any of the other vague notions that have come to be lumped under "the V-word," vitalism. We are well to be done with it, in the minds of most: real scientists do it reductively. But what if vitalism has a point, that there is, in fact, a special quality to life that cannot quite be captured under the materialist rubric, a purposeful dimension that is not illusory, but quite real, that is, indeed, the most important thing about it? If this is the case, alarm bells should go off in the minds of any scientist, because any presumptive science of life that ignores its most important feature has deliberately blinded itself. To judge whether we have done so, we must ask a number of questions. Is life qualitatively unique, and if so, what precisely makes it so? Can molecular biology tell us what an organism is, and why it, and not something else, seems to be a common evolutionary *motif*? What is the physics of the organism's seeming coherency and integrity? What "atoms of heredity" determine why certain body plans exist and others do not? What is the mathematics of life's seeming design and purpose? At the onset of the twenty-first century, after nearly a century of triumphant biological materialism, there are few answers to these questions that are not either irritatingly glib or embarrassingly tautological.

In her provocative essay,[1] "The Endurance of the Mechanism-Vitalism Controversy," Hilde Hein argued that mechanism and vitalism each represent, to their respective adherents, an internally coherent metaphysics, effectively insulating each from scrutiny by the other. Consequently, the history of biology has been an ongoing and unresolved struggle between the two for ascendancy, punctuated by episodic, and sometimes revolutionary, progress. What has made progress possible in these episodes is not one metaphysical view or the other being proved 'correct', but the discovery of common ground where both metaphysical camps can ask questions in a shared scientific discourse. These episodes of synthesis usually occur during a period of intellectual crisis, where a temporarily ascendant metaphysical view has bumped up against the limits of its explanatory power. We are, arguably, presently at such a point, where the limits of the materialist approach to life are

[1] Hein (1972).

beginning to rise dimly into view. The question becomes: where will the common ground for biology's next synthetic leap forward be?

In this essay, I argue that it will be found in the work of the great nineteenth-century French physiologist Claude Bernard (1813–1878). Bernard's reputation today rests upon two achievements. He established physiology as an experimental discipline, and so helped liberate medicine from the largely descriptive "natural history" approach that had long dominated. For this, he has been dubbed "medicine's Newton,"[2] deservedly so. And he is renowned for his famous aphorism:

> The constancy of the internal environment is the condition for a free and independent life.

to which the American physiologist Walter B Cannon gave a name: homeostasis.[3]

In the materialist metaphysics that has dominated twentieth-century biology, Bernard's insight has come to have a rather narrow operational definition: it is appropriate to speak of homeostasis of a particular property, like body temperature, and the physiological mechanisms that produce it. Bernard, however, had a much broader conception in mind. His "constancy of the internal environment" was, in fact, a bold philosophical assertion about life's unique nature, and it drew deeply from contemporary vitalist thought.[4] As the materialist juggernaut swept through twentieth century biology, this bracing vitalist metaphysics was confined, tamed, and largely forgotten. As we stand now at the leading edge of the materialist revolution in biology, homeostasis is re-emerging as something more like what Bernard himself envisioned: a cardinal law of biology that sets life apart from the materialist world,[5] what I call Biology's Second Law (the First being Darwinian natural selection).[6] I will argue in this essay that a fully coherent theory of life will remain out of reach until we recover Bernard's essentially vitalist conception of homeostasis.

2 Claude Bernard as the Vindicator of Vitalism

Claude Bernard's pioneering work in experimental physiology is well known: he laid down foundational concepts in digestion, endocrinology, circulation, nerve action and regulation of "animal heat." It is only natural, then, that he is often celebrated as the one who, through dint of rigorous experiment and heroic empiricism, drove a stake through the heart of vitalism.[7] There is a bit of Whiggish history in this story, however. That Bernard was a brilliant experimental biologist is without doubt. That he is a key founder of the experimental approach to medicine is also unquestionable.

---

[2]Normandin (2007).

[3]Cannon (1932), Wolfe et al. (2000).

[4]Oppenheimer (1948), Virtanen (1960).

[5]Varela et al. (1974), Benson (1989), Margulis (1997), Maturana (1999).

[6]Turner (2013).

[7]Bloch (1989), Gross (2008).

To claim him as a biological materialist, however, is a distortion. Far from discrediting vitalism, he vindicated it.

The roots of this contrarian conclusion are to be found in the historical development of vitalist thought in the eighteenth and early nineteenth centuries.[8] Then, as now, the basic question was this: is life a special phenomenon, and if so, what makes it so? The traditional vitalist view was essentialist in nature, dating back to the Hippocratic physicians and derived from the still more ancient doctrine of humors: life was special because it was imbued with a vital substance, a *vis essentialis*, that animated otherwise inanimate matter. By the eighteenth century, this long-standing doctrine became the subject of vigorous debate between competing schools of academic medicine in Europe.[9] What emerged from this was a transition from an "essentialist" vitalism (sometimes called "metaphysical vitalism") to a "process" vitalism (also called "physical vitalism"), that was concerned more with vital action and mechanism rather than vital substance. Initially, this emerging process vitalism drew deeply from its essentialist roots, taking its inspiration from another Hippocratic idea, the *vis mediatrix*, a substance thought to be dispersed through the nerves, and which mediated the interactions between the body's various parts. This proved to be a poison pill for essentialist vitalism, because the new operational outlook proved to be more congenial to experiment and disproof. Whereas *vis essentialis* could only offer the *fact* of the living body as evidence of its existence (or a corpse as evidence of its absence), the *vis mediatrix* invited predictions that could be tested in the laboratory. This opened the path to the distinctive "physical" form of vitalism that emerged in the nineteenth century. Much of this nascent empiricism was anatomical and observational – cataloguing the correlations between certain diseases and certain modifications of organs, for example – but some of these were experiments that called into question the very existence of the mediating "stuff." Why, for example, could an organ, like a frog's heart, survive for a time outside the body, and still be responsive to changes in its environment, even though it was removed from the influence of any supposed *vis mediatrix*? Why, if the nerves were the conduits for the *vis mediatrix*, could some organs, like the pancreas, function even though the organ had no discernible innervation? Among the most interesting of these contrary observations was Théophile de Bordeu's remarkable eighteenth-century anticipation of the superorganism idea, which drew a parallel between the coordinated behavior of a swarm of bees and a living body.[10] If bees in a swarm were separate bodies, physically disconnected from one another, how could any putative *vis mediatrix* flow between them? And if there was no *vis mediatrix*, how were the behaviors of the individuals mediated to produce the "organism-like" behavior of the swarm?

This was the stage set in 1824, when Claude Bernard came to Paris as a young man, eventually to come into the orbit of the remarkable François Magendie, the

[8] Normandin (2007), Wolfe (2008).

[9] Normandin (2007), Cheung (2008).

[10] Bordeu (1751, § CXXV), in Bordeu (1818, vol. 1, 187).

prickly scion of a revolutionary father.[11] Prior to the French Revolution, François' father, Antoine, was a "barber-surgeon," practicing at a time when physicians held that profession in some disrepute. As is often the case with upstart professionals fighting for legitimacy and respect, Antoine Magendie was aggressively anti-aristocratic, anti-clerical, skeptical and rational. When the Revolution broke out, Antoine relocated his family, including the 9-year-old François, to Paris, so he could be in the thick of the upheaval. There, Antoine became disillusioned with practicing surgery (not surprising in light of the horrific suffering inflicted on surgical patients in those pre-anaesthetic days), and quit his profession to serve the new Republic. This proved to be an impecunious decision, and despite his disillusionment with his erstwhile career, Antoine determined that his son should nevertheless follow his footsteps. So, when François came of age, he was sent as a medical apprentice to the École de Santé, a prestigious hospital founded by a group of physicians and surgeons who, unusually for the time, did not hold one another in mutual disdain. What followed was an erratic career of brilliant success interspersed with controversy and dissipation. François realized his father's upwardly mobile aspirations to become, at an early age, a physician and a member of the Academy of Medicine ("in spite of myself"). He became the beneficiary of a large inheritance only to squander it on horses, parties and drink. He fell into acrimonious disputes with colleagues within France and across Europe. He fumingly quit medical practice in frustration over his inability to cure patients and what he perceived as the blithe attitudes of his physician colleagues toward their own inabilities to do so. All came right in the end, though, for Magendie's erratic nature landed him in the very place where his energy, bluntness and irritation with received wisdom could be put to good use: as a researcher at the Academy of Medicine.

When Bernard came to him in 1832, Magendie was at his peak. The essentialist idea was gasping its last, propelled there in no small part by Magendie's pointed attacks on the endless proliferation of "vital forces" that academic physicians were wont to invoke when an unexpected phenomenon cropped up:

> Why then is it necessary in respect to every phenomenon of the living body to invent a peculiar and special vital force? Cannot one be content with a single force which one could designate vital force in a general way, while admitting that it gives rise to different phenomena depending upon the structure of the organs and tissues which function under its influence? But is not this single vital force still too much? Is it not a hypothesis pure and simple, inasmuch as we are unable to perceive it? And would it not be more advantageous if physiology began only when the phenomena of the living body became appreciable to the senses?[12]

Magendie was more than capable of delivering this criticism in the bluntest of terms: "To express an opinion [*sc.* on the existence of a vital force], to believe, is nothing else than to be ignorant ... One could with justice say to you 'You believe, therefore you don't know'".[13] One can easily imagine the attraction the brash Magendie must have had for the young Bernard, in from the provinces and looking to make his mark.

[11] Dawson (1908).

[12] Magendie (1819), in Dawson (1908, 6–7).

[13] Ibid.

Bernard's upbringing was more conservative than Magendie's.[14] His father, Jean-François, was a provincial wine-maker and landowner, and so was not as caught up in revolutionary fervor as was Antoine Magendie. Claude Bernard's early education, which bored him immensely, was mostly in local Jesuit schools, but he had an active mind, so he became an autodidact, indulging a growing interest in philosophy and the Romantic arts. One had to make a living though, so when a school friend enthused about his new career in pharmacy, Claude decided to follow his friend, and became a pharmacist's apprentice himself. This proved not to be a good experience either, so Claude surreptitiously began to plot his escape to Paris, there to follow his dream of becoming a playwright. The plot came to premature fruition when, in the aftermath of an accident in the apothecary, Bernard's pharmacist master discovered that Claude had been working on a play rather than making potions. Once in Paris, Claude's artistic aspirations quickly foundered at the hands of the critic Saint-Marc Girardin, who kindly, but firmly, advised Bernard to try his hand at medicine instead. This he did, enrolling in medical school, and to support himself, took on work as a laboratory assistant in a Parisian girl's school, where he proved to be extraordinarily skillful in dissection. Bernard's anatomical prowess soon captured the attention of Magendie, who took him on as his *préparateur*. Soon thereafter, Bernard formally became Magendie's student, setting him firmly on his own path to scientific research.

Bernard's relationship with the mercurial Magendie was both contrary and complementary. The relationship was stormy, marked by intermittent fallings-out and reconciliations. Like Magendie, Bernard brought to his work a healthy dose of skepticism, fueled in his case by the slapdash pharmaceutical formulations (which sometimes included the leavings of other potions) he had been ordered to make during his apprenticeship, and by the utter lack of consistent rationale or evidence of effectiveness for them. Bernard's skepticism was rooted in more conservative soil than Magendie's, so although Magendie's own radical skepticism bolstered Bernard's own, he was not prepared to go as far as his mentor, maintaining throughout his life the conviction that life was somehow special, an attitude that placed Bernard solidly in the mainstream of nineteenth-century physical vitalism. What Bernard got from Magendie was his distinctive belief that life's special quality, whatever it was, had to operate through a material basis, which could be probed experimentally but could only go so far: there was a limit to which materialism could explain life's special qualities. Where Magendie and Bernard were in harmony was their belief that medical science had been culturally unwilling to push to anywhere near where that limit might be. Their shared conviction, which Bernard took much further than Magendie could, was the scientist's duty to push against those limits as hard as possible.

In this sense, Bernard was indeed the nemesis of vitalism, but only of the dwindling metaphysical variety: once he was done with it, that form of vitalism would never come back. Nevertheless, Bernard's work was actually a strong empirical

[14]Foster (1899).

defense of the physical vitalist thought of his day.[15] At the time, modern cell theory was just coming to fruition, and this gave new impetus to an idea that had captivated early physical vitalists like Bordeu: organisms were assemblages of innumerable and mutually accommodating "little lives" together. This process of accommodation was, in the eyes of the physical vitalists, the core of life's unique quality, so obvious to the senses and so demanding of an explanation. If *vis mediatrix* could not do the job, as was becoming increasingly obvious to Magendie, Bernard and their contemporaries, then perhaps the key was the common interest all an organism's "little lives" had in their shared environment, the *milieu intérieur*, and their common need to make that environment suitable for the continuance of all their "little lives". Bernard's invaluable contribution to this problem was his insistence that the mediation had to occur through discernible mechanisms that were solidly grounded in known chemistry. It is worth reiterating, though, the precise phrasing of Bernard's aphorism: the constancy of the internal environment *is the condition* for a free and independent life. In short, the scientific vitalist conception of life is the antecedent for Bernard's experimental program: life does not serve the material world, the material world serves life. By showing that life's unique phenomenology was amenable to an experimental approach, Bernard did not kill vitalism, he vindicated it.

3 Bernard's Uncertain Legacy

Nevertheless, this aspect of Bernard's legacy made little impact in his lifetime, and for some time thereafter. There are many reasons for this, but in general, Bernard's vitalist speculations simply ran counter to the rising tide of materialism and rationalism that was, at that time, sweeping through the science, politics, and culture of Europe, and later, North America.[16] In biology, this tide was impelled by the increasingly powerful technical armory that enabled biologists to analyze cell chemistry in hitherto unimaginable detail and precision, a trend that continued ever stronger through most of the twentieth century. As the confidence grew that life could be explainable ultimately as a complex form of chemistry – the current conceit that "synthetic life" is within reach is the logical culmination of this trend[17] – the vitalist roots of Bernard's ideas came to acquire a whiff of disreputability. This engendered an enduring ambivalence toward them: one can find memoirs of Bernard published as recently as 1989 that do not mention Bernard's signature idea, the constancy of the *milieu intérieur*, at all.[18]

Where Bernard's ideas did persist in mainstream science, they did so in a largely materialist form, cleansed of their vitalist roots. The chemical fervor that consumed physiology in the late nineteenth-century – the analysis of living chemistry,

[15]Normandin (2007), Virtanen (1960).

[16]Gross (2008), Oppenheimer (1948).

[17]Pennisi (2010).

[18]Bloch (1989).

most notably the newly discovered class of compounds, the hormones[19] – eventually gave rise to a renewed appreciation of the phenomenon of self-regulation in physiological systems.[20] This left the physiological chemists, as they were then known, to branch off into their own discipline of biochemistry, leaving the physiologists, most notably Lawrence Joseph Henderson, to take up the Bernardian baton.[21] Henderson's work on acid-base regulation of the blood highlighted the complex participation of various parts of the body – blood cells, plasma proteins, lungs, kidneys and intestines – that were behind maintaining constant blood acidity. These were properties of organisms, not something that could effectively be captured in a cuvette. Shortly thereafter, Walter B. Cannon's discoveries of the complex web of hormonal and nervous system interactions that maintained blood sugar, body temperature, and readiness of the body for self-defense ("fight-or-flight") further bolstered the importance of the organism as a self-regulating system. Cannon went on to become Bernard's most vigorous disciple of this period: it was Cannon who can be credited with coining the term "homeostasis," first used in a *Festschrift* for Bernard's colleague, the Nobel laureate Charles Richet.[22] The problem of self-regulation proved to be a fruitful goad to fields outside physiology as well: Norbert Wiener's pioneering work in self-regulating physical systems and the seminal cybernetic concept of negative-feedback control owes its inspiration to the phenomenon in living systems.[23] (Cybernetics is, in many ways, Bernard's intellectual grandchild: I shall return to this momentarily.) In all instances, however, homeostasis was an operational outcome of the materialist conception of the organism-as-machine, to be approached strictly within the metaphysical boundaries of that metaphor.

Contemporaneously, Bernard's ideas did enjoy what we might term an "idealist renaissance" that ranked the mechanistic details of self-regulation below the phenomenon of "independent life" itself. This idealist renaissance derived in roughly equal measure from Bernard's own scientific vitalism, and from Herbert Spencer's seminal ideas about societies as "organism-like" entities.[24] In ecology, for example, Frederic Clements likened ecosystems to "superorganisms" that, like conventional organisms, maintained an "independent life" through forms of ecosystem homeostasis.[25] Homeostasis also became the conceptual core of various schools of psychology: the twin notions of "appetites" and "drives" explicitly applied homeostasis and control system theory to the phenomenon of behavior. Various philosophers, including Henri Bergson, Hans Driesch, Jan Christian Smuts and Alfred North Whitehead, also took up the phenomenon of the "independent life" as a serious

[19] Needham (1970).

[20] Cziko (2000).

[21] Henderson (1913).

[22] Cannon (1932), Wolfe et al. (2000), Cannon (1926, 1929).

[23] Rosenblueth et al. (1943).

[24] Mayr (1982).

[25] Golley (1993).

metaphysical problem that a materialist metaphysics could not quite capture.[26] This neo-vitalist renaissance was largely swept aside by the tsunami of materialism that surged through the science of life, but it persists to the present day, in disciplines that are considered "fringe" science (Claude Bernard is cited by homeopaths as an important source of scientific credibility[27]), and in more mainstream ideas, such as the re-emergent superorganism idea[28] that is at the heart of James Lovelock's Gaia theory[29] and Lynn Margulis' symbiogenic theories of evolution.[30] These ideas have faced the obstinacy of a reigning metaphysical materialism, which has criticized them as obscurantist and unscientific, at times with surprising vehemence.[31]

It is human nature to point out motes in other's eyes while ignoring the logs in one's own, so it should come as no surprise that this sanitized operational approach to homeostasis has sometimes enjoyed uncritical support beyond what evidence can support. The application of cybernetic theory to regulation of body temperature provides a good example. Since the 1950s, the dominant paradigm for body temperature homeostasis has been the "thermostat" model (i.e. constant body temperature is the operational outcome of an embodied negative feedback controller).[32] This implies there is, within the body, a cybernetic system that comprises the necessary components, namely sensors, comparators and effectors, assembled in the requisite configuration. For many years, this idea enjoyed great success. The body's "thermostat" was supposedly located in a specific part of the brain, the pre-optic hypothalamus. Temperature-sensitive nerve cells were identified there that were presumed to be the thermostat's components, including sensors and the required neural "circuitry" to make the comparator. Various effectors were also successfully identified, including modulation of cutaneous circulation, metabolic heat production, sweating, panting and so forth. There was even experimental proof of the thermostat at work: manipulating the temperature of the pre-optic "thermostat" independent of the body temperature produced results consistent with an embodied cybernetic system in operation. Warming the pre-optic hypothalamus of sheep, for example, was followed by reductions of metabolic heat production, increased cutaneous circulation, panting and sweating, all responses expected for a body temperature above the thermostatic "set point," even though the body's temperature (if not the brain's) was actually normal.

Nevertheless, the cybernetic model for body temperature regulation actually has not stood up well. In principle, one should be able to predict the behavior of any cybernetic system if one knows the system's parameters of operation: the gain of the sensors, the modulating functions that control the outputs and so forth. If body

[26] Capra (1996).
[27] Ullman (2007).
[28] Hölldobler and Wilson (2009).
[29] Lovelock (1987).
[30] Margulis (1971), Margulis (1981).
[31] Doolittle (1981), Dawkins (1982), Williams (1992).
[32] Bligh (1973).

temperature is controlled by a cybernetic controller, then a known perturbation should produce a specific response that compares well with the modeled prediction. Invariably, such comparisons show some agreement, but also significant departures from the predicted response. A more complex cybernetic model is then proposed to account for the discrepancy, which is then tested and found to exhibit deviations of its own, which requires a still more complex cybernetic model. As a result, cybernetic models of thermoregulation have proliferated into baroque assemblages of multiple controllers, sensors and effectors, distributed throughout the nervous system and body, arising *ad hoc* as explanatory needs multiplied.[33] It is a quintessentially Ptolemaic dynamic with an elaborately complicated Ptolemaic system as the result. To complicate matters further, the various components of a cybernetic system have proved elusive: rather than cells with dedicated roles being assembled into "circuits" that can behave as a cybernetic system, individual nerve cells can sense temperature, act as comparators, and even bring about a degree of self-maintenance of temperature themselves.[34]

There also is a substantial diversity of "types" of thermal homeostasis to be found amongst organisms, ranging from torpidity and hibernation, to ecological effects, even to thermoregulating plants: cybernetic models of temperature regulation have had little meaning for these.[35] The problem of fever provides an interesting example. Traditionally, fever was regarded as a pathological state and a failure of temperature homeostasis. The discovery that fever is actually a regulated state ushered in a dramatic shift in therapeutic approach (fever was now to be managed, not suppressed), as well as a cybernetic "explanation": fever is an upward adjustment of the "set point" in the pre-optic cybernetic controller.[36] While this is an attractive metaphor, it has not engendered much in the way of scientific progress. Where progress has been made, such as in the important roles of bacterial pyrogens and of prostaglandins, these have in no way derived from a cybernetic model of fever. The water is muddied further by the discovery of "behavioral" fever among poikilotherms (like "cold-blooded" lizards or other reptiles).[37] Such animals can attain a degree of steady body temperature by seeking favorable thermal microclimates. A lizard fighting off an infection exhibits a "behavioral" fever by seeking warmer microclimates than it otherwise does, producing a warmer average body temperature, and therapeutic benefits that are similar to those accruing to fever in mammals. Because the fever is behavioral, however, it now becomes a reflection not of a body temperature regulated by a cybernetic system, but by the lizard's desire to be at a certain body temperature. This introduces borderline vitalist ideas like intentionality and desire, which undercuts the materialist rationale for proposing the cybernetic system in the first place. This is

---

[33] Nagashima et al. (2000).

[34] Kobayashi (1989), Romanovsky (2007).

[35] Hudson (1973), King and Farner (1961), Kronenberg and Heller (1982), Prosser (1973), Templeton (1970).

[36] Bligh (1980), Lyons (1861), Stitt (1979).

[37] Cabanac and Rossetti (1987), Casterlin and Reynolds (1977a, b), Covert and Reynolds (1977), Kluger (1979), Reynolds and Covert (1977), McClain et al. (1988).

only the beginning, in fact. What is one to make of behavioral fever in animals, like crayfish, which operate with a markedly different neuronal architecture from vertebrates? And of "fever" in plants like the arum lily, which raise temperature to volatilize chemicals that attract insect pollinators?[38]

In short, the cybernetic model of temperature regulation has been characterized by the same proliferation of *ad hoc* complexity that so perturbed Magendie. These observations and others have even led some to argue that the core cybernetic concept of the set point has no correspondence to the reality of physiological systems.[39] In short, the cybernetic metaphor for homeostasis appears to have collapsed almost entirely. Indeed, metaphor appears to be all that is left of it.

4 Homeostasis and Evolution

Claude Bernard himself was not an evolutionist, not because he disbelieved in it but because he thought evolution to be an experimentally intractable problem. Ironically, it is in evolution that Bernard's vitalist streak holds the most interesting promise, for Bernard's metaphysical leanings shared a provenance with seminal evolutionary thinkers such as Lamarck and Cuvier. The historical revisionism that permeates modern evolutionary thought has relegated these scientists to the roles of foils, misguided thinkers who prepared the ground for Darwin to come along and finally "get it right." In fact, Darwinism, as Darwin himself originally conceived it, was steeped in the same vitalist tradition that sustained Lamarck, Cuvier and Bernard. By the late nineteenth-century, this vitalist evolutionary tradition had reached a point of crisis, from which emerged the modern form of Darwinism that we know as Neo-Darwinism. That strain of Darwinian thought was metaphysically quite alien from its "classical" Darwinian progenitor: indeed, one can argue that modern Darwinism is fundamentally an anti-Darwinian theory. The important question, of course, is which form of Darwinism, classical or modern, is the better model for what actually drives the process of evolution?

The conversion from classical Darwinism to its modern form was the outcome of a tumult that roiled evolutionary biology in the late nineteenth and early twentieth centuries.[40] The antagonists then could be broadly classed into "developmentalists," who looked to the phenomenon of adaptation to environment as the primary driver of Darwinian evolution; and "geneticists" who downplayed the role of adaptation and argued instead that evolution was constrained by heredity. It should be pointed out that the geneticists included vigorous critics of classical Darwinism, like Thomas Hunt Morgan. By that time, Darwinism had morphed into a form of Neo-Lamarckism, a development that had, for a time, the imprimatur of Darwin himself, evidenced by

[38] Raskin et al. (1987), Seymour et al. (1983).

[39] Schmidt-Nielsen (1994).

[40] Bowler (1983).

his own Lamarckian model of inheritance, pangenesis. Darwin eventually abandoned this line of thinking, because it seemed to encourage various models of "directed" or "purposeful" evolution, which Darwin came to believe were anathema to his theory. Examples of this deviationist tendency included orthogenesis, which posited a sort of "driving force" behind evolution that kept lineages evolving along certain paths, resistant to the corrective actions of natural selection. The eponymous Cope's Law (named for the American paleontologist Edward Drinker Cope), which posited a general increase of body size as lineages evolved, is one of the more durable examples of such "directed evolution" theories.

Geneticists, for their part, regarded developmentalism and adaptationism as a form of surreptitious vitalism (a not unreasonable supposition), and looked to clarify the nature of hereditary memory as the key to a coherent theory of evolution. Central to this program was clearing up the prevalent confusion over so-called "soft-inheritance," the essentially Lamarckian transmission to offspring of traits acquired during the parents' lifetimes; versus "hard-inheritance," essentially nuggets of heritable memory that were transmitted unchanged from generation to generation. Darwin's pangenesis theory was a quintessential "soft-inheritance" mechanism. Pangenesis was eventually abandoned because it could not be sustained in the face of experimental reality: it was never clear what Darwin's gemmules actually were, what structures within the cell contained them, how use and disuse of body parts could be imprinted on them and, most critically, how they could migrate to, and come to constitute, the transmissible germ-plasm.[41] Pangenesis' doom was sealed by August Weismann's famous tail-amputation experiments, from which he derived his doctrine of the radical segregation of the body into non-heritable *soma* and heritable germ-plasm (the Weismann barrier). In the geneticists' minds, this obviated any possibility of "soft-inheritance" mechanisms operating in lineages. The tumbrel was hurried along by the Dutch botanist Hugo deVries, who located hard-inheritance firmly in the nucleus and clarified its particulate nature. "Soft inheritance" was finally marched to the scaffold by the rediscovery of Gregor Mendel's "atoms of heredity." By this time, adaptationists could only babble in their defense, leaving "hard-inheritance" as the only feasible evolutionary mechanism that had any scientific support. Ever since, the Neo-Lamarckian holdouts for "soft-inheritance" have been consigned to the fringe: our modern conception of vitalism as the disreputable "V-word" is ultimately traceable to this development.

The nineteenth-century conflict between developmentalist and geneticist is one example of the many forms of Hein's ongoing conflict between vitalist and mechanist metaphysics. In this instance, mechanism ultimately won, which set biology's scientific agenda for roughly the next century. As is always the case, though, there is a trap that lurks in victory. Biology in general, and evolutionary biology in particular, has now been immersed in the mechanist metaphysic for so long that it has become easy to mistake its persistence as *disproof* of the opposing metaphysic. Yet, it was not disproof of the metaphysical correctness of vitalism that led to mechanism's triumph. It was rather the disproof of a particular operational model of it. What August

[41] Mayr (1982).

Weismann disproved by lopping the tails off generations of mice was not the Lamarckian *idea* but Darwin's particular conception of how the Lamarckian idea worked – pangenesis. To put this defeat into its proper perspective, it is therefore essential to peel away layers of caricature that have been built up around Lamarck's ideas. Of these, the most egregious is the caricature of Lamarckism as the "inheritance of acquired characteristics." This gives the mistaken impression that the kernel of Lamarck's thinking was a mechanism of inheritance. In fact, Lamarck's principal focus was on the problem of evolutionary *adaptation*: the tendency of organisms to shape themselves to function well in a fickle and unpredictable environment. Lamarck gives us much to criticize about his thinking on this problem, of course: his notions of "complexifying force" and "adaptive force" were steeped in the essentialist vitalism of his day, with all the attendant problems already described. Lamarck's particular errors should not blind us, however, to the fundamental problem he sought to confront: that quintessentially vitalist idea, the negotiation of the organism's "many little lives" into a coherent and responsive living entity. Lamarck's principal innovation was his proposal that these attributes, understood by all at the time to be operating *within* organisms during their development and adult lives, could also shape the evolutionary development of *lineages*. In short, the radical challenge inherent in Lamarck's system was to propose a fundamental unity of adaptation across all contexts in which it could occur, whether ontogenetic or phylogenetic. Cuvier's own ideas about correlation of parts and conditions for existence were similarly motivated: to explain how an organism came to exist in a well-adapted state, and how this shaped lineages of organisms in changing environments.[42] It is one of the tragic ironies of French evolutionism that these two men, who were philosophically so much in harmony, were bitter enemies.

Returning to Darwin, it is worth remembering that it was adaptation, rather than any specific inheritance mechanism, that captivated Darwin's own early thinking about evolution and adaptation, and ran as an intellectual undercurrent throughout his life. Darwin came to natural history through that peculiarly British approach to the "many little lives" problem, natural theology, with its focus on perfection and harmony in nature as evidence of the Creator's power and perfection. We remember that among the books Darwin brought as his precious cargo on the *HMS Beagle* was William Paley's exemplary tome, *Natural Theology.* It was adaptation that threw the problem of the finches and tortoises of the different islands of the Galapagos archipelago into stark relief for Darwin. And it was adaptation that was at the heart of Darwin's revelation of the Malthusian struggle for existence as a source of *natural* selection. Even when Darwin came to reject Lamarck's notion of adaptation *across* lineages as unrealistic, he retained the idea that the ongoing struggles of individual organisms to survive in whatever world they found themselves in implied a purposeful, largely unconscious, striving for survival. Even when Darwin fully turned away from adaptationism, as he did in his use of sexual selection to explain seemingly maladaptive sexual characteristics, there remained at the heart of his thinking a hard nugget of sexual striving and desire. In short, there is no escaping the deep vitalist

[42] Gayon (2000), Reiss (2009).

roots that nourished Darwin's thinking. This is why even prominent contemporaries, such as Asa Gray and Ernst Haeckel could cite Darwin as enshrining purpose within evolutionary thought, even if Darwin himself did not think so.

The materialist triumph in the late nineteenth-century effectively strangled this dimension of evolutionary thought, turning rather to a genetic theory of selection which purged adaptation of its essential intentionality and purposefulness. To see purposeful adaptation in the evolution of lineages, manifest in evolving coherency of function, emergence of biological design, evolutionary convergence and so forth, was to yield to the power of illusion and was therefore shunned as "unscientific." Here we begin to see the emergence of elaborate circumlocutions like "apparent" design and intentionality, and tortured semantics over whether hearts were "for" pumping blood (which implied purposefulness) or were organs "that" pumped blood (which supposedly was teleologically neutral), as if the organ and function just happened to coincide.[43] Where adaptation occurred, it was now the outcome of a process of selection for genetic specifiers of "apt function" which forced modern evolutionism into an inconvenient tautology from which it has never found a graceful escape – adaptation exists because genes for "apt function" are selected over genes for "inapt function," with the only criterion for being an "apt function" gene being that it is selected. For a supposedly fully coherent theory of biology, this is rather thin broth.

The problem that confronts modern evolutionism is therefore the same one that confronted Lamarck, and after him, Darwin: is there a common theory of adaptation that can explain both physiological (ontogenetic) and evolutionary (phenotypic) adaptation? Bernard opened the window to such a common theory, although Bernard himself did not make the connection, largely because he deliberately limited his conception of homeostasis to the *milieu interieur*: the environment *inside* a boundary that delimited the organism from its environment. Physiologists ever since have seen little reason to quibble with that boundary, and this has largely kept physiology from being a serious contributor to the science of evolution.[44] This is largely a philosophical choice without much scientific justification.[45] Suffice to say that the

[43] Ruse (2003, 2008).

[44] Purushotham and Sullivan (2010).

[45] In limiting his focus to the *milieu intérieur*, Bernard reveals himself to be solidly embedded in the Physicalist school of medicine. Physicalism was an outgrowth of the Cartesian division between ineffable mind and machine-like body, in which understanding the body as a machine was the principal scientific focus. A corollary of this idea was the clear division between machine-like organism and physical environment. This came to be a very powerful force in European physiology, with the German physiologists such as Hermann van Helmholtz exerting the strongest gravitational pull. Bernard himself was never quite comfortable with the strong materialist implications of this, but it was nevertheless the intellectual environment in which he lived. Political motivation also cannot be discounted, because the example of German physiology was a useful goad to the French authorities to cough up the needed resources to bolster France's relatively feeble physiological endeavors: for nearly his entire career, Bernard found himself starved for laboratory space and assistants; he did much of his pioneering work in a musty space under a staircase at the Collège de France, or out of his home, to the chagrin of his militantly anti-vivisectionist wife, who eventually left him over the matter. One cannot gainsay these efforts: for all of Bernard's considerable intellectual achievements on behalf of physiology, he also left French physiology much better off institutionally.

strong division that Bernard drew between organism and environment, and generations of physiologists continue to draw, cannot be reconciled with even the most elementary principles of conservation of mass and energy.[46] Organisms are dynamic entities, sustained by ongoing and specific flows of matter and energy through them. This means that physiology *within* an organism necessarily imparts a degree of physiology to the environment *outside* the organism as well. In other words, there is no such thing as a self-contained physiology: there is only an "extended physiology," the living organism drawing the environment into a sort of physiological conspiracy to form an "extended organism."[47] With respect to adaptation, this means that the commonly held view – that adaptation is an "in-forming" of the organism by the environment, to use Henry Plotkin's felicitous phrasing[48] – is only half the picture. The extended organism also "out-forms" the environment, shaping it and adjusting it to sustain the precarious disequilibrium of the living organism. Adaptation, and many of the ancillary phenomena that identify it – design, optimization, convergence, cognition – follow from the homeostasis of this "extended organism."[49]

The critical question for evolution is whether homeostasis of the extended organism can inform both phenotypic adaptation (where it is uncontroversial) and evolutionary adaptation (where it is more problematic). The sticking point is an enduring legacy of the triumph of "hard-inheritance" at the beginning of the twentieth century: the strong coupling of hereditary memory to the material gene. If that coupling is strong, that is to say that the material gene is the sole source of hereditary memory, physiological and evolutionary adaptation probably cannot be reconciled. If the material gene can be decoupled even partly from hereditary memory, however, physiological and evolutionary adaptation may indeed have a common mechanism. The homeostasis of the extended organism offers a means for doing just this.

Modern Darwinism is built upon what I have elsewhere called the object-gene.[50] This conceptualization has much to recommend it, but it does not provide the basis for a coherent theory of adaptation, and adherence to it has produced a body of evolutionary theory that has become ever more detached from the phenomenon it seeks to explain. Some of these are minor and untroubling, such as the old habit (now thankfully passing) of designating the non-coding regions of a genome as "junk" DNA.[51] A more worrisome example is the tendency to force difficult phenomena, like culture or intellect, into an artificial conformity to the logic of the object-gene: the reduction of cultures to collections of heritable "memes" is one example of this Procrustean tendency. It has also elevated flawed concepts to a level of generality that is unsupportable by evidence or logic. The "Weismann barrier,"

---

[46] Turner (2007c).

[47] Turner (2000).

[48] Plotkin (1993).

[49] Turner (2007a, c).

[50] I use the word in both its senses: as a material object of heredity, and as a snippet of programmed code. Turner (in press).

[51] "Junk" DNA in humans constitutes over 97 % of the nucleotide sequence of the human genome. Cf. Biemont and Vieira (2006), Wong et al. (2000).

the hermetic bulwark between germ-line and *soma*, is a good example. The Weismann barrier blocks the feedback of physiological adaptation onto the genome that is necessary for any mechanism of soft-inheritance to work. As such it has been laid down as a foundational concept in the Neo-Darwinian synthesis, and, to quote the Black Knight from *Monty Python and the Holy Grail*, "no man shall pass" beyond it. Yet it is applicable, at best, only to a small slice of the diverse forms of life on Earth, and does not even conform well to the physiological evidence in them.[52] Adherence to the object gene has deliberately closed off major problems in evolutionary thought: it is impossible, for example, to think about the origin of life during the inevitable period when there was life (or at least proto-life) without genes.[53] Finally, the object-gene has forced core evolutionary questions to be framed in some rather fantastic ways. Evolutionary fitness, for example, is measured as replication of gene-objects, which swirl about in imaginary gene "pools," carried about in bodies that serve the object-gene as disposable vehicles that power their replication and combination. Adaptation comes when these vehicles come to occupy "niches" in "adaptive landscapes" that exist in "multidimensional hyperspaces", etc....[54] This phantasmagoria, which rivals the Platonic demiurge in its dazzling mystery, is the framework that has been built up to support the object-gene as the principal driver of evolution.

Molecular genetics has steadily undermined the object-gene, however, and the emerging science of epigenetics has finally opened the door to the "soft-inheritance" that Weismann so emphatically rejected a century ago. This is a vast territory of scientific research, ably summarized elsewhere.[55] Suffice to say that the epigenetics revolution has ushered in a transformation in our conception of the gene. No longer is the gene a specifier object, its expression limited to what is embedded in an impervious sequence code. Rather, the gene is a process that is modified, even determined, by the physiological context in which it occurs. This uncouples the phenomenon of hereditary memory from the object-gene, so that hereditary memory can now exist in many forms, no longer solely as sequence information in DNA.[56] Hereditary memory can now reside in epigenetic forms such as patterns of DNA methylation, positional information of transpositional elements, and so forth, but in larger non-nucleotide forms as well: persistence of a cell's catalytic milieu, heritable cytoskeletal patterns, body plans, and even extending beyond the organism's outer boundaries. Now, the organism is not so much a phenomenon specified by material nuggets of hard-inheritance, but is the outcome of a complex negotiation between a multitude of these different forms of hereditary memory. Adaptation – the "in-forming" of the organism by environment, now emerges from a complex web of cognitive interactions among these multitudinous

[52] Ferrer et al. (2010), Franchi et al. (1962), Longo and Anderson (1974), Matthews (1962), Zuckerman et al. (1962).

[53] Cairns-Smith and Hartman (1986), Dyson (1999), Fry (2000), Shapiro (1986).

[54] Dobzhansky (1970).

[55] Jablonka and Lamb (1998a, b, 2005), Gilbert and Epel (2009).

[56] Turner (in press).

memories.[57] Fitness is now no longer the replication of object-genes, but the persistence through time of these cognitive webs. Persistence is homeostasis defined, which suggests an intriguing equivalence of homeostasis and evolutionary fitness.[58]

In short, the emerging picture from molecular genetics is leading not to a final triumph of mechanism over vitalism. It is, rather, turning inexorably back to that core concept that motivated the vitalism of the nineteenth century: the negotiation and reconciliation of innumerable "little lives" to form the complete and complex organism. Now the distinction between phenotype and genotype, drawn as a bright line for so many decades dissolves before our eyes: the "phenotype" is becoming a complex interpretive dance between multiple forms of persistent memory that produce the peculiar focus of specified disequilibrium that is life.

5 Vitalism and the Modern Science of Life

Does all this suggest that are we now poised on the cusp of a newly regnant scientific vitalism? If Hilde Hein's analysis is correct, perhaps we are, but I wish to argue here for a more ambitious proposition: that there is now, within our reach, the possibility of the final reconciliation between vitalism and mechanism that Hein thought impossible to achieve. In that reconciliation, biology may finally be established as the metaphysically distinct discipline it deserves to be. This is not only desirable, it is imperative, for the metaphysical landscape of the life sciences is today more complicated than it was in Hein's day. Presently, the life sciences are being pulled not between two metaphysical poles but three, for "vitalism" today includes not only a newly credible "scientific vitalism" but a resurgent "metaphysical vitalism" that is manifest, on the one hand, in Intelligent Design Theory (essentially a form of Neoplatonism)[59] and on the other in the essentially animist conceptions of nature that are inherent in modern environmentalism[60] (essentially a resurgent Natural Theology). The only way biology can survive as a science is to stake out a distinct metaphysical identity that does not involve a regression into spiritualism, as both intelligent design theory and environmentalism would do. Defining a new metaphysics of biology will mean engaging with and incorporating long-shunned "vitalist" concepts such as intentionality, design, cognition and intelligence as universal properties of life. Homeostasis, I propose, offers the credible scientific foundation for doing so.

Acknowledgments This chapter is part of a book project, *Biology's Second Law: Evolution, Purpose and Desire,* supported through the generosity of a grant from the John C Templeton Foundation.

---

[57] Hoffmeyer (2008), Kull (1999).

[58] Turner (2007c).

[59] Dembski (1999), Turner (2007b), Meyer (2009).

[60] Bowler (1992), Hicks (2011).

References

Benson, Keith R. 1989. Biology's "Phoenix": Historical perspectives on the importance of the organism. *American Zoologist* 29: 1067–1074.
Biemont, Christian, and Cristina Vieira. 2006. Genetics: Junk DNA as an evolutionary force. *Nature* 443(7111): 521–524.
Bligh, John. 1973. *Temperature regulation in mammals and other vertebrates*, Frontiers in biology. Amsterdam: North-Holland Publishing Company.
Bligh, John. 1980. Central neurology of homeothermy and fever. In *Central neurology of homeothermy and fever*, ed. J.M. Lipton, 81–89. New York: Raven.
Bloch, Harry. 1989. François Magendie, Claude Bernard, and the inter-relationship of science, history and philosophy. *Southern Medical Journal* 82(10): 1259–1261.
Bordeu, Théophile. 1751. *Recherches anatomiques sur la position des glandes et leur action*. Paris: Quillau père.
Bordeu, Théophile. 1818. *Œuvres complètes, précédées d'une Notice sur sa vie et ses ouvrages par Monsieur le Chevalier de Richerand*. 2 vols. Paris: Caille et Ravier.
Bowler, Peter J. 1983. *The eclipse of Darwinism. Anti-Darwinian evolution theories in the decades around 1900*. Baltimore: The Johns Hopkins University Press.
Bowler, Peter J. (ed.). 1992. *The Norton history of the environmental sciences*, The Norton history of science. New York: W.W. Norton & Company.
Cabanac, Michel, and Yves Rossetti. 1987. Fever in snails. Reflection on a negative result. *Comparative Biochemistry and Physiology* 87: 1017–1020.
Cairns-Smith, Alexander Graham, and Hyman Hartman (eds.). 1986. *Clay minerals and the origin of life*. Cambridge: Cambridge University Press.
Cannon, Walter B. 1926. Physiological regulation of normal states: Some tentative postulates concerning biological homeostatics. In *Jubilee volume to Charles Richet*, 91–93. Paris: Editions Médicales.
Cannon, Walter B. 1929. Organization for physiological homeostasis. *Physiological Reviews* 9(3): 399–431.
Cannon, Walter B. 1932. *The wisdom of the body*. New York: W W Norton & Company.
Capra, Fritjof. 1996. *The web of life. A new scientific understanding of living systems*. New York: Anchor Books.
Casterlin, Martha E., and William W. Reynolds. 1977a. Behavioral fever in anuran amphibian larvae. *Life Sciences* 20: 593–596.
Casterlin, Martha E., and William W. Reynolds. 1977b. Behavioral fever in crayfish. *Hydrobiologia* 56: 99–101.
Cheung, Tobias. 2008. Regulatory agents, functional interactions, and stimulus-reaction schemes. The concept of "organism" in the organic system theories of Stahl, Bordeu and Barthez. *Science in Context* 21(4): 495–519.
Covert, Jerry B., and William W. Reynolds. 1977. Survival value of fever in fish. *Nature* 267: 43–45.
Cziko, Gary. 2000. *The things we do. Using the lessons of Bernard and Darwin to understand the what, how and why of our behavior*. Cambridge, MA: The MIT Press.
Dawkins, Richard. 1982. *The extended phenotype*. Oxford: W H Freeman & Co.
Dawson, Percy Millard. 1908. *A biography of François Magendie*. New York: Huntington.
Dembski, William A. 1999. *Intelligent design. The bridge between science and theology*. Downer's Grove: Intervarsity Press.
Dobzhansky, Theodosius. 1970. *Genetics of the evolutionary process*. New York: Columbia University Press.
Doolittle, W. Ford. 1981. Is nature really motherly? *The CoEvolution Quarterly* Spring 1981: 58–65.
Dyson, Freeman. 1999. *Origins of life*, 2nd ed. Cambridge: Cambridge University Press.
Ferrer, Aurelie, Anthony F. Dixon, Marc Gibernau, and Jean-Louis Hemptinne. 2010. Ovarian dynamics and specialisation in ladybirds. *Ecological Entomology* 35(1): 100–103.

Foster, Michael. 1899. *Claude Bernard.* New York: Longman's Green & Co.
Franchi, L.L., A.M. Mandl, and Solly Zuckerman. 1962. The development of the ovary and the process of oogenesis. In *The ovary*, vol. I, ed. S. Zuckerman, A.M. Mandl, and P. Eckstein, 1–88. New York: Academic.
Fry, Iris. 2000. *The emergence of life on earth. A historical and scientific overview.* New Brunswick: Rutgers University Press.
Gayon, Jean. 2000. History of the concept of allometry. *American Zoologist* 40: 748–758.
Gilbert, Scott F., and D. Epel. 2009. *Ecological developmental biology: Integrating epigenetics, medicine, and evolution.* Sunderland: Sinauer Associates.
Golley, Frank Benjamin. 1993. *A history of the ecosystem concept in ecology.* New Haven: Yale University Press.
Gross, Charles G. 2008. Claude Bernard and the constancy of the internal environment. *The Neuroscientist* 4(5): 380–385.
Hein, Hilde. 1972. The endurance of the mechanism – vitalism controversy. *Journal of History of Biology* 5(1): 159–188. doi:10.1007/bf02113490.
Henderson, Lawrence J. 1913. *The fitness of the environment. An inquiry into the biological significance of the properties of matter.* New York: Macmillan.
Hicks, Stephen R.C. 2011. *Explaining postmodernism. Skepticism and socialism from Rousseau to Foucault.* Roscoe: Ockham's Razor Publishing.
Hoffmeyer, Jesper. 2008. *Biosemiotics: An examination into the signs of life and the life of signs.* Scranton: University of Scranton Press.
Hölldobler, Bert, and Edward O. Wilson. 2009. *The superorganism: The beauty, elegance, and strangeness of insect societies.* New York: W. W. Norton.
Hudson, Jack W. 1973. Torpidity in mammals. In *Comparative physiology of thermoregulation*, vol. 3, ed. G. Causey Whittow, 98–166. New York: Academic.
Jablonka, Eva, and Marion J. Lamb. 1998a. Epigenetic inheritance in evolution. *Journal of Evolutionary Biology* 11: 159–183.
Jablonka, Eva, and Marion J. Lamb. 1998b. Genic neo-Darwinism – Is it the whole story? *Journal of Evolutionary Biology* 11: 243–260.
Jablonka, Eva, and Marion J. Lamb. 2005. *Evolution in four dimensions: Genetic, epigenetic, behavioral and symbolic variation in the history of life.* Cambridge, MA: MIT Press.
King, J.R., and D.S. Farner. 1961. Energy metabolism, thermoregulation and body temperature. *Biology and Comparative Physiology of Birds* II: 215–288.
Kluger, M.J. 1979. Phylogeny of fever. *Federation Proceedings* 38: 30–34.
Kobayashi, Shigeo. 1989. Temperature-sensitive neurons in the hypothalamus: A new hypothesis that they act as thermostats, not as transducers. *Progress in Neurobiology* 32: 103–135.
Kronenberg, Fredi, and H. Craig Heller. 1982. Colonial thermoregulation in honeybees (*Apis mellifera*). *Journal of Comparative Physiology* 148: 65–76.
Kull, Kalevi. 1999. Biosemiotics in the twentieth century: A view from biology. *Semiotica* 172(1/4): 385–414.
Longo, F.J., and E. Anderson. 1974. Gametogenesis. In *Concepts of development*, ed. J. Lash and J.R. Whittaker, 3–47. Stamford: Sinauer Associates.
Lovelock, James E. 1987. *Gaia. A new look at life on earth.* Oxford: Oxford University Press.
Lyons, Robert D. 1861. *A treatise on fever.* London: Longman, Green, Longman and Roberts.
Magendie, François. 1819. *Recherches physiologiques et chimiques sur l'emploi de l'acide prussique ou hydrocyanique dans le traitement des maladies de poitrine et particulièrement dans celui de la phtisie pulmonaire.* Paris: Méquignon-Marvis.
Margulis, Lynn. 1971. Symbiosis and evolution. *Scientific American* 225: 49–57.
Margulis, Lynn. 1981. *Symbiosis in cell evolution. Life and its environment on the early earth.* San Francisco: W.H. Freeman.
Margulis, Lynn. 1997. Big trouble in biology: Physiological autopoiesis versus mechanistic neo-Darwinism. In *Slanted truths: Essays on Gaia, symbiosis and evolution*, ed. L. Margulis and D. Sagan, 265–282. New York: Copernicus.
Matthews, L.H. 1962. The structure of the ovary. A. Invertebrates. In *The ovary*, vol. I, ed. S. Zuckerman, A.M. Mandl, and P. Eckstein, 89–120. New York: Academic.

Maturana, Humberto R. 1999. The organization of the living: A theory of the living organization. *Journal of Human-Computer Studies* 51(51): 149–168.

Mayr, Ernst. 1982. *The growth of biological thought: Diversity, evolution, and inheritance*. Cambridge, MA: Belknap/Harvard University Press.

McClain, E., P. Magnuson, et al. 1988. Behavioural fever in a Namib desert tenebrionid beetle, *Onymacris plana*. *Journal of Insect Physiology* 34: 279–284.

Meyer, Stephen C. 2009. *Signature in the cell. DNA and the evidence for intelligent design*. New York: Harper One.

Nagashima, Kei, Sadamu Nakai, Mutsumi Tanaka, and Kazuyuki Kanosue. 2000. Neuronal circuitries involved in thermoregulation. *Autonomic Neuroscience* 85(1–3): 18–25. doi:10.1016/s1566-0702(00)00216-2.

Needham, Joseph. 1970. *The chemistry of life: Eight lectures on the history of biochemistry*. Cambridge: Cambridge University Press.

Normandin, Sebastian. 2007. Claude Bernard and an Introduction to the study of experimental medicine: "Physical vitalism", dialectic, and epistemology. *Journal of the History of Medicine and Allied Sciences* 62(4): 495–528. doi:10.1093/jhmas/jrm015.

Oppenheimer, J. 1948. Book review. *La Philosophie de Claude Bernard* by A.-D. Sertillanges. *The Quarterly Review of Biology* 23(4): 337–338.

Pennisi, Elizabeth. 2010. Synthetic genome brings new life to bacterium. *Science* 328(5981): 958–959. doi:10.1126/science.328.5981.958.

Plotkin, Henry. 1993. *Darwin machines and the nature of knowledge*. Cambridge, MA: Harvard University Press.

Prosser, Clifford Ladd. 1973. Temperature. In *Comparative animal physiology*, ed. C.L. Prosser, 362–428. Philadelphia: W.B. Saunders.

Purushotham, Anand D., and Richard Sullivan. 2010. Darwin, medicine and cancer. *Annals of Oncology* 21(2): 199–203. doi:10.1093/annonc/mdp537.

Raskin, Ilya, Axel Ehmann, Wayne R. Melander, and Bastiaan J.D. Meeuse. 1987. Salicylic acid: A natural inducer of heat production in Arum lilies. *Science* 237: 1601–1602.

Reiss, John O. 2009. *Not by design: Retiring Darwin's watchmaker*. Berkeley: University of California Press.

Reynolds, W.W., and J.B. Covert. 1977. Behavioral fever in aquatic ectothermic vertebrates. In *Drugs, biogenic amines & body temperature*, ed. K.E. Cooper, P. Lomax, and E. Schönbaum, 108–110. Banff: S Karger.

Romanovsky, Andrej A. 2007. Thermoregulation: Some concepts have changed. Functional architecture of the thermoregulatory system. *American Journal of Physiology. Regulatory, Integrative and Comparative Physiology* 292: R37–R46. doi:10.1152/ajpregu.00668.2006.

Rosenblueth, Arturo, Norbert Wiener, and Julian Bigelow. 1943. Behavior, purpose, and teleology. *Philosophy of Science* 10: 18–24.

Ruse, Michael. 2003. *Darwin and design. Does evolution have a purpose?* Cambridge, MA: Harvard University Press.

Ruse, Michael. 2008. Purpose in a Darwinian world. In *The deep structure of biology. Is convergence sufficiently ubiquitous to give a directional signal?* ed. S. Conway Morris, 178–194. West Conshohocken: Templeton Foundation Press.

Schmidt-Nielsen, Knut. 1994. How are controlled systems controlled? *American Scientist* 82: 38–44.

Seymour, Roger S., G.A. Bartholomew, and M.C. Barnhart. 1983. Respiration and heat production by the inflorescence of Philodendron sellorum Koch. *Planta* 157: 336–343.

Shapiro, Robert. 1986. *Origins. A skeptics guide to the creation of life on earth*. London: Heinemann.

Stitt, J.T. 1979. Fever vs hyperthermia. *Federation Proceedings* 38: 39–43.

Templeton, J.R. 1970. Reptiles. In *Comparative physiology of thermoregulation*, Invertebrates and nonmammalian vertebrates, vol. 1, ed. G. Causey Whittow, 167–221. New York: Academic.

Turner, J. Scott. 2000. *The extended organism. The physiology of animal-built structures*. Cambridge: Harvard University Press.

Turner, J. Scott. 2007a. Homeostasis, complexity and the problem of biological design. In *Explorations in complexity thinking: Pre-proceedings of the 3rd international workshop on complexity and philosophy*, ed. K.A. Richardson and P. Cilliers, 131–147. Mansfield: ISCE Publishing.

Turner, J. Scott. 2007b. Signs of design. *The Christian Century* 124(12): 18–22.

Turner, J. Scott. 2007c. *The tinkerer's accomplice. How design emerges from life itself.* Cambridge, MA: Harvard University Press.

Turner, J. Scott. 2013. Biology's second law. Homeostasis, purpose and desire. In *Beyond mechanism. Putting life back into biology*, ed. G. Henning and A. Scarfe, 183–203. Lanham: Lexington Books.

Turner, J. Scott. In press. The semiotics of a superorganism. In *Approaches to semiosis of evolution*, eds. K. Kull and Jesper Hoffmeyer. Heidelberg: Springer.

Ullman, Dana. 2007. The philosophical and historical roots of the holistic approaches to health. A review of the first three volumes of "Divided Legacy" by Harris L. Coulter, Ph.D. Homeopathic Educational Services, Berkeley.

Varela, Francisco J., Humberto R. Maturana, and Ricardo Uribe. 1974. Autopoiesis: The organization of living systems, its characterization and a model. *Bio Systems* 5: 187–196.

Virtanen, Reino. 1960. *Claude Bernard and his place in the history of ideas*. Lincoln: University of Nebraska Press.

Williams, George C. 1992. *Gaia*, nature worship and biocentric fallacies. *The Quarterly Review of Biology* 76(4): 479–486.

Wolfe, Charles T. 2008. Introduction. Vitalism without metaphysics? Medical vitalism in the enlightenment. *Science in Context* 21(4): 461–463.

Wolfe, Elin L., A. Clifford Barger, and Saul Benison. 2000. *Walter B. Cannon, science and society*. Cambridge, MA: Harvard University Press.

Wong, Gane Ka-Shu, Douglas A. Passey, Ying-zong Huang, Zhiyong Yang, and Yu Jun. 2000. Is "Junk" DNA mostly intron DNA? *Genome Research* 10(11): 1672–1678. doi:10.1101/gr.148900.

Zuckerman, Solly, A.M. Mandl, and P. Eckstein. 1962. *The ovary*, vol. I. New York: Academic.

Chapter 12
Unanticipated Trends Stemming from Initial Events in the History of Cell Culture: Vitalism in 2013?

Carlos Sonnenschein, David Lee, Jonathan Nguyen, and Ana M. Soto

Abstract During the period 1907–1912, tissue culture pioneers developed the basic techniques that, with modifications, have been adopted by experimental biologists worldwide to resolve a variety of scientific and technological questions. Because their immediate pragmatic concern was the "growth" of the cells, these pioneers may have inadvertently ignored the theoretical underpinnings of why those cells grew in the artificial conditions they imposed on them. By theoretical underpinnings we mean what premises they adopted to interpret the fact that cells grew outside the organism from where they were explanted, i.e., did they favor *proliferation* or *quiescence* as their default state? Here, we argue that the premises adopted and the interpretation of the data they collected introduced important misconceptions that still remain in place. The crucial one has been the notion that *quiescence,* instead of *proliferation,* is the default state of cells in metazoan. Later on, this notion led to the claim that there were "signals," so-called growth factors, that would stimulate those passively quiescent cells to undergo proliferation. Additionally, the notion that *quiescence* as the default state of cells in metazoa is inimical to evolutionary theory because it implies the intervention of some external, undefined entity that instruct cells to enter their cycle of reproduction. Probably unintentionally, this mistaken conclusion carrying a specific command may be considered as the core of a sort of a naïve physicalism that hinders the understanding of biological organization.

Keywords Default state • Emergence • Growth factors • Proliferation • Tissue culture • Senescence

---

C. Sonnenschein • D. Lee • J. Nguyen • A.M. Soto (✉)
Department of Anatomy and Cellular Biology, Tufts University School of Medicine,
136 Harrison Ave., 02111 Boston, MA, USA
e-mail: ana.soto@tufts.edu

S. Normandin and C.T. Wolfe (eds.), *Vitalism and the Scientific Image in Post-Enlightenment Life Science, 1800–2010*, History, Philosophy and Theory of the Life Sciences 2, DOI 10.1007/978-94-007-2445-7_12,

1 Introduction

Much ink has been spilled on the merits of vitalism and physicalism (and related stances such as materialism and mechanicism), i.e., the major competing stances upheld by biologists from the Scientific Revolution to the "triumph" of materialistic views.[1] An analysis of these views reveals that neither vitalism nor physicalism was monolithic.[2] Since the Scientific Revolution, biologists wondered whether physics provided a sound approach to the study of phenomena inherent to biology such as reproduction, embryogenesis, and death. Mayr remarked that postulating special invisible biological forces was inspired by Newtonian physics and gravitation, and thus, when it began, vitalism was totally materialistic. In this vein, the physicists Niels Bohr, Erwin Schrödinger and Walter Elsasser professed a materialistic vitalism as they contemplated the possibility of additional laws operating in biology. They did so at a time when biologists had already abandoned vitalistic ideas. We will leave the analysis of this nuanced range of stances, sometimes of ontological nature, sometimes epistemological, sometimes heuristic, to historians and philosophers.

However, central to the disputes between vitalists and mechanicists was (a) reductionism, the core of physicalism, and (b) emergence, a main issue in vitalism. This conflict was superseded in the twentieth century by organicism, which accepts both bottom up causality (the only one accepted by reductionism) and top-down causalities, which accounts for the ability of the organism to act upon its parts.[3] According to Mayr, the vitalist/mechanicist controversy was settled when the "entelechy," "vis vitalis" and similar notions were replaced by the concept of "program."[4] Ironically, "program" and "information" are not physical entities, but instead, they derive from a human artifact, the computer, and the mathematical theories of computer science.[5] Proponents of the idea of a developmental "program" consider that DNA is the material that contains the "information." From this view emerged the metaphor that the genome is the "book of life," which triggered the loaded question "who wrote the book of life?,"[6] that in turn introduced a most dreaded visitor, a non-material actor. Interestingly, the chasing of vitalism out of scientific discourse brought about a nebulous sort of agency closer to creationism that is lurking behind the "information" discourse. The "program" and "information" detour is increasingly being viewed as inadequate.[7]

These introductory remarks are relevant to the visit we will pay to episodes that took place at the beginning of last century that relate to the momentous introduction

[1] Mayr (1982, 1996), Allen (2005).

[2] Wolfe (2013), Allen (2005).

[3] Gilbert and Sarkar (2000), Soto and Sonnenschein (2005), Soto et al. (2008).

[4] Mayr (1996).

[5] Longo et al. (2012).

[6] Kay (2000).

[7] Longo et al. (2012).

of an experimental tool that has affected our understanding of biology. This tool is cell and tissue culture. Admittedly, however, experimental biologists at large and the pharmaceutical industry have productively used this tool when addressing questions in the fields of developmental, cellular and molecular biology, prevention of diseases (antibodies, vaccines), and therapeutics.

2 The Initial Steps in Tissue Culture

For over a century, tissues and cells have been explanted from multicellular organisms and placed in an artificial environment where they continue to function as living matter. In these conditions, cells proliferate, move and do much more. Groundbreaking techniques introduced at the beginning of the twentieth century provided the means to elucidate developmental processes in those organisms while studying their tissues and cells in an external environment observable through transparent glass.[8] During recent decades, industrial applications of cell and tissue culture techniques have been crucial for the manufacturing of drugs and biological reagents. These beneficial outcomes, however, simultaneously introduced a series of surely unintended misconceptions that have been overlooked for a century. This chapter deals with the impact of those misconceptions in current biology.

In the beginning...

Our interest in the historical aspects of tissue culture stems from our research program initiated in the 1970s focused on explaining why cells proliferated, with particular emphasis on how ovarian estrogens regulated the proliferation of their target cells. Everyone – us included – was then persuaded that estrogens directly *stimulated* the proliferation of their target cells. All along, it remained clear that when estradiol was administered to ovariectomized rodents, epithelial cells of the uterine and vaginal mucosa proliferated. However, *bona fide* estrogen target cells proliferated in culture conditions regardless of whether the standard medium[9] was supplemented or not with estrogens. Meanwhile, when inoculated into suitable animal hosts, these same target cells required estrogens to proliferate at the site of inoculation and to form a tumor. This represented a paradox.

Eventually, when trying to reconcile the above-mentioned paradox, we adopted a premise that *a posteriori* we recognized as being compatible with evolutionary theory, namely, that *proliferation* was the default state of *all* cells. Indeed, microbiologists had for long accepted this concept with regard to the unicellular organisms that they dealt with, be they prokaryotes or eukaryotes; it was axiomatically clear that they proliferated when there were sufficient nutrients in the culture medium.[10] That

---

[8] Landecker (2004), Willmer (1965).

[9] By standard medium, we mean a mixture of salts, sugars, amino acids, vitamins and a few miscellaneous ingredients in a buffered solution plus variable amounts of fetal calf or horse serum (between 5 and 10 % by volume).

[10] Luria (1975).

is, they did not require “positive (stimulatory) signals” to enter into the cell cycle. The vast majority of biologists studying multicellular organisms took for granted, instead, that cells in those organisms did not proliferate unless they were stimulated by “specific” growth factors. To our knowledge, no publication addressed this radical switch of premises until 1980.[11] Meanwhile, however, the premise that *quiescence* was the default state in multicellular organisms was being openly acknowledged, for instance, in the 1994 edition of *Molecular Biology of the Cell* that claimed, but did not explain, why such a switch in default state was necessary: “Thus, while a well-fed yeast cell proliferates unless it gets a negative signal (such as a mating factor) to halt, an animal cell halts unless it gets a positive signal to proliferate.”[12] The 2008 edition of this same popular textbook carries the following comment on the subject: “…for an animal cell to proliferate, it must receive extracellular signals, in the form of mitogens, from other cells, usually their neighbors. Mitogens overcome intracellular mechanisms that block progress through the cell cycle.”[13]

Our bibliographic search provided no plausible explanation for such a radical switch of default state (from *proliferation* to *quiescence*). Thus, starting in 1977, we concluded that the change in default state never happened during evolution. Instead, we reckoned that (a) cells placed *in culture* conditions exercised their constitutive ability to proliferate, and (b) in animals, estrogen administration lead to the proliferation of its target cells because this hormone cancelled the inhibition under which they were actively kept by a plasma-borne inhibitor.[14] Later on, others reached a comparable conclusion regarding the proliferation of estrogen-target cells.[15] Also, a couple of decades later, it was experimentally shown that quiescence in lymphocytes and in hemopoietic cells is an actively induced state, and that proliferation is the default state of embryonal stem cells.[16] Again, the universality of the premise that *proliferation* was the default state of cells made sense from an evolutionary perspective, because it is difficult to imagine the emergence of life, and its maintenance, if cells were not endowed with a dominant and constitutive ability to proliferate.

The above-mentioned conclusion represents a significant departure from the accepted perspective adopted by researchers in the last 100 years. During those years, researchers working with whole animal models approached proliferation according to the perceptions of their discipline (i.e. endocrinologists were prone to see hormones as stimulatory agents of proliferation, while pathologists who focused on organ regeneration, perceived quiescence as an induced state).[17] For

---

[11] Sonnenschein and Soto (1980).

[12] Alberts et al. (1994).

[13] Alberts et al. (2008, 1102).

[14] Soto and Sonnenschein (1987), Sonnenschein et al. (1996).

[15] Lykkesfeldt and Briand (1986), Laursen et al. (1990), Sirbasku and Moreno-Cuevas (2000).

[16] Yusuf and Fruman (2003), Yusuf et al. (2008), Passegué and Wagers (2006), Ying et al. (2008), Casanova (2012).

[17] Weiss and Kavanau (1957).

their part, mathematical modelers of liver regeneration concluded that the kinetics of cell proliferation during regeneration could be best explained if hepatocytes secreted their own plasma-borne inhibitor.[18]

The advent of cell culture provided an opportunity to explore these alternative modes by bypassing organismal complexity in a two-dimensional phase-space. Within this context, we will now explore the pioneers' motivations and the assumptions that culminated in the birth of a technique that has so heavily influenced a broad variety of biological fields, biotechnology included.

3 When, Where and How Were Fundamental Premises on Proliferation Adopted?[19]

Historians have delved into diverse aspects of the birth of tissue culture.[20] An aspect that remains incompletely addressed has been the presumed motivations of the pioneers, the *Zeitgeist* at the time of their contributions (1907–1912) and the impact of the pioneers' narrative on today's experimental biology. We hasten to add that those here identified as pioneers were not the first who attempted to use tissue culture; the prestigious names of Wilhelm Roux and Leo Loeb are the most prominent among those who initiated efforts in this direction, but because according to historians of science their efforts were short-lived and for the most part unsuccessful, we will not deal with them.[21]

The acknowledged pioneers in this field were Ross Granville Harrison, the man actually credited with the invention of tissue culture, as well as Alexis Carrel, Montrose Burrows, and Margaret and Warren H. Lewis. During the critical period between 1907 and 1912, the possible applications of tissue culture seemed limitless and the excitement generated by its development was at its peak. Within these few years it was argued that a new form of life was being created, that internal developmental processes in tissues could be observed directly, that the autonomous powers of cells and tissues 'freed' from the organism could be tested, and that the controls of cellular proliferation could be defined, explored and even manipulated for human benefit. Amidst the novelty and excitement generated by tissue culturing, these biologists, particularly Alexis Carrel, thought themselves to be god-like in their ability to create, control, and exploit living cells.[22] Alexis Carrel's

[18]Bard (1978).

[19]Our current narrative is the result of our bibliographic search and interpretation of the content of the papers cited. We have tried to incorporate all papers that we considered relevant while trying to avoid a "Whiggish" view of past history. Nevertheless, we trust that there is room for a more extended, rigorous scholarly treatment of this period and ideas on the default state of proliferation and motility that prevailed from 1907 to the present.

[20]Oppenheimer (1966), Landecker (2004), Maienschein (1983), Witowski (1979).

[21]Landecker (2004).

[22]Ibid.

contributions to the field of tissue culture deserve a more extended analysis given their sociological connotations (eugenics); however, this is beyond the purview of this inquiry.

4 Milestones in Tissue Culture

The origin of tissue culturing can be attributed to Ross Granville Harrison's landmark paper, 'Observations of a Living Developing Nerve Fiber' read before the Society for Experimental Biology and Medicine and published in the *Proceedings of the Society for Experimental Biology and Medicine* in May 1907.[23] At that time, there were conflicting explanations as to how the nerve fiber developed. One of the competing hypotheses asserted that nerve fibers did not grow out from nerve centers, but rather they differentiated from pre-existing protoplasmic threads in the organism.[24] The other hypothesis, by His and Ramon y Cajal, proposed that nerve fibers grew out from nerve centers toward the periphery.[25] Harrison described a tense debate reaching "…a certain culmination in the controversy between Held and Ramon y Cajal…in which it became clear that the evidence for and against the two hypotheses respectively, rested upon such minute histological details that a decision to which all would subscribe was impossible of attainment."[26]

After a series of experiments that failed to conclusively demonstrate the His-Cajal outgrowth hypothesis, Harrison finally developed the only method that could definitively end the debate: tissue culture. Harrison took a piece of a frog embryo "known to give rise to nerve fibers,"[27] immersed it in clotted frog lymph and inverted the drop on a cover slip into a hollow glass tube. Through this method, Harrison observed that the explanted embryonic nerve cells produced developing nerve fibers that grew out into the lymph-clot. Most importantly, he was able to directly view the development of the nerve fibers as they grew into the medium. This process, which could previously only be inferred by His and Cajal through a series of two-dimensional histological specimens, could finally be observed first-hand through Harrison's technique. This was the irrefutable proof that Harrison simultaneously vindicated the His-Cajal hypothesis and ruled out Hensen's. Harrison triumphantly declared the debate over because "these observations show beyond question that the nerve fiber develops by the outflowing of protoplasm from the central cells."[28]

[23] Abercrombie (1961), Harrison (1907).

[24] Harrison (1910).

[25] Abercrombie (1961).

[26] Held's theory is described by Harrison as a "compromise" theory between Hensen and Cajal, but typically classed as favoring Hensen's theory. Harrison (1910).

[27] Harrison (1907).

[28] Ibid.

In 1910, Harrison restated his results and justified his conclusions regarding the development of the nerve fiber to remaining skeptics. In addition, recognizing the power and potential of tissue culture, Harrison expanded on the possible uses of this new technique:

> This method, which obviously has many possibilities in the study of the growth and differentiation of tissues, has two very distinct advantages over the methods of investigation usually employed. It not only enables one to study the behavior of cells and tissues in an unorganized medium free from the influences that surround them in the body of the organism, but it also renders it possible to keep them under direct continuous observation, so that all such developmental processes as involve movement and change of form may be seen directly instead of having to be inferred from a series of preserved specimens taken at different stages.[29]

From this quotation, it is evident that Harrison primarily viewed tissue culture as a means to study the "growth and differentiation of tissues."[30]

Alexis Carrel and Montrose Burrows took tissue culturing in a completely different direction. Burrows, an assistant to the surgeon Alexis Carrel then at the Rockefeller Institute, joined Harrison's lab in the spring of 1910.[31] Burrows' objective was "...to adapt, if possible, his [Harrison's] method to the investigation of the growth of the tissues of warm-blooded adult animals ..." at which he was successful.[32] Burrows modified Harrison's techniques by first, using plasma, rather than clotted lymph, because of its relative abundance and uniform firmness; next, using chicken embryos rather than frog embryos to apply his technique to a warm blooded animal, and last, Burrows utilized his modified tissue culture method to culture several different types of tissues.[33] Burrows and Carrel spent the next few years feverishly exploring and expanding the realm of tissue culture.

In their first joint publication, Carrel and Burrows cultured various tissues and organs of adult mammals and they commented on the relative ease and simplicity of culturing tissues remarking that: "The cultivation of normal cells would appear to be no more difficult than the cultivation of many microbes."[34] Aside from achieving 'growth' of mammalian tissues in plasma, this publication also described the first efforts to prolong the survival of tissues in culture by placing them in fresh plasma after 6–7 days marking the earliest attempt to artificially manipulate the growth of cells outside the body and established the basis on which to test the presumed 'immortality' of cells.[35]

---

[29] Harrison (1910).

[30] All along this manuscript, we have tried to avoid the use of the word "growth" because of its vague meaning. For over a century, many authors have alluded to this lack of precision in defining "growth" and therefore, to remedy this situation we use the word 'hyperplasia' when referring to increase in cell number and 'hypertrophy' when there is an increase in cell mass of a cell or a tissue. When quoting others, we cannot vouch for the intention of the authors.

[31] Burrows (1910).

[32] Ibid.

[33] Ibid.

[34] Carrel and Burrows (1910a).

[35] Ibid.

Carrel and Burrows also cultivated cancerous tissues in a plasma-based medium derived from animals with or without cancer and concluded that the rate at which the tumor masses grew correlated with the type of medium used (i.e. plasma from cancerous or non-cancerous animal). They could only speculate as to why plasma taken from a cancerous animal and plasma taken from a normal one produced different growth rates in both normal and cancerous cells. While they considered the notion of the "stimulation" or "inhibition" of cell proliferation, they elected not to follow up on this aspect of their work.[36]

In 1911, Carrel aimed at suppressing the factors that may have caused cellular death and proposed to circumvent this cell death by aseptically rinsing the tissue and placing it in fresh plasma, a process that could be repeated continuously.[37] He speculated that the explanted tissues stopped growing because of a buildup of waste materials or the exhaustion of the nutritive factors present in the plasma and concluded that "… under the conditions and within the limits of the experiments, senility and death are not a necessary (in the explanted cells), but merely a contingent, phenomenon."[38] In hindsight, Carrel's views in this regard could have implied that once cells were prevented from dying, they could have adopted a sustained and unquenchable ability to proliferate. This was Carrel's first attempt to create an 'immortal' form of life; later, he claimed to have kept a fragment of heart tissue from a chicken embryo alive in his lab for the next 28 years until his retirement from the Rockefeller Institute in 1939.[39] His recognition that cells removed from their natural milieu were very much alive, and his use of embryonal extracts and body fluids, rather than an arbitrary mixture of defined nutrients could in part be explained by his professed intention to study cell function in a milieu that mimicked as much as possible that of a tissue *in situ*. In this regard, we infer that he was most likely a vitalist who embraced methodological reductionism. His frame of reference was physiology and the notion of internal milieu as the cell proper environment in a Bernardian context.[40] At no point does Carrel take evolutionary theory into consideration to explain why cells proliferate.

Research conducted in the second half of the twentieth century exposed the complexity of the issues of cellular senescence and immortality. As a sample of such complexity it can be mentioned that the probability of establishing a cell line at will (a manifestation of immortality) is highly dependent on the species and tissue of origin; for instance, (a) no established cell line of chicken is known to exist, (b) human "established" cell lines from fibroblasts or epithelial cells are rare while they are easily derived from white blood cells, (c) it is easy to establish cell lines from Chinese hamster tissues, while it is practically impossible to obtain them

---

[36]Carrel and Burrows (1910b, c).

[37]It should be remembered that no antibiotic was then available to reduce bacterial contamination of these cultures. Carrel (1911).

[38]Ibid.

[39]This claim has been a subject of reinterpretation. Ebling (1942).

[40]Carrel (1931).

from Armenian hamsters. A detailed analysis of this subject is beyond the scope of this chapter.

Following the successes of Harrison, Carrel and Burrows on tissue culture using clotted lymph and plasma, in April 1911, Margaret Reed Lewis and Warren H. Lewis of Johns Hopkins University claimed in a landmark paper that the medium in which they cultured embryonic chicken tissue was of known composition. Margaret Reed Lewis was probably influenced by her previous experience cultivating amoeba in nutrient agar and she adapted this approach to the cultivation of mammalian cells. They reported encouraging results with the artificial media, although the "growths have, as a rule, not been so extensive as those in the plasma."[41] The introduction of a partially defined artificial medium to culture tissues was a significant step away from the use of plasma. The Lewises aimed at using a 'very specific' medium when compared to the undefined and complex media used by Harrison, Carrel, and Burrows. Today, it is acknowledged that plasma and serum represent a complex mixture of proteins, sugars, hormones, lipids, and a variety of other components that so far have remained surely undefined. Understandably, the Lewises were attempting to use media with more clearly defined ingredients to reduce the variables known to affect cell proliferation outside of the body.

In sum, Harrison answered the first question '*if/whether or not*' tissues and cells could be kept alive outside of the body, Carrel and Burrows examined '*how*' tissues and cells could be kept alive outside the body, and the Lewises directed their experimental protocol at '*what*' allowed cells to live outside the body. The Lewises went on to explore tissue culture using increasingly, but not truly, defined media. Upon the discovery that embryonic tissue would 'grow readily' in Locke's solution in combination with various salts, the Lewises "attempt(ed) to cultivate such tissues in media all the constituents of which are (were) known,"[42] thus setting the stage for later studies in which other factors ("stimulators of cell proliferation") in the media could be manipulated, isolated and observed for their effect on cells and tissues in culture conditions.

5 The Influence of Claude Bernard on Experimental Biology and Tissue Culture

The rationale for experimental biology at large drew heavily from the contributions of Claude Bernard, the French scientist who is regarded as the greatest physiologist of the nineteenth century, and according to Denis Noble, the first systems biologist.[43] Bernard wanted to resolve the questions of physiology through experimental approaches in a manner equivalent to those then used by chemists and physicists.[44]

[41] Lewis and Lewis (1911a).

[42] Lewis and Lewis (1911b).

[43] Noble (2007).

[44] Lafollette and Shanks (1994).

He realized, however, that biology dealt with complex living beings and that the search for the basic cause-and-effect relationship in living matter was a difficult task without killing the organism or the cells under study. Bernard proposed the concept of the internal milieu in an effort to distinguish living things from the cold and "dead" external world while creating a framework from which to experimentally analyze and eventually "manipulate" organisms.[45]

Bernard also unambiguously talked of "dominating nature and mastering phenomena, an aim that guided experimental sciences since the times of Bacon and Descartes."[46] "Though we do not know the essence of phenomena," he wrote:

> … We can produce or prevent their appearance, because we can regulate their physico-chemical conditions. We do not know the essence of fire, of electricity, of light, and still we regulate their phenomena to our own advantage. We know absolutely nothing of the essence even of life; but we nevertheless regulate vital phenomena as soon as we know enough of their necessary conditions. Only in living bodies these conditions are much more complex and difficult to grasp than in inorganic bodies; that is the whole difference.[47]

While tissue culture disproved to a certain extent the dependence of individual cells on the internal milieu, Bernard's core ideals of experimental attack and mastery over nature were clearly imprinted during the discovery and advancement of tissue culture techniques.[48]

6 Motivations to Pursue Tissue Culture

Though the contributions of Harrison, Carrel, Burrow, and the Lewises to our current understanding of development and to the response of cells to alternate environments are irrefutable, it would be illuminating to retrospectively analyze their thoughts and motivations now with added emphasis on the lack of an evolutionary perspective in their interpretation of data. Thrust in the middle of the ongoing debate regarding the morphogenetic origin of nerve fibers, in the short run, Harrison sought to definitively prove the His-Cajal hypothesis.[49] As an embryologist, however, Harrison primarily saw the uses of tissue culture as a means to study tissue development.[50] Under a Cartesian and Bernardian approach, he sought merely to isolate pieces of the organism in order to better understand the tissue as it related to the organism as a whole, and summed up this sentiment with this analogy: "Logically, then, this method of isolation of cells or pieces of tissue is but the application of the method

[45] Wasserstein (1996).

[46] Ibid.

[47] Bernard (1957).

[48] For an expanded analysis on Bernard's leanings on the vitalism/physicalism controversy cf. Wolfe (2013).

[49] Harrison (1907).

[50] Harrison (1910).

of the physiologist when an organ is isolated in order to find out its function, or that of the experimental embryologist when he isolates the blastomere of the segmenting egg to determine its developmental potencies."[51]

Alternatively, Carrel and Burrows aimed at culturing tissue to "determine some of the laws of cellular physiology." However, other than finding that "adult tissues and organs of mammals can be cultivated outside of the animal body,"[52] they largely failed to find those laws. Later, alone and with the help of Burrows, Carrel continued to report techniques for "cultivating a large quantity of tissue"[53] and "pure cultures of cells,"[54] and especially aimed at prolonging the lifespan of cells in culture.[55] Carrel's motive also appears to have been the exploration of the "immortal" attribute that cells gained outside of the body and towards this aim he developed techniques to find ideal environments in which to facilitate a new, permanent, form of life. Carrel explicitly acknowledged that cells in culture were "liberated" and thus were able to achieve this new form of (immortal) life, an idea that readily fits with the premise that *proliferation* is the default state of cells.[56] However, this notion was not pursued further by him or by the overwhelming majority of experimental biologists who adopted cell culture as a tool during the last and current centuries (see above). Carrel and Burrows shared Harrison's view that tissue culture could be used as a means of exercising control over diverse forms of living matter (embryonic tissue, adult tissue, cancerous tissue).[57]

Lastly, Margaret and Warren Lewis aimed at culturing tissues in completely known, "defined" media.[58] The Lewises concluded that "the most important advantage ... consists in the fact that we are dealing with solutions of known chemical constitution and that our picture is uncomplicated by structures except those which have grown out from the original piece."[59] The Lewises' contributions point toward the isolation and reduction of variables in both the tissue and the media. Though never explicitly stated, based on the design of many of their experiments with supposedly chemically defined media, it can be inferred that the Lewises were searching for factors that operationally allowed or stimulated the 'growth' of cells or "growth factors," much likely in the spirit of the earlier empirical search for suitable conditions to grow bacteria.[60] Landecker described this period as a time when the field of biology was shifting away from its stance as a natural science, to one where biologists saw themselves as designers and

[51] Harrison (1913).

[52] Ibid.

[53] Carrel (1912a).

[54] Carrel (1912b).

[55] Carrel and Burrows (1910b), Carrel (1911).

[56] Carrel (1912c).

[57] Carrel and Burrows (1910a).

[58] Lewis and Lewis (1911a, b).

[59] Lewis and Lewis (1912a, b).

[60] Sonnenschein and Soto (1999).

inventors of new things, and when the significance of 'nature' in experimental biology began to evaporate.[61] Although no obvious intellectual connection could be established between the tissue-culture pioneers and like-minded groups, the 1907–1912 period coincides with the pioneering efforts of Morgan's genetics lab at Columbia University that also transported "nature" to the well-regulated environment of the bench-laboratory.

Regardless of intentions, the use of tissue culture by the pioneers appears as a reflection of their view that life, within the sphere of the experiment, could be magically controlled by humans. Even though no reference is made in the pioneers' papers to Darwin's ideas, it is unlikely that they would have been unaware of their existence and of their impact on society at large. In fact, both Bernard[62] and Harrison[63] acknowledged the value of evolution; however, they appear to consciously avoid discussing it in the context of their experimental work. So, what component of the "big picture" did the pioneers miss in their attempt to highlight their methodological contribution to experimental biology?

7 Concepts Not Challenged and Questions Not Asked

By the time the pioneers undertook their respective aim to culture cells in glass flasks, the Darwinian blueprint of evolution had gone through five decades of discussions and a decade from the rediscovery of Mendel's original experiments (in 1865) during which their merits were assessed. As mentioned above, the pioneers did not show much concern about those momentous discoveries and neither were they concerned with the question… *why* cells 'grew' (proliferated) in culture. Had the pioneers framed this latter question within an evolutionary perspective, they could have speculated about having at least two plausible alternatives to choose from. The first one would have been that metazoan cells proliferated as soon as they were removed from inhibitory influences prevalent in the whole intact body, a thought hinted at, but not pursued, by Carrel (see above). *Proliferation* would have been then declared as the default state of all cells. In fact, Darwin made explicit the concept that living entities reproduce following a geometrical progression ("There is no exception to the rule that every organic being naturally increases at so high a rate, that, if not destroyed, the earth would soon be covered by the progeny of a single pair"). Implicit in Darwin's theory is the concept of a dominant and constitutive ability of organisms to reproduce, which directly leads to the concept of cell proliferation as a dominant and constitutive property of cells and thus to *proliferation* as their default state. The second alternative would have been that, once removed from whole organisms where they were subject to stimulation by

[61] Landecker (2004).

[62] Wolfe (2013).

[63] Maienschein (1983).

endogenous signals, cells would have become *quiescent* in culture conditions, and thus the "growth factors" presumably present in either plasma or serum would have stimulated their proliferation. The influential paper by Eagle and Piez,[64] published in 1960, consolidated the notion that the default state of cells was *quiescence* when they concluded that plasma proteins were not used by cells as nutrients but as alleged cell proliferation stimulators.[65] Imperceptibly but effectively, the second alternative prevailed and the operational definition of "growth factors" *à la* Lewises eventually morphed into the meaning adopted in the second half of the twentieth century and still prevalent today.[66]

From a methodological perspective, during the period 1907–1912, the evaluation of the role of ingredients included in the culture medium either as plain nutrients or as alleged stimulators of cell proliferation was rather limited because no rigorous way to measure increases or decreases in cell mass, DNA content or cell numbers were then available. Accurate cell number estimates in culture were incorporated into routine laboratory practices after Moscona re-introduced trypsin treatment (a follow-up of an original observation by Peyton Rous in 1916 (Moscona 1952)). Also, particle-counting machines were only introduced in the 1960s and 1970s.

What has been the impact on experimental biology and biomedicine of not having explicitly chosen, or even discussed, the alternatives regarding what we have called the *default state* of cells? Aware of the impropriety of adopting a 'Whiggish' historical approach, and with a century of hindsight on what has been productive and unproductive, one may divide the answers into epistemological and pragmatic ones, the latter loosely defined as "what worked" (after all, regardless of which was the cells' default state, experimental results were publishable using cells in culture). The epistemological answers require that those cell and tissue culture pioneers would have had a solid background in evolutionary theory, which would have likely prevented them from accepting without challenge the premise that *quiescence* was the default state of cells in metazoa.

Regardless of motivations, the adoption of *quiescence* as the default state of cells in metazoa generated fields of biology whose relevance are now being increasingly questioned. The two most obvious are (a) the introduction of growth factors and oncogenes,[67] operational notions that lack reliable quantitative support to explain cell proliferation in metazoa,[68] and indirectly, (b) the acceptance that cancer is a disease of the control of cell proliferation. Those siding with the somatic mutation theory of carcinogenesis, which was hatched in 1914 by Theodor Boveri, tacitly incorporated into this theory the premise that *quiescence* was the default state of cells in metazoa. Six decades later (in the 1970s), this arbitrary decision facilitated the creation of the notion of oncogenes (i.e. mutated genes that in a dominant

[64] Eagle and Piez (1960).

[65] For an extended discussion of this subject cf. Sonnenschein and Soto (1999).

[66] Alberts et al. (2008), Alberts (2010).

[67] Bishop (1991).

[68] Sonnenschein and Soto (1999, 2008).

fashion and through as yet-to-be defined ways overcome the alleged quiescent state of cells in metazoa which causes unrestrained cell proliferation).[69] Critical assessments of this notion have been published.[70]

8 Looking Back…and Conclusions

The pioneers of cell and tissue culture did not explicitly state the philosophical stances behind their research. One could interpret Harrison's aims as compatible with both vitalism and mechanicism, given that he tried to keep his explants in conditions as close as possible to those *in situ* and tried to observe a phenomenon that could not be observed *in vivo*, but did not try to apply to it an external "causal agent." At the same time, he separated a tissue from the embryo, and thus, his approach was not holistic (a prevalent stance among some vitalists who at that time thought that removal from the organism deprived organs and tissues of the forces that made them alive). In fact, Harrison believed that separating a tissue from the influences of the organism was advantageous because it made possible the direct observation of some vital phenomena, and because it provided added information about how the "parts" behave when isolated from the whole.

Carrel's stance regarding vitalism/mechanicism is also complex. On the one hand, he was a surgeon who wanted to preserve tissues outside the organism. On the other, he thought that keeping cells outside the organism suppressed senescence and death, two characteristics of multicellular organisms, while keeping intact the ability of cells to proliferate as if they were bacteria. He did not elaborate on the meaning of this phenomenon that he related to the ability of microbes to thrive and reproduce. His unrealized purpose of discovering laws of cell biology, as well as his stated adherence to physiology and the role of the internal milieu suggest that he was a vitalist who practiced methodological reductionism. The idea of dominating Nature, in this case, avoiding aging and death, is a recurrent theme of the mechanicists, from Descartes to Jacques Loeb.

For their part, the Lewises seem to have operated in a totally pragmatic way, that of creating artificial conditions of life, and of having control over these cells. This aim, as well as their search for "growth factors," seems to be on the side of physicalism, if not mechanicism. Obviously, the main difference between the world of the living and that of inanimate matter is that the latter is in inertia (if placed in a vacuum where there is not friction) and it is passive (it does nothing by itself and moves only when external forces act upon it). Living organisms, instead, are active, they move, they proliferate, they generate heat. In the Lewises' scheme of things, these cells needed to be "stimulated" to grow and proliferate; for them, they probably were as passive as stones. Should we attribute to the Lewises physicalist thinking?

[69] Hanahan and Weinberg (2000), Soto and Sonnenschein (2004).

[70] Bizarri et al. (2008), Sonnenschein and Soto (2011), Soto and Sonnenschein (2011).

Or, did they consider that agency, a property of the living so central to vitalism, should have been transferred to the will of the person who placed the cells in a dish? Thus, cells multiplied because the researcher (the cell/tissue culture person) was the agency that commanded them to do so. The addition in the second half of the twentieth century, of the idea of a "program" in biology led to the incongruent state of affairs whereby cells needed to receive "information" or "signals" in order to do something that they were inherently endowed to do (i.e. to proliferate and to move). Can this be considered as a covert form of neo-creationism?

From an organicist view, a perspective that we embrace, cell culture represents a state of de-emergence, whereby the cells that form part of an organism are liberated from the constraints imposed by it. Under extra-organismic conditions, these cells regain properties that mimic those of the unicellular organisms from which the multi-celled organism evolved. This brings up the relevance of placing cell and tissue culture in an evolutionary framework. Neither Bernard nor Harrison recognized a need to apply evolutionary theory to the practice of organismal biology (Bernard) or when venturing into quasi-artificial life (Harrison). In hindsight, this was a squandered opportunity to recognize that in the quasi-artificial life of the culture flask, metazoan cells behave as unicellular organisms, and thus exert their constitutive ability to proliferate and move, properties that enabled the last universal common ancestor (LUCA) to generate the diversity of life on earth as we know it today.

Acknowledgements This work was supported by grants from the Parsemus Foundation and the NIH (ES0150182, ES012301 and ES08314). We are grateful to Cheryl Schaeberle and Michael Askenase for their excellent editorial assistance.

References

Abercrombie, M. 1961. Ross Granville Harrison 1870–1959. *Biographical Memoirs of Fellows of the Royal Society* 7: 111–126.

Alberts, B. 2010. Model organisms and human health. *Science* 330: 1724.

Alberts, B., et al. 1994. *Molecular biology of the cell*. New York: Garland.

Alberts, B., et al. 2008. *Molecular biology of the cell*. London: Garland Science.

Allen, Garland E. 2005. Mechanism, vitalism and organicism in late nineteenth and twentieth-century biology: The importance of historical context. *Studies in History and Philosophy of Biological and Biomedical Sciences* 36: 261–283.

Bard, J.B.L. 1978. A quantitative model of liver regeneration in the rat. *Journal of Theoretical Biology* 73: 509–530.

Bernard, Claude. 1957. *An introduction to the study of experimental medicine*. New York: Dover Press.

Bishop, J.M. 1991. Molecular themes in oncogenesis. *Cell* 64: 235–248.

Bizzarri, M., A. Cucina, F. Conti, and F. D'Anselmi. 2008. Beyond the oncogene paradigm: Understanding complexity in cancerogenesis. *Acta Biotheoretica* 56: 173–196.

Burrows, Montrose. 1910. The cultivation of tissues of the chick-embryo outside the body. *Journal of the American Medical Association* 20: 2057–2058.

Carrel, Alexis. 1911. Rejuvenation of cultures of tissues. *Journal of the American Medical Association* 57: 1611.

Carrel, Alexis. 1912a. Technique for cultivating a large quantity of tissue. *The Journal of Experimental Medicine* 15: 393–396.

Carrel, Alexis. 1912b. Pure cultures of cells. *The Journal of Experimental Medicine* 16: 165–168.

Carrel, Alexis. 1912c. On the permanent life of tissues outside the body. *The Journal of Experimental Medicine* 15: 516–528.

Carrel, Alexis. 1931. The new cytology. *Science* 73: 297–303.

Carrel, Alexis, and Montrose Burrows. 1910a. Cultivation of adult tissues and organs outside of the body. *Journal of the American Medical Association* 55: 1379–1381.

Carrel, Alexis, and Montrose Burrows. 1910b. Cultivation of sarcoma outside of the body: A second note. *Journal of the American Medical Association* 55: 1554.

Carrel, Alexis, and Montrose Burrows. 1910c. Artificial stimulation and inhibition of the growth of normal and sarcomatous tissues. *Journal of the American Medical Association* 56: 32–33.

Casanova, J. 2012. Stemness as a cell default state. *EMBO Reports* 13: 396–397.

Eagle, H., and K.A. Piez. 1960. The utilization of proteins by cultured human cells. *Journal of Biological Chemistry* 235: 1095–1097.

Ebling, A.H. 1942. Dr. Carrel's immortal chicken heart: Present, authentic facts about the oft-falsified scientific "celebrity". *Scientific American* 22–24.

Gilbert, Scott F., and Sahotra Sarkar. 2000. Embracing complexity: Organicism for the 21st century. *Developmental Dynamics* 219: 1–9.

Hanahan, D., and R.A. Weinberg. 2000. The hallmarks of cancer. *Cell* 100: 57–70.

Harrison, R.G. 1907. Observations of the living developing nerve fiber. *Anatomical Record* 1: 116–128.

Harrison, R.G. 1910. The outgrowth of the nerve fiber as a mode of protoplasmic movement. *The Journal of Experimental Zoology* 9: 787–846.

Harrison, R.G. 1913. The life of tissues outside the organism from the embryological standpoint. *Transactions of the Congress of American Physicians and Surgeons* 9: 63–75.

Kay, Lily E. 2000. *Who wrote the book of life?: A history of the genetic code*. Stanford: Stanford University Press.

LaFollette, H.S., and N. Shanks. 1994. Animal experimentation: The legacy of Claude Bernard. *International Studies in the Philosophy of Science* 8: 195–209.

Landecker, H. 2004. *Culturing life*. Cambridge, MA: Harvard University Press.

Laursen, I., P. Briand, and A.E. Lykkesfeldt. 1990. Serum albumin as a modulator on growth of the human breast cancer cell line, MCF-7. *Anticancer Research* 10: 343–351.

Lewis, M.R., and W.H. Lewis. 1911a. The growth of embryonic chick tissues in artificial media, agar, and bouillon. *Johns Hopkins Hospital Bulletin* 22: 126–127.

Lewis, M.R., and W.H. Lewis. 1911b. The cultivation of tissues from chick embryos in solutions of NaCl, $CaCl_2$, KCl, and $NaHCO_3$. *Anatomical Record* 5: 277–293.

Lewis, M.R., and W.H. Lewis. 1912a. The cultivation of sympathetic nerves from the intestine of chick embryos in saline solutions. *Anatomical Record* 6: 7–31.

Lewis, W.H., and M.R. Lewis. 1912b. The cultivation of chick tissues in media of known chemical constitution. *Anatomical Record* 6: 207–211.

Longo, Giuseppe, Paul-Antoine Miquel, Carlos Sonnenschein, and Ana M. Soto. 2012. Is information a proper observable for biological organization? *Progress in Biophysics and Molecular Biology* 109: 108–114.

Luria, Salvador Edward. 1975. *36 lectures in biology*. Cambridge, MA: MIT Press.

Lykkesfeldt, A.E., and P. Briand. 1986. Indirect mechanism of oestradiol stimulation of cell proliferation of human breast cancer cell lines. *British Journal of Cancer* 53: 29–35.

Maienschein, Jane. 1983. Experimental biology in transition: Harrison's embryology 1885–1910. *Studies in the History of Biology* 6: 107–127.

Mayr, Ernst. 1982. *The growth of biological thought: Diversity, evolution, and inheritance*. Cambridge, MA: Belknap.

Mayr, Ernst. 1996. *This is biology*. Cambridge, MA: Harvard University Press.

Moscona, A. 1952. Cell suspensions from organ rudiments of chick embryo. *Experimental Cell Research* 3: 535–539.

Noble, Denis. 2007. Claude Bernard, the first systems biologist, and the future of physiology. *Experimental Physiology* 93: 16–26.

Oppenheimer, J. 1966. Ross Harrison's contributions to experimental embryology. *Bulletin of the History of Medicine* 40: 525–543.
Passegué, E., and A.J. Wagers. 2006. Regulating quiescence: New insights into hematopoietic stem cell biology. *Developmental Cell* 10: 415–417.
Sirbasku, D.A., and J.E. Moreno-Cuevas. 2000. Estrogen mitogenic action. II. Negative regulation of the steroid hormone-responsive growth of cell lines derived from human and rodent target tissue tumors and conceptual implications. *In Vitro Cellular & Developmental Biology* 36: 428–446.
Sonnenschein, Carlos, and Ana M. Soto. 1980. But ... are estrogens per se growth-promoting hormones? *Journal of the National Cancer Institute* 64: 211–215.
Sonnenschein, Carlos, and Ana M. Soto. 1999. *The society of cells: Cancer and control of cell proliferation*. New York: Springer.
Sonnenschein, Carlos, and Ana M. Soto. 2008. Theories of carcinogenesis: An emerging perspective. *Seminars in Cancer Biology* 18: 372–377.
Sonnenschein, Carlos, and Ana M. Soto. 2011. The death of the cancer cell. *Cancer Research* 71: 4334–4337.
Sonnenschein, Carlos, Ana M. Soto, and C.L. Michaelson. 1996. Human serum albumin shares the properties of estrocolyone-I, the inhibitor of the proliferation of estrogen-target cells. *Journal of Steroid Biochemistry and Molecular Biology* 59: 147–154.
Soto, Ana M., and Carlos Sonnenschein. 1987. Cell proliferation of estrogen-sensitive cells: The case for negative control. *Endocrine Reviews* 8: 44–52.
Soto, Ana M., and Carlos Sonnenschein. 2004. The somatic mutation theory of cancer: Growing problems with the paradigm? *BioEssays* 26: 1097–1107.
Soto, Ana M., and Carlos Sonnenschein. 2005. Emergentism as a default: Cancer as a problem of tissue organization. *Journal of Biosciences* 30: 103–118.
Soto, Ana M., and Carlos Sonnenschein. 2011. The tissue organization field theory of cancer: A testable replacement for the somatic mutation theory. *BioEssays* 33: 332–340.
Soto, Ana M., Carlos Sonnenschein, and Paul-Antoine Miquel. 2008. On physicalism and downward causation in developmental and cancer biology. *Acta Biotheoretica* 56: 257–274.
Wasserstein, A.G. 1996. Death and the internal milieu: Claude Bernard and the origins of experimental medicine. *Perspectives in Biology and Medicine* 39: 313–326.
Weiss, P., and J.L. Kavanau. 1957. A model of growth control in mathematical terms. *Journal of General Physiology* 41: 1–47.
Willmer, E.N. 1965. *Cells and tissues in culture: Methods, biology, and physiology*. Cambridge, MA: Academic.
Witowski, J.A. 1979. Alexis Carrel and the mysticism of tissue culture. *Medical History* 23: 279–296.
Wolfe, Charles T. 2013. Vitalism and the resistance to experimentation on life in the eighteenth century. *Journal of the History of Biology* 46: 255–282.
Ying, Q.L., J. Wray, J. Nichols, L. Batlle-Morera, B. Doble, J. Woodgett, P. Cohen, and A. Smith. 2008. The ground state of embryonic stem cell self-renewal. *Nature* 453: 519–523.
Yusuf, I., and D.A. Fruman. 2003. Regulation of quiescence in lymphocytes. *Trends in Immunology* 24: 380–386.
Yusuf, I., M.G. Kharas, J. Chen, R.Q. Peralta, A. Maruniak, P. Sareen, V.W. Yang, K.H. Kaestner, and D.A. Fruman. 2008. KLF4 is a FOXO target gene that suppresses B cell proliferation. *International Immunology* 20: 671–681.

[illegible]

[illegible]
[illegible] Review [illegible]

[illegible]
[illegible]

[illegible]
[illegible]
[illegible]

Chapter 13
Varieties of Living Things: Life at the Intersection of Lineage and Metabolism

John Dupré and Maureen A. O'Malley

Abstract We address three fundamental questions: What does it mean for an entity to be living? What is the role of inter-organismic collaboration in evolution? What is a biological individual? Our central argument is that life arises when lineage-forming entities collaborate in metabolism. By conceiving of metabolism as a collaborative process performed by functional wholes, which are associations of a variety of lineage-forming entities, we avoid the standard tension between reproduction and metabolism in discussions of life – a tension particularly evident in discussions of whether viruses are alive. Our perspective assumes no sharp distinction between life and non-life, and does not equate life exclusively with cellular or organismal status. We reach this conclusion through an analysis of the capabilities of a spectrum of biological entities, in which we include the pivotal case of viruses as well as prions, plasmids, organelles, intracellular and extracellular symbionts, unicellular and multicellular life forms. The usual criterion for classifying many of the entities of our continuum as non-living is autonomy. This emphasis on autonomy is problematic, however, because even paradigmatic biological individuals, such as large animals, are dependent on symbiotic associations with many other organisms. These composite individuals constitute the metabolic wholes on which selection acts. Finally, our account treats cooperation and competition not as polar opposites but as points on a continuum of collaboration. We suggest that competitive relations are a transitional state, with multi-lineage metabolic wholes eventually outcompeting selfish competitors, and that this process sometimes leads

J. Dupré (✉)
ESRC Centre for Genomics in Society (Egenis), University of Exeter,
Byrne House, St. German's Road, Exeter EX4 4PJ, UK
e-mail: j.a.dupre@ex.ac.uk

M.A. O'Malley
Department of Philosophy, University of Sydney, Sydney, NSW, Australia
e-mail: m.a.o'malley@ex.ac.uk

S. Normandin and C.T. Wolfe (eds.), *Vitalism and the Scientific Image in Post-Enlightenment Life Science, 1800–2010*, History, Philosophy and Theory of the Life Sciences 2, DOI 10.1007/978-94-007-2445-7_13,

to the emergence of new types or levels of wholes. Our view of life as a continuum of variably structured collaborative systems leaves open the possibility that a variety of forms of organized matter – from chemical systems to ecosystems – might be usefully understood as living entities.

Keywords Living • Non-living • Collaboration • Lineage formation • Metabolic whole • Autonomy

1 Foreword: 'Varieties of Living Things' and Its Relation to Vitalism

This chapter[1] was not written with vitalism in mind, and we do not use the word even once in our lengthy discussion of lineage and metabolism. However, we did write about the concept of life, and vitalism is a theme intimately entwined with questions about how life should be understood. We suggest vitalism is relevant to our discussion in two ways. The first is indirectly, as a heuristic that stimulates productive inquiries into the nature of living and non-living things, and how the former may or may not become the latter. The vitalism heuristic makes certain claims about the relationship between life and non-life that can be addressed philosophically and sometimes scientifically. Whatever the outcome for vitalism as a theoretical movement, addressing the questions it raises has a long and intriguing history, as this volume shows. We recognize that the legacies of this historical debate may well be precisely the stimulus for the sorts of questions we address in our discussion of the varieties of living things.

The second connection between our paper and vitalism could be considered a more direct one. The earlier work of one of us on reductionism and pluralism (Dupré 1993) argued for a non-reductive materialism. Materialism, of course, rules out any kind of substance vitalism. However, a non-reductive materialism claims only that everything is composed of the same (material) stuff, not that the properties of that stuff are sufficient to explain the properties of the complex entities that it composes. So it is entirely consistent with this kind of materialism that there should be distinctive principles that apply only to material when it is arranged into living processes.

But our ultimate rejection of any sharp boundary between life and non-life might make it hard to see how distinctive principles could apply to all and only instances of the former. Note, however, that we also stress a particular kind of dynamic organization that emerges from systems of molecules up to and beyond complex consortia of organisms. This focus, and the way investigations of this dynamic are instantiated in many modern molecular life sciences, may bring resolution to the issues that motivate many varieties of vitalism.

We are agnostic and inclined to be doubtful as to whether the philosophical developments that we see as important for twenty-first century philosophy of

[1]An earlier version of which appeared in the online journal *Philos Theor Biol* (2009) 1:e003. Thanks to *Philosophy and Theory in Biology* for their permission to reprint.

biology will best be served by recovering ideas from the vitalist tradition. But regardless of whether vitalism redux guides future trajectories of philosophy of biology, the vitalism heuristic and the attention it draws to the complex issues surrounding distinctions between the living and non-living will be enduringly valuable.

2 Introduction

> It would seem that 60 years after Erwin Schrödinger wrote his book 'What is Life?' we should be able to answer the question. However, Nature never ceases to challenge the limits of our imagination.
>
> —M. Y. Galperin (2005, 149)

This essay will not attempt to provide a definition that answers Schrödinger's question. We shall instead address it by describing a spectrum of biological entities that illustrates why no sharp dividing line between living and non-living things is likely to be useful. The more positive goal of these reflections will be to offer a flexible view of life that does in fact make good sense of why particular organizations of matter can be described as living. By identifying the different capacities exhibited by the various entities constituting our spectrum, especially problem cases such as viruses, we hope to address at least some of the issues that lie behind Schrödinger's question and its many earlier precursors and subsequent echoes. Such concerns have been raised in a striking way by recent attempts under the rubric of 'synthetic biology' to synthesize life from basic chemical building blocks.

In this chapter we shall highlight a tension in standard discussions of characteristics of life, which tend to prioritize one or other of two fundamental but very different features of living things: the capacity to form lineages by replication and the capacity to exist as metabolically self-sustaining wholes. We suggest that this tension can best be resolved by seeing life as something that arises only at the intersection of these two features: matter is living when lineages are involved – directly or indirectly – in metabolic processes. But also crucial to our argument and, we suggest, to many of the difficulties that have confronted attempts to comprehend life, is the observation that the entities that form lineages are not always, or even usually, the same as those that form metabolic wholes. Metabolism, the transformative biochemical reactions that sustain life processes, we shall argue to be a collaborative affair. Life, we claim, is typically found at the collaborative intersections of many lineages, and we even suggest that collaboration should be seen as a central characteristic of living matter – a claim that also has implications for how we understand the origins of life. Further corollaries of this non-coincidence of parts of lineages with metabolic wholes are, first, that we cannot assume the identification of living things with organisms (at least as standardly conceived), and nor, second, can we assert traditional organisms to be 'the' biological individuals on which selection operates.

3 Collaboration and the Diversity of Life

The collaborative nature of living entities and processes is our essential starting point. Darwin's theory of natural selection has, quite appropriately, focused a great deal of theoretical interest on questions of competition. This focus, however, has had the less salutary consequence of diverting attention from the equally important topic of cooperation and has culminated in the assumption that altruism, understood as the conferral of a benefit by one biological entity on another, is a profound theoretical problem. Although this is generally seen as a problem pertaining to organisms, a similar argument has notoriously been applied to the topic of genes. Richard Dawkins (1976) made famous the idea that genes are fundamentally selfish entities in competition with one another. From this point of view, it is truly remarkable that the whole consortium of genes in an organism's genome can nevertheless manage to collaborate on a task as momentous as development.

In this chapter we place selfishness in a wider context and emphasize the broader perspective of life as a collaborative enterprise. We are not arguing that interpretations of selfishness are invalid but that, at best, they can only provide a limited perspective on life and evolution. Rather than reducing cooperation to selfishness, we suggest selfishness and cooperation might better be understood within a framework of collaboration. By collaboration, we mean interactions between components of a system that lead to different degrees of stability, maintenance or transformation of that system. As in scientific collaborations, there may be some strongly selfish interests involved in such interactions (Hull 1988) but these selfish activities can only operate in a collaborative context. Defecting from collaboration is only possible if collaboration is the general default.

In every domain of organismal life, there are extensive sets of organisms that are problematic for standard evolutionary understandings of selfish individuals (Roughgarden 2009). Shared interests can lead to highly cooperative 'team' behaviour, described by Joan Roughgarden as 'cooperative teamwork' (2009, 13). Evolutionary payoffs for such team members may not be equal, but are distributed across the whole team. Collaboration, however, may also include the 'mere' coincidence of individual interests, and it is often in the interest of any individual to collaborate – at least to some extent. Collaboration from this point of view covers a range of interactive processes that may include both cooperative and competitive activities. At one end of this continuum the goals of participants may be completely aligned, while at the other end of the continuum, relationships may be largely or wholly hostile. We will try particularly to understand the evolutionary persistence of apparently 'parasitic' or selfish interactions between organisms, and the nature of the entities formed by what are usually conceived as separate biological individuals.

Part of cooperation is merely interactive combination. Thus atoms combine to produce molecules, and the latter have properties that are not found in any of the atoms of which they are composed. But certainly more than this is required to count as collaboration in the sense we are elaborating. In common with most who have considered the question of how living entities are constituted, we assume that there is one necessary condition for being a living thing that most combinations of atoms and molecules lack: the ability to reproduce. Though we take this to be a necessary condition, it is less obvious that it is sufficient. Living entities have also to be understood in relation to their capacity to sustain themselves through biochemical transformations. Metabolism in our account can be engaged in autonomously (this is the usual understanding) or collaboratively, through interactions with other biological entities. At any rate, as the microbial and microbe-like entities that we shall describe below illustrate, a very diverse group of things both reproduce and participate in metabolic systems.

Our empirically informed investigation of living matter will not be based on the animal, fungi or plant life that has been the main concern of philosophers and scientists working on these issues; nobody questions the status of these as living things, and the problem is only one of deciding which of their characteristics confer the status of living on them. We shall focus instead on the realm loosely referred to as microbial, which includes some entities only contentiously afforded living status. Microbes are a group of organisms biologically and conceptually diverse to the point of incoherence, but then so are the macroorganisms or macrobes that loom large in most perspectives on life (O'Malley and Dupré 2007). The category of microbes includes at least protists (unicellular eukaryotes, which have membrane-bound nuclei and other organelles), prokaryotes (which do not have such compartments but are highly organized in other ways), and viruses.

Viruses are the biological objects that are the pivot of our discussion because many biologists deny that they are living organisms. In fact, they are frequently considered to be prime examples of the boundary between life and non-life, organism and non-organism, and biology and chemistry (e.g., Stanley 1957; Wimmer 2006). They are most often assigned to the second of each of these pairs of categories. Viruses are often deemed not to be alive on the grounds that they cannot reproduce themselves autonomously, nor can they metabolize. They can, however, carry out such biologically impressive activities as entering cells, co-opting the transcription and translation machinery of the cell, and picking up and moving about DNA from the organisms with which they interact. And by exploiting or collaborating with cellular organisms in these ways, they very effectively reproduce themselves and have no need of autonomous metabolism.

Thinking about viruses and their relegation to the realms of non-living and non-organismal entities necessitates a consideration of whether organism and living entity are identical categories, and whether a minimal account of life has to begin with cells. Such thoughts then invite further reflection on other biological entities that seem to have some autonomy but are almost never described as living organisms. Joshua Lederberg, a pioneer in molecular biology who first formulated the term 'plasmid' (Grote 2008), places these biological entities in the same category of 'symbiotic organisms' as he does mitochondria and chloroplasts. For him, they comprise part of 'the organic whole' (Lederberg 1952, 403). He argues more broadly that any scheme of life has to work out where to place prions, plasmids, integrons (gene capture and integration systems) and transposons – mobile genetic elements in a genome, sometimes called 'jumping genes' (Lederberg 1998).[2]

We will take our cue from Lederberg and start our examination of life with a discussion of some of the biological entities that inhabit this grey area between living and non-living, specifically prions, plasmids, organelles, endosymbionts and reduced extracellular symbionts. As we move along this continuum of biological organization to entities whose living status is never questioned (micro- and macro-organisms), we will investigate whether these instances of entities possess some of the most frequently cited life-endowing characteristics, such as spatial boundedness, reproduction, metabolism and evolvability, and how our criterion, collaborativity, relates to these characteristics. We will also argue that our account of cellular and sub-cellular entities fits very well with origin-of-life scenarios that stress chemical collaboration and community. Our bottom-up perspective, starting at the microscopic level of biology, rather than top-down from its most complex and undisputed exemplars, will suggest that much standard thinking is based on quite restricted and even covertly normative conceptions of what life is. This perspective will ultimately challenge the view that entities such as viruses are not alive and that the minimal definition of life must be cellular.

4 A Spectrum of Biological Entities

4.1 Prions

Once thought of as 'slow viruses', prions are now commonly understood to be self-propagating proteins that are able to convert normal proteins of the same type into the pathogenic prion conformation (Weissmann 2004; Prusiner 1998; Soto and Saborio 2001).[3] They have a life cycle from induction[4] to self-perpetuation (the conversion of another protein). Prions are very robust and persisting entities, because their conformation makes them highly resistant to inactivation by chemical, heat and irradiation treatments.

The central oddity of prions is that they propagate autocatalytically in a protein-only form, without DNA involvement.[5] For this reason, they are frequently referred

[2]Lederberg also includes in his 1998 list heterokaryon cells that have a diversity of nuclei in a common cytoplasm (cf. Rayner 1997 for details). We will leave these interesting entities out of our discussion.

[3]PrP^C is the generic protein, and PrP^{Sc} is the pathogenic protein isoform. The designation of prion is made in relation to the pathogenic form's still hypothesized function (Weissmann 2004, 863). Cf. Manuelidis (2004) for an argument against the protein-only understanding of prions and in favour of their viral status.

[4]Induction can occur spontaneously, through vertical inheritance and by lateral infection.

[5]The gene encoding the prion protein has to be expressed, of course, but the same nucleotide sequence can express either the pathogenic or non-pathogenic conformation of the protein.

to as protein-based genes (Wickner et al. 2004; Uptain and Lindquist 2002). Although best known as non-Mendelian hereditary elements[6] in diseased sheep, cattle and humans, prions exist in unicellular organisms too. Yeast and other fungal prions share no amino acid sequence similarities with mammal prions, and they function and are transmitted very differently (Bousset and Melki 2002; Uptain and Lindquist 2002; Weissmann et al. 2002). Nevertheless, experimental work on yeast prions has provided deep insights into conformational change in proteins and their transmission (Wickner et al. 2007).

The Modern Synthesis does not cope well with prions, and this has led some commentators to propose that a more comprehensive theory of inheritance is needed for prions to be properly understood evolutionarily (Jablonka and Lamb 2005; Chernoff 2001). The prion-forming potential of the implicated yeast proteins is evolutionarily conserved, implying that it is adaptive (Chernoff et al. 2000). Diverse functions have been identified or proposed for prions in a range of taxa. There is some evidence that prions are associated with epigenetically enabling yeast cells to cope with fluctuating environments, and that they play a role in memory formation in sea slugs (Shorter and Lindquist 2005). The non-pathogenic isoform of human prion proteins (the functions of which are still largely mysterious) is linked to the prevention of Alzheimer's disease (Parkin et al. 2007).

These capabilities and characteristics do not give a ready answer to the question of whether the self-propagational status of prions gives them the status of being alive. Although genes are frequently given a special 'informational' role in accounts of heredity (e.g. Hood and Galas 2003), the conferral of a similar status on proteins – as information-bearing molecules – does not simultaneously make them into living entities. Genes and proteins are not classified as alive in their own right,[7] despite the widespread 'selfish DNA' thesis that seems to confer autonomy on nucleotides (Doolittle and Sapienza 1980; Orgel and Crick 1980), and despite the recognition of the absolute centrality of enzymes to life processes (Kornberg 1989; Lezon et al. 2006).

Prions exhibit collaborative behaviours that benefit themselves, as a class of protein isoforms, as well as their hosts. When low amounts of the non-pathogenic isoform are produced, the prion conversion process halts, and when high amounts of the former are produced, it may stimulate spontaneous prion formation in the previously prion-free cell (Chernoff et al. 2000; Derkatch et al. 2001). Prion propagation in yeast requires the involvement of chaperon proteins. Moreover, prions in yeast are associated with greater adaptability in yeast because they increase protein variation – a factor that may prove advantageous in variable environments and eventually be genetically assimilated (True and Lindquist 2000; Pál 2001; Masel and Bergman 2003). It is these abilities to interact with biological processes at different levels of organization that presumably explain evolutionarily the prion's powers of persistence.

[6]Prions are described as 'non-Mendelian hereditary elements' because they self-propagate by transmitting their conformational characteristics in a lineage-forming manner, but do not form Mendelian patterns of inheritance (Liebman and Derkatch 1999).

[7]Cf. H. J. Muller (1966, 512) for an older view that the gene is a uniquely living material because of its capacity for reproduction, mutation and enzyme production.

4.2 Plasmids

Plasmids are small, stably inherited and self-replicating molecules of DNA (sometimes RNA) independent of the chromosomal DNA in bacterial, archaeal and eukaryotic cells. Plasmids are prolific and diverse; they may be larger than some prokaryote genomes (del Solar et al. 1998). Many are mobile genetic elements that direct their own transmission to new host cells during conjugation (the unicellular equivalent of sex), thereby spreading themselves to closely related and evolutionarily distant prokaryotes (Thomas 2000, 2006; Sørensen et al. 2005). They are then transmitted vertically, from mother to daughter cells.

Plasmids have a two-stage life cycle of establishment and proliferation followed by a steady state that matches the cell cycle (del Solar and Espinosa 2000). Neighbouring plasmid-free cells are often killed by plasmids, and this leads to a very high rate of successful infection (Gerdes et al. 1986; Eberhard 1990; Bingle and Thomas 2001). The complexities of plasmid characteristics have led some biologists to describe them as 'subcellular organisms' or endosymbionts with distinct autonomy from their host (Perlin 2002, 508). Because of their many talents, plasmids have become a mainstay of laboratory genetic manipulation as vectors of gene transfer.

Plasmids are often described as selfish in the same way that other genetic elements are because they encode genes that are not essential for the host and may impose fitness costs (Kado 1998). Importantly, however, they also play cooperative roles in cells (Wegrzyn 2005). Plasmids often encode and express genes of a variety of functions apart from those for their own mobility and replication, such as antibiotic resistance, virulence, environmental protection (including biofilm formation), DNA repair and supplementary metabolic pathways (Barton et al. 1995; Ghigo 2001). They can thus be seen as collaborative elements that enhance the functionality and adaptiveness of their host cells. The fact that these features favour plasmid survival has allowed these phenomena to be interpreted as instances of selfishness (Kado 1998), but in our framework they could equally well be interpreted as examples of (sometimes mutualistic) collaboration.

4.3 Organelles

Organelles are diverse membrane-bound compartments in eukaryote cells.[8] They carry out highly specialized biochemical functions and communicate between themselves to achieve this division of labour (Lowe and Barr 2007;

[8]There are increasing reports of a variety of compartments in prokaryote cells and the rising use of 'organelle' to describe these structures (e.g. Niftrik et al. 2004; Seufferheld et al. 2003; Kerfeld et al. 2005; Komeili et al. 2006).

Munro 2004). Major organelles include mitochondria and plastids (including chloroplasts, the organelles enabling photosynthesis in plants), as well as peroxisomes (compartments involved in metabolic activities that include the oxidative metabolism of fatty acids and the breakdown of hydrogen peroxide), Golgi complexes and endoplasmic reticula. Apart from the nucleus, most organelles are primarily involved in energy generation, transport and storage. They are often highly dynamic, mobile structures that react to relevant features of the environment to maintain cell function (Cutler and Ehrhardt 2000; Braun and Schleif 2007; Collings et al. 2002).

Organelles are often considered to be 'autonomous structures' because of their semi-independent inheritance strategies (Warren and Wickner 1996, 398; Nunnari and Walter 1996). Organelles reproduce within cells and a complete set is passed on to the daughter cells during cell division. However, because most membranes have to be inherited from pre-existing membranes and are usually not constructed de novo,[9] organelles are templated from pre-existing organelles. They self-assemble on the basis of the information their membranes carry about membrane polarity, type and location (Cavalier-Smith 2000; Lowe and Barr 2007).

Two of the most evolutionarily fascinating organelles were once free-living bacteria. Mitochondria and plastids functioned first as intracellular symbionts until most of their DNA migrated to the nucleus of the host over a billion years ago – a process that profoundly shaped the structure and content of the eukaryote genome and cell (Timmis et al. 2004; Martin 2003).[10] Now, to obtain the proteins they need for many functions, including their own metabolic activities, mitochondria and plastids rely on a protein import mechanism provided by the host's cellular machinery (Thiessen and Martin 2006; Cavalier-Smith and Lee 1985). This loss of genetic autonomy is not total, however, because plastids and mitochondria retain genes for translation and transcription machinery as well as metabolic function. They divide and grow independently of the cell cycle, although mitochondria gain some division assistance from the host cell (Osteryoung and Nunnari 2003). As well as inheriting their membranes directly, both organelles inherit their own organelle-specific DNA.

Mitochondria and plastids are not only essential to their cellular hosts, but are defining characteristics of them: there are no eukaryotes without mitochondria or plants without plastids.[11] Again, it is obvious that collaboration is happening here in ways that benefit – and make dependent – both organelles and the cells they inhabit. Indeed, the eukaryote cell could no more survive without its mitochondrial residents than the latter could survive in natural circumstances outside the cell.

[9]Peroxisomes and Golgi bodies can sometimes be reconstituted from other membrane types (Cavalier-Smith 2000; Lowe and Barr 2007).

[10]Mitochondria were incorporated into early eukaryote cells before plastids. As well as primary plastids, obtained in a single endosymbiotic event, there are also secondary and tertiary plastids gained from endosymbioses of plastid-carrying organisms (Archibald 2007).

[11]Subsequent loss or dysfunction notwithstanding. Cf. Embley and Martin (2006) for a demolition of the 'amitochondriate eukaryotes' hypothesis.

4.4 Viruses

Viruses are typically very small packages of single- or double-stranded DNA or RNA[12] (often just a few genes), wrapped up in a coating of protein and sometimes an additional lipid envelope.[13] They are prolific, highly diverse and ancient, although there is incomplete agreement about their evolutionary origins (as we shall see below). Viruses are generally excluded from organismal status because although they can synthesize some of their own proteins, they do not metabolize or reproduce independently (Van Regenmortel 2007). They either use their hosts, which probably include every organism past and present, or occasionally work in collaboration with other viruses to make necessary enzymes. Viruses do not reproduce by division but by self-assembly of the components that they manufacture with the help of the host cell. Some viruses influence host behaviour quite significantly by, for example, conferring either protection against other viruses or virulence properties (e.g. diphtheria or cholera toxins).

Viruses have well defined life cycles that are often described as consisting of 'developmental' stages (e.g. Luria et al. 1978). The cycle begins with virions, the inert form of viruses, which are transformed into the next stage of adsorption, when viruses or phages (the viruses with affinities for prokaryotes rather than eukaryotes) 'dock' onto the outer cell membrane of their hosts and either enter the cell or have their DNA absorbed into it. Their protein coats dissolve or are discarded, after which the viruses co-opt the host's cellular machinery to express genes that lead to genome replication, maturation (in which the new genomes are wrapped in freshly synthesized protein) and, finally, release from the intact or lysed cell. A number of plant viruses move actively from cell to cell, using virus-encoded movement proteins (Boevink and Oparka 2005). Some viruses have an extra developmental stage in which they remain dormant in the host cell or genome as prophages or proviruses and are inherited (Casjens 2003; Bannert and Kurth 2004). Endogenous retroviruses, which are viruses that have integrated permanently into the host chromosomes and are inherited vertically, have left their mark on many organismal genomes, including our own (Griffiths 2001; Hamilton 2006). Included amongst these viruses are those that are crucial for the development of the placenta in mammals (Mallet et al. 2004).

The diversity and mutability of viruses makes them difficult to classify, although both genome sequence and protein structure analyses are constantly refining viral groupings, which were once based primarily on pathogenic effect (Bamford et al. 2005). The term 'species' is often applied, with many caveats, to subgroups of virus divisions (Lawrence et al. 2002; Hendrix et al. 1999; Van Regenmortel 2007). The

[12] Single-stranded RNA viruses can be divided into positively and negatively stranded (sense and anti-sense) genomes, and retroviruses, which make DNA copies of themselves with their own reverse transcriptase before entering the host chromosome and being transcribed back to RNA (Ahlquist 2006).

[13] Viroids, which are tiny RNA viruses that infect plants, have no protein coat.

aim of such language is to 'bring the definition of virus species into line with the species definitions of cellular organisms' (Gibbs and Gibbs 2006, 1419). One earlier and another more recent division of life into superkingdoms give viruses a superkingdom (domain) of their own: the Acytota or Akamara, both of which are categories for acellular organisms possessing genomes (Jeffrey 1971; Hurst 2000; Weinbauer 2004). These domain-level classification schemas have the potential to identify viruses as genuine forms of life but have yet to gain many adherents.

There are three main hypotheses about the origins of viruses: primeval pre-cellular life (the virus-first or primordial hypothesis), degenerate intracellular parasites (the reduction or regression hypothesis), and as renegade prokaryote genes (the escape hypothesis). The most popular is currently the third one, which is that viruses are actually genetic elements that opted out of cellular organization and are thus true instantiations of 'selfish' genetic material (Campbell 2001; Hendrix et al. 2000). However, new versions of the primordial hypothesis are also being promoted. They shift the discussion back to the pre-cellular 'unselfish' gene pool and give viruses major roles as evolutionary innovators (e.g., Forterre 2006; Koonin et al. 2006; Hendrix 2002; Hendrix et al. 2000; Claverie et al. 2006). Whatever their origins, viruses have made extraordinary contributions to the evolution of non-viral life through their proclivity for mutation and recombination, and their ability to pick up and move genes from one organism to another (transduction) and integrate their own and other genetic material into host genomes (Weinbauer and Rassoulzadegan 2004; Lawrence et al. 2002; Karam 2005; Villarreal 2004; Hambly and Suttle 2005). Moreover, their role as carbon regulators in the global oceans, for example (Suttle 2005), shows how a broader conception of collaboration is necessary to understand the evolutionary, biogeochemical and ecosystemic contributions of viruses to all living systems.

The recently discovered Mimivirus (short for 'mimicking microbe') provides an additional challenge to some prevalent ideas about viruses and their capabilities. Mimiviruses are huge (larger in volume and genome size – over 900 protein-coding genes – than many of the smallest bacteria, some of which are described below) and, most surprisingly, they carry genes that are known to encode translation, DNA repair and metabolic activities (Raoult et al. 2004).[14] They do not seem to have picked these genes up from their hosts.[15] Although these viruses cannot synthesize their own ribosomes and do not metabolize (their metabolic pathways are incompletely coded), they can easily be conceived of as entities in transition from viruses to free-living organisms (Forterre 2006; Raoult 2005; Claverie et al. 2006). Mimiviruses certainly exhibit more independence than organelles and, moreover, seem to be in an 'evolutionary steady state' with no apparent signs of genome reduction (Claverie et al. 2006, 142).

Microbiologists and other biologists are highly ambivalent about the biological status of viruses. Although a strong line of thinking throughout much of the history

[14]While some other large viruses also carry translation and metabolic genes, Mimivirus greatly extends the known repertoire of these genes in viruses (Koonin 2005).

[15]Although cf. Moreira and López-Gárcia (2005) for the opposite claim.

of virus research and microbiology has advocated that viruses are alive and at least proto-organismal (Burnet 1945; Stanley 1941, 1957; Luria et al. 1978; van Helvoort 1992), the dominant view of viruses is still fixed by the assumption that only cellular entities are appropriately designated as living (Moreira and López-García 2009). According to virologist Marc Van Regenmortel,

> Only unicellular and multicellular organisms possess the property of being alive while the organelles, macromolecules and genes found in cells are not themselves considered to be alive. The differences between viruses [which are not alive] and various types of organisms is quite obvious when the functional roles of the proteins found in viruses and organisms are compared (Van Regenmortel 2007, 133).

Other microbiologists, however, believe that there are numerous reasons to give viruses the status of living matter. Because they 'have the intrinsic ability to mediate their own transfer from one host to another,' say Salvador Luria and co-authors,

> viruses are independent genetic systems. They are not accidentally separated fragments of a cell genome. They are endowed with genetic continuity and mutability, and contain sets of genes working in concert to make more virus. They have their own evolution, which is independent, to some extent at least, of the evolution of organisms in which they reproduce (Luria et al. 1978, 481).

Some virologists go even further and argue that viruses exhibit the same primary features common to all life forms, such as internal homeostatic controls that enable survival in changing environments, organization that is based on heritable nucleic acids, reproduction, exploitation of environmental resources, diversity of components and their functions, and the capacity to adapt and evolve (Mindell and Villarreal 2003, 1677; Mindell et al. 2003; Stanley 1941, 1957). Discovery of the debilitating effects of a minute 'virophage' on a huge virus has been argued as evidence for the aliveness of viruses: if they can be infected themselves, and respond in various ways to these infections, then the 'imaginary boundary' between viruses and true organisms seems to have been crossed (Claverie, Koonin, in Pearson 2008).

A further stream of reflection sees no contradiction in regarding viruses as *alternating* between living and non-living phases.

> Outside the host cell, poliovirus is as dead as a ping-pong ball. It is a chemical that has been purified ... and crystallized ... with its physical and chemical properties largely determined ... and its three-dimensional structure solved. Just like a common chemical, poliovirus has been synthesized in the test-tube. Once poliovirus, the chemical, has entered the cell, however, it has a plan for survival. Its proliferation is then subject to evolutionary laws: heredity, genetic variation, selection towards fitness, evolution into different species and so forth–that is, poliovirus obeys the same rules that apply to living entities (Wimmer 2006, 56).

The inertness of virions outside the cell leads us to think that viruses are similar to prokaryotes with spore stages as well as to plant seeds and fungal spores. In our conclusion we shall (cautiously) endorse this perspective, and also suggest that it is helpful to distinguish the developmental cycle, which includes both active and inert stages, from the life cycle, which should be applied only to metabolically active phases of lineage-forming systems.

Historical echoes of the discussion of the status of viruses are amplified by recent practical achievements of creating synthetic viral genomes. Several of these have now been synthesized from scratch and used successfully to infect cells (e.g. Tumpey et al. 2005; Smith et al. 2003; Cello et al. 2002). Some of these researchers claim their achievements are the final nails in the coffin of vitalism, because their virus 'chemical' was resurrected in a cellular extract and not a living cell (e.g. Cello et al. 2002). However, those who do *not* see viruses as organisms perceive synthetic viral genomes as further proof that 'true' (cellular) life – still resistant to synthesis from the top down or bottom up – is something fundamentally different from the much more easily created biology of viruses or plasmids.

It is clear to us that leaving viruses out of evolutionary, ecological, physiological or conceptual studies of living entities, would allow only an incomplete understanding of life at any level (Weinbauer and Rassoulzadegan 2004; Wilhelm and Suttle 1999; Suttle 2005). This deep and extensive interaction is too biologically important, from our perspective, to be considered as purely parasitic. Conceived of collaboratively, cellular life is constantly 'bathing in a virtual sea of viruses', within and without every cell, with evolutionarily significant consequences for the past, present and future of all cellular lifeforms (Bamford 2003, 232). In fact, says virologist Dennis Bamford (2003, 235), it is time to consider dividing life into two realms: the cellular realm and the viral one. He believes that only by dealing more thoroughly with a concept of life fully cognizant of the role of viruses will we be able to achieve an adequate view of life even as it applies to its cellular manifestations.

4.5 Endosymbionts

Endosymbionts are entities that live inside the cells of other organisms. Some are mutualists while others are more parasitic. Parasites are generally distinguished from other symbionts by their mode of collaboration with their hosts. While endosymbionts have a mutual give-take relationship with their hosts, obligate endoparasites are generally viewed primarily as receivers of benefits and not givers. Increasingly, however, these are being understood as more fluctuating and complex relationships (Valdivia and Heitman 2007). Numerous bacteria are obligate parasites that have reduced genomes and depleted cellular function. *Rickettsia*, *Chlamydia*[16] and microsporidia are well known examples. Microsporidia have lost so many genomic, biochemical and morphological features that they were once thought to be the most primitive eukaryotes (Keeling and Fast 2002). Now, however, they are deemed to be fungi that are highly adapted to their parasitic lifestyles. Rather than relinquishing their genes to the host genome (as have organelles), obligate endoparasites have simply lost the genes that have become redundant due to reliance on host provisions (Timmis et al. 2004; Tamas et al. 2001). These are usually metabolic and mobility genes, although some of these symbionts retain capacities for intra- and intercellular mobility (Gouin et al. 2004).

[16]Inert *Chlamydia* 'spores' (elementary bodies) exist outside cells but the 'live' form of the organism conducts all its activities intracellularly. Note the parallels with the developmental cycle of viruses, which *Chlamydia* was once thought to be.

Despite the ongoing reduction of their genomes, some of these parasites also acquire and exchange DNA via conjugation and transduction (Darby et al. 2007). Obligate bacterial parasites can be vertically as well as horizontally transmitted, and transmission between mammals and other animals often involves vector organisms such as ticks or fleas (Darby et al. 2007). Another form of symbiosis, 'reproductive parasitism' (Wernegreen 2004), is employed by *Wolbachia*. These are widespread hereditary endosymbionts of insects, crustaceans, spiders and nematodes. The hosts do not depend on their endosymbionts for metabolism or defence,[17] but the bacteria significantly influence host lives and may induce speciation events by reproductively isolating insect lineages (Charlat et al. 2003; Weeks et al. 2002). *Wolbachia* control the reproduction and development of many of their hosts by biasing sex ratios and reproductive strategy (asexual rather than sexual), as well as feminizing genetic males (Werren 1997; Stouthamer et al. 1999). In addition to being inherited vertically via maternal transmission, *Wolbachia* spread themselves laterally, sometimes to evolutionarily distant insect hosts. Their genes are also transferred laterally (in one case the entire genome!) into insect host genomes (Dunning-Hotopp et al. 2007).

Many mutualist endosymbionts cannot live without their hosts and the hosts are frequently just as dependent on their endosymbionts. They are almost always transmitted vertically from host to host through the maternal line (Wernegreen 2002). Numerous insects are involved in obligate intracellular mutualisms with bacteria, to the extent that separate insect and bacterial lineages are fused into single, highly coordinated metabolic systems (Wu et al. 2006). These endosymbionts frequently live in specialized cells (bacteriocytes) created within the host organism and their primary endosymbioses are quite commonly associated with secondary endosymbioses (Douglas and Raven 2003; Baumann 2005).

One of the most intensively studied mutualist endosymbionts is *Buchnera aphidicola*, which lives in tight association with its aphid hosts (about a million *Buchnera* cells per aphid) and produces essential amino acids for them. It is vertically inherited from one generation of aphids to the next and its few regulatory genes appear to control its life cycle in relation to its aphid host (Moran and Degnan 2006). *Buchnera* have tiny genomes due to gene loss and no uptake of mobile genetic elements. They are about one-seventh the size of *E. coli* (although *Buchnera* cells are actually larger and contain many copies of the genome), with which they shared a common ancestor about 200 million years ago (Moran and Degnan 2006). Aphids and *Buchnera* coevolve and codiversify, meaning the phylogenies of associated lineages map onto each other (Moran 2006). *Buchnera* are commonly classed as endosymbionts but the depth of their dependence on their host means that some biologists see these bacteria as closer in status to organelles (e.g. Andersson 2000; Douglas and Raven 2003).

One key difference often said to distinguish endosymbionts from organelles is that endosymbiont genomes encode most of their essential proteins whereas in

---

[17] However, *Wolbachia* in nematodes do provide host-related metabolic and other physiological functions (Fenn and Blaxter 2006). There is also increasing evidence of insect host benefits from *Wolbachia* infections (Iturbe-Ormaetxe and O'Neill 2007).

organelles, many of the genes for organelle function have shifted to the host genome and been replaced by a protein import apparatus (Cavalier-Smith and Lee 1985, 378; Thiessen and Martin 2006). Not everyone accepts this distinction, however, and other commentators see variable degrees of biochemical and cellular integration between host and endosymbiont/organelle (e.g. Bhattacharya and Archibald 2006; Bodyt et al. 2007). There certainly appear to be numerous endosymbionts making the transition from organism to organelle status[18] and any definition of either will have to be based on a continuum of collaborative strategies rather than clear categories of distinct entities (Bodyt et al. 2007; Rodríguez-Ezpeleta and Phillipe 2006). Concomitant with observations about the occurrence of these evolutionary transitions from free-living organism to endosymbiont to organelle appears to be a shift in the language used by biologists: from autonomous 'invaders' to domesticated 'servants' to 'captives' or 'slaves' that have almost totally lost their bacterial identity (e.g. Dyall et al. 2004; Baumann 2005).

4.6 Reduced Extracellular Symbionts

A plethora of bacteria and other microbes live in intimate extracellular liaison with plants, animals and fungi (sometimes these arrangements are called ectosymbioses or episymbioses). Cyanobacteria, as well as being ancestral to plastids, live in close symbioses with eukaryotes, providing nitrogen fixing and photosynthesizing capabilities through a variety of mechanisms (Douglas and Raven 2003). Some are vertically transmitted and a few free-living cyanobacteria exhibit trends towards genome reduction very similar to those in endosymbionts (Marais et al. 2007). Some ultimately obligate symbiotic arrangements have free-living stages, such as the *Rhizobium* bacteria that colonize plant roots and fix nitrogen for their partners.

Fascinating as many of these symbiotic arrangements are (e.g. bacteria that provide 'legs' for ciliates; others that oxidize sulphur for tube worms that lose their mouths and guts as juveniles when colonized by these ectosymbionts), we will focus here on 'transitional' organisms that seem to be on the very edge of 'independent' living. One example is *Nanoarchaeum equitans*, an exceedingly tiny archaeon, which is always described as an organism despite its extremely reduced genome and consequent inability to metabolize, grow and reproduce independently of another archaeon, *Ignicoccus hospitalis* (Huber et al. 2002). A better-known example is the genus *Mycoplasma*, which consists of very small obligate parasites that are notable for having no cell walls (almost all bacteria do, as do plants and fungi but not animals or most protists).[19] They are usually regarded as the smallest

---

[18] Or from an endosymbiont with increasingly limited function to extinction. Cf. Pérez-Brocal et al. (2006).

[19] All organisms have cell membranes, of course, but not the more rigid cell walls that plants and bacteria possess.

free-living cell[20] although they are heavily dependent on their hosts for amino acid and co-factor biosynthesis, and fatty acid metabolism, especially sterols for membrane maintenance (Fraser et al. 1995; Rottem and Naot 1998). They have lost large numbers of their genes and are considered to have 'little adaptive capability' (Glass et al. 2006, 425).[21]

Because of this reduced genome and restricted function, *Mycoplasma* (*M. genitalium* in particular) have been popular candidates for minimal cell research, in which synthetic biologists attempt to recreate the simplest cellular form of life from synthetic or engineered components. One of the recent breakthroughs in synthetic biology involved 'rebooting' a *Mycoplasma* cell with a genome from a different *Mycoplasma* taxon (Lartigue et al. 2007). Although the experiment was successful, doubts were raised about the transferability of the technique to less closely related organisms and to those with cell walls (Pennisi 2007).

In none of this research, however, is it doubted that *Mycoplasma* is a living organism, so its dependent nature and restricted function are apparently insufficient reasons to consider it in the same light as a virus. One of the characteristics tending to confer organismal status on it is genetic autonomy, or the capability of a biological entity to initiate and complete its own reproduction. This status does not obtain for plastids or organelles, however, which are usually perceived as mere parts of the cell in which they are found. The additional biosynthetic and metabolic capabilities of endosymbionts and exosymbionts, no matter how reduced, seem to be essential to the conferral of organismal status. Given the complete dependence of these processes on contributions from the host cell, however, the grounds for this sharp distinction between viruses and (other) symbionts is far from clear.

4.7 Unicellular Organisms and Single Cells

It might seem a strange turn in our discussion to interrogate unicellular organisms for whether they are alive or not, when nobody has questioned that status. Our point here, however, is to continue to press the question of whether the boundaries of life are clear cut and, in particular, whether cellularity is enough in itself to confer 'aliveness'. Certainly, a single mammalian cell on a petri dish, for example, is not normally considered a living entity in its own right,[22] in part because of the highly technical requirements for keeping this cell and its descendants alive (Bhardwaj et al. 2006). This ambiguous status is, we believe, the same ambiguity that bedevils our understanding of prions, plasmids, organelles and viruses. Single animal or plant cells are only truly alive when they are collaborating with other cells. Whether

[20]There is increasing evidence, however, that *Mycoplasma* are also intracellular symbionts (e.g. Meseguer et al. 2003). And the genome of *N. equitans* mentioned above is smaller than that of any mycoplasma.

[21]Mycoplasmas do, however, have multifunctional enzymes that have taken on unusual roles.

[22]See however, Theodore Puck (1972) for an argument about the autonomy of mammalian cells.

prokaryote or eukaryote, microorganismal or macroorganismal, cells work together in a great variety of ways, collectively structuring their activities through numerous mechanisms. In the same way that cellular life forms are only fully functional when collaborating with other cells, so are viruses, plasmids and prions. Is there a hard line worth drawing between different modes of cellular and subcellular collaboration – between collaboration and exploitation? We think not.

Moreover, even when single cells are considered in isolation, each cell is a complex of collaborating parts. In the case of eukaryote cells, those parts – as we saw in the discussion of organelles – may include once free-living cellular entities. A eukaryote cell, in the minds of some biologists, "can be likened to a society composed of a nucleus and a crowd of subcellular organelles in which all members cooperate for the common good" (Eberhard 1980, 231). This is a complex collaboration, however, because competitive reproductive relationships may also exist between organelles or plasmids in a cell (for examples of such competition, cf. Walsh 1992; Eberhard 1980; Paulsson 2002). Such competition can also occur between cells in clones, as when somatic mutations occur in the meristems of vegetatively reproducing plants (Klekowski 2003; Pineda-Krch and Fagerström 1999). Although the philosophy of biology has directed considerable attention to the problem of conflict in the transition from single cells to multicellularity (e.g. Okasha 2004), it has not extended a similar level of scrutiny to intracellular cooperation and competition. We believe this is worth doing for a better understanding of these collaborative relationships between biological entities at multiple levels.

4.8 Multicellular Organisms

Multicellular organisms, particularly plants and animals, and most notably ourselves, are considered to be 'paradigmatic' examples of living entities (Wilson 2000). Again, we think that this is far from clear, and that whatever aliveness consists of for an animal, for example, it is a much less autonomous state than is usually recognized in discussions of life (especially, but not only, philosophical discussions). The evolution of eukaryotes has largely been driven by microorganismal interactions, and a variety of modes of dependence between eukaryotes and prokaryotes endures and diversifies in every existing eukaryotic organism. Vast numbers of eukaryotes cannot reproduce, develop or metabolize without their prokaryote partners. We noted earlier that achieving organismal status is often understood to be the achievement of autonomy. This interpretation can easily mislead our understanding of life and what it is to be alive. Traditionally conceived biological entities are systems elaborated around unique genomes, but to consider them as autonomous individuals is a mistake, we argue: functional wholeness, the basis of any attribution of autonomy, is a characteristic of collaborative interactions, almost always involving diverse entities.

Not only are paradigmatic multicellular organisms more multicellular than is usually supposed (in that a multicellular organism should be understood as including all the entities that interact to achieve shared metabolic and reproductive goals), but even 'simple' prokaryotes could be thought to qualify for multicellular status on this basis. Take, for example, magnetotactic bacteria, which have organelles of magnetic crystals (magnetosomes) that line up inside the cell and are attached to the flagella of the bacteria. The magnetosomes function as compasses and guide the bacteria along local magnetic field lines (preferentially north or south, depending on which hemisphere the bacteria live in). As if this were not astonishing enough, some magnetotactic bacteria live in strictly multicellular arrangements. The individual cells form a spherical group of up to 40 bacteria, constructing an empty compartment in the middle of the group. As well as sensing magnetic lines together and moving in a fully coordinated manner, the groups reproduce together by coordinated cell division. They grow at the same rate (increasing in volume, not cell number) and then simultaneously divide into a new multicellular organism that swims off immediately after separation (Keim et al. 2004, 2007; Abreu et al. 2007). Most multicellular organisms have a unicellular stage, whereas these magnetotactic bacteria have a strong claim to be exclusively multicellular throughout their life cycle.[23]

More variable in their organization than magnetotactic bacteria and other specialized multicellular structures of unicellular organisms (such as the well known aggregating examples of *Dictyostelium* and myxobacteria) are other collaborative arrangements known as communities. Prokaryotes and other microbes seldom live as isolated single cells but cohabit in a variety of communal organizations such as biofilms. Microorganisms that live as parts of biofilms express genes very differently from free-floating (planktonic) microbes, and in patterns that are structured at each stage of the biofilm's development (Stoodley et al. 2002; Costerton et al. 1995). Communities such as biofilms (which may be single or multi-taxa), as well as some populations of unicellular organisms, exhibit well-defined cell organization and a functional division of labour that includes specialized cell-to-cell interactions, the suppression of cellular autonomy and competition, metabolic collaboration, combined defence and attack strategies, and the coordination of movement, growth and reproduction (Cho et al. 2007; Aguilar et al 2007; Kaiser 2001; Shapiro 1998; Kolenbrander 2000; Crespi 2001; Dworkin 1997). Many of these are activities that no individual microbe can accomplish on its own, and the collective behaviour is often achieved with a cost for individual 'altruistic' microorganisms (if they are perceived through the lens of selfishness).

Some biologists and philosophers may prefer to define multicellularity in ways derived from reflection on animals and plants, and thereby exclude these microbial communities from that category. But certainly any general account of the varieties of biological organization will need to take account of them and explain how they conform to concepts such 'multicellularity', 'individuality' and 'autonomy'. Do humans, for example, stop at their skin and have to be conceived of as tubular rather than solid in order to avoid incorporating large internal populations of gut microbes?

[23] Abreu et al. (2007) assign the name *Candidatus Magnetoglobus multicellularis* to this organism ('*Candidatus*' indicates that it has not been cultured).

Lederberg, with his concept of 'symbiome', raises the question of whether organisms are necessarily monogenomic or whether a multi- or metagenomic state is the usual state of organismal organization (Lederberg, in Hooper and Gordon 2001; Dupré and O'Malley 2007). Discussions of life and its organization have to take into account the fact that symbiotic relationships are ubiquitous and all organisms, when conceived as the functional wholes that interact with their surroundings, are multilineal and multigenomic.

All multicellular organisms function with the inherited assistance of endosymbiotic partners in interplay with numerous other forms of partnership. All unicellular organisms are infected with phages and other unicellular organisms, and even viruses have their own phages, 'virophages' (La Scola et al. 2008). Although viruses are generally thought of as strictly parasitic, this view may owe more to preconception than to biological fact. The functions of micro-alliances with viruses are only beginning to be investigated, and one early investigative success has been delineating the contribution of cyanophages to cyanobacterial photosynthesis (Lindell et al. 2005). Similarly, the phages that infect the anthrax bacterium, *Bacillus anthracis*, play major roles in the bacterium's capacity to build communities and to produce the long-lived spores that ensure the perpetuation of the cycle of anthrax infections in animals (Schuch and Fischetti 2009). More broadly, the role of viruses as facilitators of genetic variation in multi-lineage communities and as fundamental agents in biogeochemical cycles (Suttle 2005) means they cannot be assumed to be exclusively parasitic and self-serving.

Overall, deep and extensive collaborations between biological entities blur – at the very least – any distinction between so-called individual organisms and these larger organismal groupings of which they are parts (Moran 2006; Dyer 1989). They also call attention to the non-discrete and highly dynamic nature of biological individuals (Rayner 1997). Although this is not our present focus, we should note that great evolutionary significance has been attributed to symbiosis (e.g. Sapp 1994). Symbioses have constituted innovations that have made possible some of the most significant transitions in evolutionary history, as our discussion of mitochondria and plastids made clear.

5 Characteristics of Living Biological Entities

5.1 Common Criteria of Life

How is it usually decided which of these diverse entities is alive? All the definitions of life in current circulation emphasize particular life-bestowing properties. Some of these definitions take functional criteria (such as reproductive autonomy) to be the most important, whereas others emphasize evolutionary criteria, such as continuity or evolvability, or foreground metabolic or organizational characteristics (Popa 2004; Koshland 2002; Pályi et al. 2002; Zhuravlev and Avetisov 2006; Szathmáry 2006). The most inclusive range of criteria for deciding whether entities are alive or not is derived from exemplars already regarded as unquestionably alive. Animal characteristics often dominate these criteria, which are then modified to include plants, fungi and unicellular organisms but to exclude entities such as fire and crystals (Chyba and Hand 2005).

Spatial boundedness is widely assumed to be a fundamental criterion of living entities, and is one reason larger biological systems, such as ecosystems, are seldom classified as living entities in their own rights. Boundaries usually consist of enclosing materials such as membranes, cell walls and skin, which separate internal from external environments and enable internal activities such as metabolism (Popa 2004). Associated with spatial boundedness, and again almost inescapably connected to the project of distinguishing coherent subunits from encompassing systems, are stability and the ability to maintain a buffer against fluctuating environments. However, the interconnectedness of the diverse entities discussed in the preceding text points to obvious dangers in assuming that spatial boundaries can be straightforwardly and uniquely identified (Rayner 1997). The boundaries of a plant and animal are precisely the sites where complex interactions occur between entities generally considered distinct, but these interactions are so closely coupled that we are strongly tempted to see them as parts of the same system.[24]

Perhaps the most widely agreed criteria for being a living thing are *metabolism,* or energy transformation, and *reproduction*, the capacity of entities to make more of themselves. Biochemical transformation of energy from the environment, first to maintain their own structural and functional integrity, and second to reproduce themselves, is a plausible general account of what living things most fundamentally do. Metabolism, then, is a basic means of survival for anything alive.[25] For many biologists, this is the most fundamental biological process and the true demarcator of living and non-living entities (Gánti 1997; Luisi 1998). An internal capacity for self-sustainability on the basis of the processing of external resources is a common understanding of organismal function (Luisi 1998).

Our reservations about this criterion are not about whether metabolism is a basic characteristic of living systems, but whether it can effectively be deployed to make the kinds of distinctions into discrete living entities that are generally expected by theorists of biology. The reason for this is that metabolism is typically a collaborative activity involving many of the things that are generally supposed to be discrete living entities. It is generally supposed, for example, that a human, qua discrete biological entity, consists of a lineage of cells deriving in a series of divisions from an original zygote. But a functional human consists also of very large numbers of symbiotic bacteria, in fact amounting to 90 % of the cells in the total human system. These microbial cells are deeply involved in the metabolic processes, most obviously

[24]Of course we accept the importance of membranes in the origins and maintenance of life in general, as well as the epistemological necessity of imposing boundaries for both theoretical and experimental biologists. This epistemic function does not, however, require that such boundaries be uniquely and unequivocally identifiable.

[25]The necessity of biochemical transformation rules out phenomena such as computer viruses as candidates for life, because they do not sustain themselves through biochemical means.

digestion, that maintain the functioning of the system (Gill et al. 2006; Hooper and Gordon 2001). Hence, a human, conceived in the way just described, is not capable of performing *autonomously* the metabolic processes essential for its survival. If it is considered sufficient merely to carry out independently some metabolic processes, but not all those necessary for the survival of the entity, then organelles and endosymbionts will count as living entities.

While we noted earlier that reproduction is a necessary feature of life, we also mentioned its inadequacies for a full understanding of life. As we have shown, viruses, organelles and even prions reproduce themselves. The reproduction criterion is sometimes tied to autonomous reproduction, so that viruses and the like, though they are very effective replicators, are often taken to fail this criterion because they do not reproduce independently and must use 'true' organisms from different lineages to achieve their reproduction. However, it is doubtful whether even paradigmatic multicellular organisms can meet the criterion of lineage-exclusive autonomous reproduction. Those insects in which reproduction is substantially under the control of endosymbiotic *Wolbachia* are one obvious counterexample. But more generally, insofar as reproduction requires the deployment of metabolic processes, as it surely must, it depends also on endo- and exosymbiotic microbes.

Another criterion sometimes proposed as definitive of life is *evolvability* (Ruiz-Mirazo et al. 2004). One highly cited definition of evolvability is that provided by Marc Kirschner and John Gerhart, in which "evolvability is an organism's capacity to generate heritable phenotypic variation" (1998, 8420). A consequence of taking this as criterial for living entities is that it would include all the entities we have described down to viruses and prions. Our interest in the concept, however, is rather to make a much more general point which, we think, cuts to the heart of the difficulty in defining a living entity. Evolvability, in the sense of Kirschner and Gerhart at any rate, is a characteristic of lineages. Viruses, prions, organelles, unicellular organisms, and multicellular organisms conceived as monogenomic wholes, all form the appropriate kinds of lineages. So, although we agree that these criteria of spatial boundedness, metabolism, reproduction and evolvability are truly important to understand life, we believe that they are being understood within a framework that misconceives living entities in a fundamentally important way.

5.2 *Reframing the Criteria of Life*

Our reservations about the above criteria arise from the fact that none of the entities we discuss are the functional entities that interact with their environment and whose success or failure in such interaction determines the success or failure of these lineages. These functional entities are, rather, associations of a variety of such lineage-forming entities. A typical large eukaryote, for instance, is constituted by entities of all the kinds we have distinguished above. We might invoke here David Hull's (1980) well-known distinction between replicators and interactors, but in a very different way from that originally supposed by Hull. Interactors, in our view, are complex systems involving the collaboration of many highly diverse lineage-forming entities. This sort of interactor, we also suggest, is the most fundamental unit of selection. This perspective has radical implications for the way we think about evolution. It would entail but obviously go beyond contemporary concepts of group selection in multi-level selectionism (Sober and Wilson 1998).

Amongst those implications is the importance of the notion of collaboration, which is seldom proposed as a criterion of life (although cf. e.g. Lezon et al. 2006, for an emphasis on cooperation). It is hard to imagine life that is not collaborative, in the sense described above, both at the intracellular and the intercellular level, and we suggest that collaboration is, therefore, one of the central characteristics of life. To treat it as such it will be necessary to specify more carefully what the relevant sense of collaboration entails. We do not want to rule out automatically even simple chemical systems. Some chemical aggregations exhibit growth, reproduction (leading to lineage formation of varying persistence), error correction and environmental sensitivity (Schulman and Winfree 2007; Weber 2007). It would be surprising if these features were not to be found in the chemical world, because otherwise it would be hard to imagine how life would have originated. Our continuum view of life is open to chemical systems being sometimes describable as living systems, though perhaps it is likely that they will meet the relevant criteria only transiently. Because biological entities in our conception are series of dynamic and diverse collaborations, boundaries are flexible and unfixed. Any claim that something is a living thing needs to be assessed in relation to the general characteristics we describe.[26]

But more important than any attempt to specify limits on what is or is not alive will be to emphasize the contrast between this perspective on life as collaborative and the much more familiar assumption that life is fundamentally selfish and entails competition between reproductively and metabolically autonomous organisms. The outlines of our response to the view that only selfish entities will win out in the battle of all against all should by now be clear: The unit of selection, the entity in which selfishness may perhaps be expected as the norm, is a collaboration of many different lineage-forming entities.

The context in which the latter evolve, then, is quite typically one of collaboration. We said that the collaborative whole may perhaps be expected typically to display selfishness. But, of course, this assumes that there is some natural terminus to the process of collaboration. We hypothesize that competitive activity is a transitional rather than a terminal state and that such temporarily competitive wholes will exhibit a strong tendency ultimately to compete most successfully by engaging in new levels of collaboration with similar or different entities. We see the emergence of sociality as an instantiation of such a process as are, more generally, the evolutionary transitions that have been highlighted by the work of John Maynard Smith and Eörs

[26]Interactors thus conceived also rule out the whole planet as a candidate for evolving life, because although the planet could be conceived as metabolizing (in a highly collaborative way), it does not interact with other such wholes. It also lacks any means of reproducing itself. We are more ambivalent about ecosystems, which may frequently interact, but a case would have to be made for ecosystems forming a lineage that was more than mere continuity (as a substitute for replication).

Szathmáry (1995). Our spectrum of biological entities exemplifies a number of different forms of collaboration that are central to such an evolutionary schema. As we have emphasized, our concept of collaboration assumes no sharp boundary between selfish and cooperative interactions, something surely to be expected if the former is inclined to evolve into the latter.

We have certainly not exhausted the criteria that have been proposed as characteristic of living entities. We have not explicitly considered, for example, environmental responsiveness, the ability to detect and respond appropriately to salient features of environments, or development, the recurrent production of the characteristic stages of a life cycle (although much that we have said has addressed these criteria implicitly). We do not mean to minimize the significance of these, and perhaps other, distinctive characteristics of living systems. What we do argue is that a focus on metabolism and reproduction, widely agreed to be fundamental features of life, has the additional virtue of drawing attention to a characteristic that has been greatly underemphasized: that of collaboration. That this has been downplayed is a readily intelligible consequence of the importance that has been attached by biological theorists to competition (Roughgarden 2009). But for this very reason giving collaboration proper emphasis could provide important fresh insight into the nature of evolutionary processes because it affects how we conceptualize the entities and activities central to evolution.

6 Autonomy and the Origins of Life

Our collaborative interpretation of life suggests that it is possible to sidestep the usual problems associated with defining life. Although we do not claim to have provided a definition of life, we do believe we have offered a view of living matter that offers a flexible resource for understanding the many ways in which life can be organized. The tension between replicating lineages as one criterion of life, and metabolic self-sustainability as the other, can be reconciled by taking a much more interactive view of metabolic processes and by reconceiving cooperation and competition within a broader framework of collaboration. Life, according to our analysis, occurs at the intersection of lineage formation and (typically collaborative) involvement in metabolism. Entities that are problem cases, such as viruses, can be understood as alive when actively collaborating. When not collaborating, they have at most a potential for life. We invite our readers to apply our framework further along the spectrum than we have gone, to various chemical and physical systems and to ecosystems.

What of the autonomous individual organism, often the conceptual target of attempts to define life, and the thing that is assumed by models of evolution through competition and selection? To the extent that such individual autonomy requires just an individual life or life history, then it surely applies much more broadly than is generally intended by biological theorists. Countless non-cellular entities have individual life histories, which they achieve through contributing to the lives and life histories of the larger entities in which they collaborate, and this collaboration constitutes their claim to life. But – and this is our central point – no more and no less could be said of the claims to individual life histories of paradigmatic organisms such as animals or plants; unless, that is, we think of these as the collaborative focus of communities of entities from many different reproductive lineages. In much the same way, whatever sense we might try to make of the Dawkinsian idea of selfish genes, molecular replication is *always*, and has always been from the pre-cellular molecular community to the present, the achievement of ensembles of molecules, not of individual molecules (Segré and Lancet 2000).

It is entirely reasonable to think of autonomy as centrally exhibited in collaboration rather than just rugged independence.[27] Assuming that this kind of autonomy is what is needed to be a living thing, our account therefore includes viruses as not only living matter, but as full-blown living entities when they enter cells and interact with the cell's metabolic capacities. As virions, they are still lineage elements but are temporarily disengaged from metabolic collaboration (likewise bacteria such as *Chlamydia* in their inert spore-like state and perhaps even many plant seeds and fungal spores). This is why we suggested above that viruses should strictly be described as having developmental cycles rather than life cycles.

Taking this perspective not only renders unproblematic the idea of an entity being sometimes living and at other times non-living but also reinforces the idea of life and the evolution of life as a continuum of collaborativity. Given the acceptance that life has evolved from a chemical context, ruling out self-replicating complexes of chemicals and molecules on the grounds that they are not cells seems misguided. A commitment to life as exclusively cellular and monogenomically organismal would mean that the origins of life must involve a single leap from fully non-living to fully living, something that is conceptually difficult to accept and, for that matter, provides a natural target for creationists to insist on the need for supernatural intervention. The spectrum of biological entities we have described shows that an inflexible dichotomy of life and non-life is, in any case, highly problematic, even for making sense of the entities that now exist. Our more generous framework can encompass a range of theories about the organization and evolution of pre-cellular life that give prions, plasmids and viruses important roles, as well as other macromolecular complexes (e.g. Rode et al. 1999; Lupi et al. 2006, 2007; Eberhard 1990; Kado 1998; Koch 1995).

We also think that our thesis of multi-modal, interconnected and overlapping life processes suggests a more continuous vision of evolutionary history. Many discussions of early life posit a radical transition from a community of genetic exchange to one of restricted vertical inheritance, cellular autonomy and stable genealogy (Woese 2005; Dawkins 2008). Although some biologists believe that pre-cellular

[27] It is tempting here to invoke Kant's analysis of autonomy as the possibility of conformity to duty, an essentially socially defined concept, and his rejection of the notion of autonomy imagined as following no more than the pursuit of contingent individual interests or desires.

life is best conceived as 'unselfish' communality in which genetic resources are shared (Woese et al. 2000), others such as Dawkins presume that pre-cellular life was driven by selfish replication, and that promiscuous horizontal exchange simply *extends* the opportunities for selfishness (Dawkins 2008). Rather than restricting some evolutionary processes to a discontinued past, we prefer to incorporate them into a schema that allows for the continuity of lateral gene transfer as an important characteristic of today's collaborative evolution.

We find here the reflections of Norris et al. (2007) and Hunding et al. (2006) very helpful. They argue that life evolved as a 'diverse interacting community of molecules' – a 'pre-biotic ecology' that implies a more ecological and community-based view of any biological entity, pre- or post the evolution of cells. Their model describes

> the emergence of life as a functional ecological system through a process of integration from diverse components, not as a single entity ... there is no identifiable point at which life emerged. Rather, [it is] a continuous *process* by which increasingly complex, integrated, self-replicating, autocatalytic, module systems evolve new properties in tandem with their environments (Hunding et al. 2006, 409–410).

We believe this sort of dynamic system-based scenario fits more appropriately what we know of the rest of the evolution of life.[28] Evolutionary history suggests that life involves a range of coevolving hierarchies, and that non-life and life share a huge and biologically significant territory that buffers and makes more complex any account of either. Ecology presents us with scenarios of collaboration at least ascompelling as those that highlight competition, and the former are rapidly increasing our understanding of the macrobial and microbial world (Dupré and O'Malley 2007). Thinking of life as the result of the intersection of lineage-forming, metabolically collaborative matter, organized within different interacting levels, allows a smooth transition from the earliest living matter to standard examples of life and beyond them all the way up to contemporary ecosystems. A general account such as ours is not, and need not be, definitional. It is, however, sufficient to encompass what is known about an ever more striking variety of biological entities and their evolutionary histories, and to reorient approaches to life around a biologically realistic interpretation of collaborativity.

Acknowledgements We wish to thank Mark Bedau, the Egenis Philosophy of Biology Journal Club, our two anonymous referees and our editor for very constructive comments and advice. We found our second referee's summary of our paper especially helpful and incorporated it into a revised abstract. We also gratefully acknowledge funding from the Economic and Social Research Council (UK). The research in this chapter was part of the programme of the ESRC Centre for Genomics in Society (Egenis).

[28] Dynamic system-based scenarios of life are a general idea more precisely formulated in work by theoretical biologists such as Tibor Gánti, Robert Rosen, Stuart Kauffman, and Humberto Maturana and Francisco Varela. Our emphasis on collaborative interaction suggests congruences with this sort of work, but we have yet to explore these in any specific way.

References

Abreu, Fernanda, Juliana Lopes Martins, Thaís Souza Silveira, Carlina Neumann Keim, Lins de Barros, G.P. Henry, Federico J. Gueiros-Filho, and Ulysses Lins. 2007. 'Candidatus Magnetoglobus multicellularis', a multicellular, magnetotactic prokaryote from a hypersaline environment. *International Journal of Systematic and Evolutionary Microbiology* 57: 1318–1322.

Aguilar, Claudio, Hera Vlamakis, Richard Losick, and Roberto Kolter. 2007. Thinking about *Bacillus subtilis* as a multicellular organism. *Current Opinion in Microbiology* 10: 638–643.

Ahlquist, Paul. 2006. Parallels among positive-strand RNA viruses, reverse-transcribing viruses and double-stranded RNA viruses. *Nature Reviews Microbiology* 4: 371–382.

Andersson, Jan O. 2000. Evolutionary genomics: Is *Buchnera* a bacterium or an organelle? *Current Biology* 10: R866–R868.

Archibald, John M. 2007. Nucleomorph genomes: Structure, function origin and evolution. *BioEssays* 29: 392–402.

Bamford, Dennis H. 2003. Do viruses form lineages across different domains of life? *Research in Microbiology* 154: 231–236.

Bamford, Dennis H., Jonathan M. Grimes, and David I. Stuart. 2005. What does structure tell us about virus evolution? *Current Opinion in Structural Biology* 15: 655–663.

Bannert, Norbert, and Reinhard Kurth. 2004. Retroelements and the human genome: New perspectives on an old relation. *Proceedings of the National Academy of Sciences USA* 101(Suppl. 2): 14572–14579.

Barton, Bret M., Gordon P. Harding, and Anthony J. Zuccarelli. 1995. A general method for detecting and sizing large plasmids. *Analytical Biochemistry* 226: 235–240.

Baumann, Paul. 2005. Biology of bacteriocyte-associated endosymbionts of plant sap-sucking insects. *Annual Review of Microbiology* 59: 155–189.

Bhardwaj, Urvashi, Yu-Hao Zhang, Zakir Rangwala, and Edward R.B. McCabe. 2006. Completely self-contained cell culture system: From storage to use. *Molecular Genetics and Metabolism* 89: 168–173.

Bhattacharya, Debashish, and John M. Archibald. 2006. The difference between organelles and endosymbionts – Response to Theissen and Martin. *Current Biology* 16: R1017–R1018.

Bingle, Lewis E.H., and Christopher M. Thomas. 2001. Regulatory circuits for plasmid survival. *Current Opinion in Microbiology* 4: 194–200.

Bodył, Andrzej, Paweł Mackiewicz, and John W. Stiller. 2007. The intracellular cyanobacteria of *Paulinella chromatophora*: Endosymbionts or organelles? *Trends in Microbiology* 15: 295–296.

Boevink, Petra, and Karl J. Oparka. 2005. Virus-host interactions during movement processes. *Plant Physiology* 138: 1815–1821.

Bousset, Luc, and Ronald Melki. 2002. Similar and divergent features in mammalian and yeast prions. *Microbes and Infection* 4: 461–469.

Burnet, Frank Macfarlane. 1945. *Virus as organism: Evolutionary and ecological aspects of some human virus diseases*. Cambridge: Harvard University Press.

Campbell, Alan. 2001. The origins and evolution of viruses. *Trends in Microbiology* 9: 61.

Casjens, Sherwood R. 2003. Prophages and bacterial genomics: What have we learned so far? *Molecular Microbiology* 49: 277–300.

Cavalier-Smith, Thomas. 2000. Membrane heredity and early chloroplast evolution. *Trends in Plant Science* 5: 174–182.

Cavalier-Smith, Thomas, and John J. Lee. 1985. Protozoa as hosts for endosymbioses and the conversion of symbionts into organelles. *Journal of Protozoology* 32: 376–379.

Cello, Jeronimo, Aniko V. Paul, and Eckard Wimmer. 2002. Chemical synthesis of poliovirus cDNA: Generation of infectious virus in the absence of natural template. *Science* 297: 1016–1018.

Charlat, Sylvain, George D.D. Hurst, and Hervé Merçot. 2003. Evolutionary consequences of Wolbachia infections. *Trends in Genetics* 19: 217–223.

Chernoff, Yury O. 2001. Mutation processes at the protein level: Is Lamarck back? *Mutation Research* 488: 39–64.
Chernoff, Yury O., Alexey P. Galkin, Eugene Lewitin, Tatiana A. Chernova, Gary P. Newnam, and Slava M. Belenkiy. 2000. Evolutionary conservation of prion-forming abilities of the yeast Sup35 protein. *Molecular Microbiology* 35: 865–876.
Cho, Ho.Jung., Henrik Jönsson, Kyle Campbell, Pontus Melke, Joshua W. Williams, Bruno Jedynak, Ann M. Stevens, Alex Groisman, and Andre Levchenko. 2007. Self-organization in high-density bacterial colonies: Efficient crowd control. *PLoS Biology* 5(11): e302.
Chyba, Christopher F., and Kevin P. Hand. 2005. Astrobiology: The study of the living universe. *Annual Review of Astronomy and Astrophysics* 43: 31–74.
Claverie, Jean-Michel, Hiroyuki Ogata, Stéphane Audic, Chantal Abergel, Karsten Suhre, and Pierre-Edouard Fournier. 2006. Mimivirus and the emerging concept of 'giant' virus. *Virus Research* 117: 133–144.
Collings, David A., John D.I. Harper, Jan Marc, Robyn L. Overall, and Robert T. Mullen. 2002. Life in the fast lane: Actin-based motility of plant peroxisomes. *Canadian Journal of Botany* 80: 430–441.
Costerton, J.William, Zbigniew Lewandowski, Douglas E. Caldwell, Darren R. Korber, and Hilary M. Lappin-Scott. 1995. Microbial biofilms. *Annual Review of Microbiology* 49: 711–745.
Crespi, Bernard J. 2001. The evolution of social behaviour in microorganisms. *Trends in Ecology & Evolution* 16: 178–183.
Cutler, Sean, and David Ehrhardt. 2000. Dead cells don't dance: Insights from live-cell imaging in plants. *Current Opinion in Plant Biology* 3: 532–537.
Darby, Alistair C., Nam-Hyuk Cho, Hans-Henrik Fuxelius, Joakim Westberg, and Siv G.E. Andersson. 2007. Intracellular pathogens go extreme. *Trends in Genetics* 23: 511–520.
Dawkins, Richard. 1976. *The selfish gene*. Oxford: Oxford University Press.
Dawkins, Richard. 2008. Life: A gene-centric view: Craig Venter and Richard Dawkins. A conversation in Munich (Transcript). *Edge*, 235, Feb 6. www.edge.org/documents/archive/edge235.html#life (also www.edge.org/documents/life/life_index.html). Retrieved from 24 Feb 2008.
del Solar, Gloria, and Manuel Espinosa. 2000. Plasmid copy number control: An ever-growing story. *Molecular Microbiology* 37: 492–500.
del Solar, Gloria, Rafael Giraldo, Maria Jesus Ruiz-Echevarría, Manuel Espinosa, and Ramón Díaz-Orejas. 1998. Replication and control of circular bacterial plasmids. *Microbiology and Molecular Biology Reviews* 62: 434–464.
Derkatch, Irina L., Michael E. Bradley, Joo Y. Hong, and Susan W. Liebman. 2001. Prions affect the appearance of other prions: The story of [PIN+]. *Cell* 108: 171–182.
Doolittle, W.Ford, and Carmen Sapienza. 1980. Selfish genes, the phenotype paradigm and genome evolution. *Nature* 284: 601–603.
Douglas, Angela E., and John A. Raven. 2003. Genomes at the interface between bacteria and organelles. *Philosophical Transactions of the Royal Society of London B: Biological Sciences* 358: 5–18.
Dunning-Hotopp, Julie C., Michael E. Clark, Deodoro C.S.G. Oliveira, Jeremy M. Foster, Peter Fischer, Mónica C. Muñoz-Torres, Jonathan D. Giebel, Nikhil Kumar, Nadeeza Ishmael, Shiliang Wang, Jessica Ingram, Rahul V. Nene, Jessica Shepard, Jeffrey Tomkins, Stephen Richards, David J. Spiro, Elodie Ghedin, Barton E. Slatko, Hervé Tettelin, and John H. Werren. 2007. Widespread lateral gene transfer from intracellular bacteria to multicellular eukaryotes. *Science* 317: 1753–1756.
Dupré, John. 1993. *The disorder of things: Metaphysical foundations of the disunity of science*. Cambridge: Harvard University Press.
Dupré, John, and Maureen O'Malley. 2007. Metagenomics and biological ontology. *Studies in History and Philosophy of Biological and Biomedical Sciences* 38: 834–846.
Dworkin, Martin. 1997. Multiculturism versus the single microbe. In *Bacteria as multicellular organisms*, ed. James A. Shapiro and Martin Dworkin, 3–13. New York: Oxford University Press.
Dyall, Sabrina D., Mark T. Brown, and Patricia J. Johnson. 2004. Ancient invasions: From endosymbionts to organelles. *Science* 304: 253–257.

Dyer, Betsey Dexter. 1989. Symbiosis and organismal boundaries. *American Zoologist* 29: 1085–1095.
Eberhard, William G. 1980. Evolutionary consequences of intracellular organelle competition. *The Quarterly Review of Biology* 55: 231–249.
Eberhard, William G. 1990. Evolution in bacterial plasmids and levels of selection. *The Quarterly Review of Biology* 65: 3–22.
Embley, T.Martin, and William F. Martin. 2006. Eukaryotic evolution, changes and challenges. *Nature* 440: 623–630.
Fenn, Katelyn, and Mark Blaxter. 2006. *Wolbachia* genomes: Revealing the biology of parasitism and mutualism. *Trends in Parasitology* 22: 60–65.
Forterre, Patrick. 2006. The origin of viruses and their possible roles in major evolutionary transitions. *Virus Research* 117: 5–16.
Fraser, Claire M., Jeannine D. Gocayne, Owen White, Mark D. Adams, Rebecca A. Clayton, Robert D. Fleischmann, Carol J. Bult, Anthony R. Kerlavage, Granger G. Sutton, Jenny M. Kelley, Janice L. Fritchman, Janice F. Weidman, Keith V. Small, Mina Sandusky, Joyce Fuhrmann, David Nguyen, Teresa R. Utterback, Deborah M. Saudek, Cheryl A. Phillips, Joseph M. Merrick, Jean-François Tomb, Brian A. Dougherty, Kenneth F. Bott, Ping-Chuan Hu, Thomas S. Lucier, Scott N. Peterson, Hamilton O. Smith, Clyde A. Hutchison III, and J. Craig Venter. 1995. The minimal gene complement of *Mycoplasma genitalium*. *Science* 270: 397–403.
Galperin, Michael Y. 2005. Life is not defined just in base pairs. *Environmental Microbiology* 7: 149–152.
Gánti, Tibor. 1997. Biogenesis itself. *Journal of Theoretical Biology* 187: 583–593.
Gerdes, Kenn, Poul B. Rasmussen, and Søren Molin. 1986. Unique type of plasmid maintenance function: Postsegregational killing of plasmid-free cells. *Proceedings of the National Academy of Sciences USA* 83: 3116–3120.
Ghigo, Jean-Marc. 2001. Natural conjugative plasmids induce bacterial biofilm development. *Nature* 412: 442–445.
Gibbs, Adrian J., and Mark J. Gibbs. 2006. A broader definition of 'the virus species'. *Archives of Virology* 151: 1419–1422.
Gill, Steven R., Mihai Pop, Robert T. DeBoy, Paul B. Eckburg, Peter J. Turnbaugh, Buck S. Samuel, Jeffrey I. Gordon, David A. Relman, Claire M. Fraser-Liggett, and Karen E. Nelson. 2006. Metagenomic analysis of the human distal gut microbiome. *Science* 312: 1355–1359.
Glass, John I., Nacyra Assad-Garcia, Nina Alperovich, Shibu Yooseph, Matthew R. Lewis, Mahir Maruf, Clyde A. Hutchison III, Hamilton O. Smith, and J.Craig Venter. 2006. Essential genes of a minimal bacterium. *Proceedings of the National Academy of Sciences USA* 103: 425–430.
Gouin, Edith, Coumaran Egile, Pierre Dehoux, Véronique Villiers, Josephine Adams, Frank Gertler, Rong Li, and Pascale Cossart. 2004. The RickA protein of *Rickettsia conorii* activates the Arp2/3 complex. *Nature* 427: 457–461.
Griffiths, David J. 2001. Endogenous retroviruses in the human genome sequence. *Genome Biology* 2(6): 1017.1–1017.5.
Grote, Mathias. 2008. Hybridizing bacteria, crossing methods, cross-checking arguments: The transition from episomes to plasmids (1961–1969). *History and Philosophy of the Life Sciences* 30: 407–430.
Hambly, Emma, and Curtis A. Suttle. 2005. The viriosphere, diversity and genetic exchange within phage communities. *Current Opinion in Microbiology* 8: 444–450.
Hamilton, Garry. 2006. Virology: The gene weavers. *Nature* 441: 683–685.
Hendrix, Roger W. 2002. Bacteriophages: Evolution of the majority. *Theoretical Population Biology* 61: 471–480.
Hendrix, Roger W., Margaret C.M. Smith, R.Neil Burns, Michael E. Ford, and Graham F. Hatfull. 1999. Evolutionary relationships among diverse bacteriophages and prophages: All the world's a phage. *Proceedings of the National Academy of Sciences USA* 96: 2192–2197.
Hendrix, Roger W., Jeffrey G. Lawrence, Graham F. Hatfull, and Sherwood Casjens. 2000. The origins and ongoing evolution of viruses. *Trends in Microbiology* 8: 504–508.

Hood, Leroy, and David Galas. 2003. The digital code of DNA. *Nature* 421: 444–448.
Hooper, Lora V., and Jeffrey I. Gordon. 2001. Commensal host-bacterial relationships in the gut. *Science* 292: 1115–1118.
Huber, Harald, Michael J. Hohn, Reinhard Rachel, Tanja Fuchs, Verena C. Wimmer, and Karl O. Stetter. 2002. A new phylum of Archaea represented by a nanosized hyperthermophilic symbiont. *Nature* 417: 63–67.
Hull, David L. 1980. Individuality and selection. *Annual Review of Ecology and Systematics* 11: 311–332.
Hull, David L. 1988. *Science as a process*. Chicago: University of Chicago Press.
Hunding, Axel, François Képès, Doron Lancet, Abraham Minsky, Vic Norris, Derek Raine, K. Sriram, and Robert Root-Bernstein. 2006. Compositional complementarity and prebiotic ecology in the origin of life. *BioEssays* 28: 399–412.
Hurst, Christon J. 2000. An introduction to viral taxonomy and the proposal of Akamara, a potential domain for the genomic acellular agents. In *Viral ecology*, ed. Christon J. Hurst, 41–62. San Diego: Academic/Elsevier.
Iturbe-Ormaetxe, Iñaki, and Scott L. O'Neill. 2007. *Wolbachia*-host interactions: Connecting phenotype to genotype. *Current Opinion in Microbiology* 10: 221–224.
Jablonka, Eva, and Marion J. Lamb. 2005. *Evolution in four dimensions*. Cambridge: MIT Press.
Jeffrey, C. 1971. Thallophytes and kingdoms – a critique. *Kew Bulletin* 25(2): 291–299.
Kado, Clarence I. 1998. Origin and evolution of plasmids. *Antonie van Leeuwenhoek International Journal of General and Molecular Microbiology* 73: 117–126.
Kaiser, Dale. 2001. Building a multicellular organism. *Annual Review of Genetics* 35: 103–123.
Karam, Jim D. 2005. Bacteriophages: The viruses for all seasons of molecular biology. *Virology Journal* 2: 19.
Keeling, Patrick J., and Naomi M. Fast. 2002. Microsporidia: Biology and evolution of highly reduced intracellular parasites. *Annual Review of Microbiology* 56: 93–116.
Keim, Carolina N., Juliana L. Martins, Fernanda Abreu, Alexandre Soares Rosado, Henrique Lins de Barros, Radovan Borojevic, Ulysses Lins, and Marcos Farina. 2004. Multicellular life cycle of magnetotactic prokaryotes. *FEMS Microbiology Letters* 240: 203–208.
Keim, Carolina N., Marcos Farina, and Ulysses Lins. 2007. *Magnetoglobus*, magnetic aggregates in anaerobic environments. *ASM News* 2: 437–445.
Kerfeld, Cheryl A., Michael R. Sawaya, Shiho Tanaka, Chau V. Nguyen, Martin Phillips, Morgan Beeby, and Todd O. Yeates. 2005. Protein structures forming the shell of primitive bacterial organelles. *Science* 309: 936–938.
Kirschner, Marc W., and John C. Gerhart. 1998. Evolvability. *Proceedings of the National Academy of Sciences USA* 95: 8420–8427.
Klekowski, Edward J. 2003. Plant clonality, mutation, diplontic selection and mutational meltdown. *Biological Journal of the Linnean Society* 79: 61–67.
Koch, Arthur L. 1995. The origin of intracellular and intercellular pathogens. *The Quarterly Review of Biology* 70: 423–437.
Kolenbrander, Paul E. 2000. Oral microbial communities: Biofilms, interactions, and genetic systems. *Annual Review of Microbiology* 54: 413–437.
Komeili, Arash, Zhuo Li, Debra K. Newman, and Grant J. Jensen. 2006. Magnetosomes are cell membrane invaginations organized by actin-like protein mamK. *Science* 311: 242–245.
Koonin, Eugene V. 2005. Virology: Gulliver among the Lilliputians. *Current Biology* 15: R167–R169.
Koonin, Eugene V., Tatiana G. Senkevich, and Valerian V. Dolja. 2006. The ancient virus world and evolution of cells. *Biology Direct* 1: 29. doi:10.1186/1745-6150.1-29.
Kornberg, Arthur. 1989. *For the love of enzymes: The odyssey of a biochemist*. Cambridge, MA: Harvard University Press.
Koshland Jr., Daniel E. 2002. The seven pillars of life. *Science* 295: 2215–2216.
Lartigue, Carole, John I. Glass, Nina Alperovich, Rembert Pieper, Prashanth Parmar, Clyde A. Hutchison III, Hamilton O. Smith, and J.Craig Venter. 2007. Genome transplantation in bacteria: Changing one species to another. *Science* 317: 632–638.

Lawrence, Jeffrey G., Graham F. Hatfull, and Roger W. Hendrix. 2002. Imbroglios of viral taxonomy: Genetic exchange and failings of phenetic approaches. *Journal of Bacteriology* 184: 4891–4905.
Lederberg, Joshua. 1952. Cell genetics and hereditary symbiosis. *Physiological Reviews* 32: 403–430.
Lederberg, Joshua. 1998. Personal perspective: Plasmid (1952–1997). *Plasmid* 39: 1–9.
Lezon, Timothy R., Jayanth R. Banavar, and Amos Maritan. 2006. The origami of life. *Journal of Physics: Condensed Matter* 18: 847–888.
Liebman, Susan W., and Irina L. Derkatch. 1999. The yeast [PSI+] prion: Making sense of non-sense. *Journal of Biological Chemistry* 274: 1181–1184.
Lindell, Debbie, Jacob D. Jaffe, Zachary I. Johnson, George M. Church, and Sallie W. Chisholm. 2005. Photosynthesis genes in marine viruses yield proteins during host infection. *Nature* 438: 86–89.
Lowe, Martin, and Francis A. Barr. 2007. Inheritance and biogenesis of organelles in the secretory pathway. *Nature Reviews Molecular Cell Biology* 8: 429–439.
Luisi, Pier-Luigi. 1998. About various definitions of life. *Origins of Life and Evolution of the Biosphere* 28: 613–622.
Lupi, Omar, Paula Dadalti, Eduardo Cruz, and Paul R. Sanberg. 2006. Are prions related to the emergence of early life? *Medical Hypotheses* 67: 1027–1033.
Lupi, Omar, Paula Dadalti, and Clyde Goodheart. 2007. Did the first virus self-assemble from self-replicating proteins prion proteins and RNA? *Medical Hypotheses* 69: 724–730.
Luria, Salvador Edward, James E. Darnell Jr., David Baltimore, and Allan Campbell. 1978. *General virology*, 3rd ed. New York: Wiley.
Mallet, François, Olivier Bouton, Sarah Proudhomme, Valerie Chaynet, Guy Oriol, Bertrand Bonnaud, Gerard Lucotte, Laurent Duret, and Bernard Mandrand. 2004. The endogenonous retroviral locus ERVWE1 is a bona fide gene involved in hominoid placental physiology. *Proceedings of the National Academy of Sciences USA* 101: 1731–1736.
Manuelidis, Laura. 2004. A virus behind the mask of prions? *Folia Neuropathology* 42(Supplement B): 10–23.
Marais, Gabriel A.B., Alexandra Calteau, and Olivier Tenaillon. 2007. Mutation rate and genome reduction in endosymbiotic and free-living bacteria. *Genetica*. doi:10.1007/s1079-007-9226-6.
Martin, William R. 2003. Gene transfer from organelles to the nucleus: Frequent and in big chunks. *Proceedings of the National Academy of Sciences USA* 100: 8612–8614.
Masel, Joanna, and Aviv Bergman. 2003. The evolution of the evolvability of the yeast prion [PSI+]. *Evolution* 57: 1498–1512.
Maynard Smith, John, and Eörs Szathmáry. 1995. *The major transitions in evolution*. New York: W. H. Freeman.
Meseguer, María A., Alberto Álvarez, T. Rejas María, Carlos Sánchez, José C. Pérez-Díaz, and Fernando Baquero. 2003. *Mycoplasma pneumoniae*: A reduced-genome intracellular bacterial pathogen. *Infection, Genetics and Evolution* 3: 47–55.
Mindell, David P., and Luis P. Villarreal. 2003. Don't forget about viruses. *Science* 302: 1677.
Mindell, David P., Joshua Rest, and Luis P. Villarreal. 2003. Viruses and the tree of life. In *Assembling the tree of life*, ed. Joel Cracraft and Michael J. Donoghue, 107–118. New York: Oxford University Press.
Moran, Nancy A. 2006. Symbiosis. *Current Biology* 16: R866–R871.
Moran, Nancy A., and Patrick H. Degnan. 2006. Functional genomics of *Buchnera* and the ecology of aphid hosts. *Molecular Ecology* 15: 1251–1261.
Moreira, David, and Purificación López-Gárcia. 2005. Comment on 'The 1.2-megabase genome sequence of Mimivirus'. *Science* 308: 1114a.
Moreira, David, and Purificación López-Gárcia. 2009. Ten reasons to exclude viruses from the tree of life. *Nature Reviews Microbiology* 7: 306–311.
Muller, Hermann Joseph. 1966. The gene material as the initiator and the organizing basis of life. *American Naturalist* 100: 493–517.
Munro, Sean. 2004. Organelle identity and the organization of membrane traffic. *Nature Cell Biology* 6: 469–472.
Norris, Vic, Axel Hunding, François Képès, Doron Lancet, Abraham Minsky, Derek Raine, Robert Root-Bernstein, and K. Sriram. 2007. Question 7: The first units of life were not cells. *Origins of Life and Evolution of the Biosphere* 37: 429–432.

Nunnari, Jodi, and Peter Walter. 1996. Regulation of organelle biogenesis. *Cell* 84: 389–394.
O'Malley, Maureen, and John Dupré. 2007. Size doesn't matter: Towards a more inclusive philosophy of biology. *Biology and Philosophy* 22: 155–191.
Okasha, Samir. 2004. Multi-level selection and the major transitions in evolution. In *Proceedings PSA 19th Biennial meeting – PSA 2004: PSA contributed papers*. Austin: PSA.
Orgel, Leslie E., and Francis H.C. Crick. 1980. Selfish DNA: The ultimate parasite. *Nature* 284: 604–607.
Osteryoung, Katherine W., and Jodi Nunnari. 2003. The division of endosymbiotic organelles. *Science* 302: 1698–1704.
Pál, Csaba. 2001. Yeast prions and evolvability. *Trends in Genetics* 17: 167–169.
Pályi, Gyula, Claudia Zucchi, and Luciano Caglioti (eds.). 2002. *Fundamentals of life*. Paris: Elsevier.
Parkin, Edward T., Nicole T. Watt, Ishrut Hussain, Elizabeth A. Eckman, Christopher B. Eckman, Jean C. Manson, Herbert N. Baybutt, Anthony J. Turner, and Nigel M. Hooper. 2007. Cellular prion protein regulates β-secretase cleavage of the Alzheimer's amyloid precursor protein. *Proceedings of the National Academy of Sciences USA* 104: 11062–11067.
Paulsson, Johan. 2002. Multileveled selection on plasmid replication. *Genetics* 161: 1373–1384.
Pearson, Helen. 2008. 'Virophage' suggests viruses are alive. *Nature* 454: 677.
Pennisi, Elizabeth. 2007. Replacement genome gives microbe new identity. *Science* 316: 1827.
Pérez-Brocal, Vicente, Rosario Gil, Silvia Ramos, Araceli Lamelas, Marina Postigo, José Manuel Michelena, Francisco J. Silva, Andrés Moya, and Amparo Latorre. 2006. A small microbial genome: The end of a long symbiotic relationship? *Science* 314: 312–313.
Perlin, Michael H. 2002. The subcellular entities a.k.a. plasmids. In *Modern microbial genetics*, 2nd ed, ed. Uldis N. Streips and Ronald E. Yasbin, 507–560. New York: Wiley.
Pineda-Krch, Mario, and Torbjörn Fagerström. 1999. On the potential for evolutionary change in meristematic cell lineages through interorganismal selection. *Journal of Evolutionary Biology* 12: 681–688.
Popa, Radu. 2004. *Between necessity and probability: Search for the definition and origin of life*. Berlin: Springer.
Prusiner, Stanley B. 1998. Prions. *Proceedings of the National Academy of Science USA* 95: 13363–13383.
Puck, Theodore Thomas. 1972. *The mammalian cell as a microorganism: Genetic and biochemical studies in vitro*. San Francisco: Holden-Day.
Raoult, Didier. 2005. The journey for *Rickettsia* to Mimivirus. *ASM News* 71: 278–284.
Raoult, Didier, Stéphane Audic, Catherine Robert, Chantal Abergel, Patricia Renesto, Hiroyuki Ogata, Bernard La Scola, Marie Suzan, and Jean-Marie Claverie. 2004. The 1.2-megabase genome sequence of Mimivirus. *Science* 306: 1344–1350.
Rayner, Alan D.M. 1997. *Degrees of freedom: Living in dynamic boundaries*. London: Imperial College Press.
Regenmortel, Van, and H.V. Marc. 2007. Virus species and virus identification: Past and current controversies. *Infection, Genetics and Evolution* 7: 133–144.
Rode, Bernd M., Wolfgang Flader, Christoph Sotriffer, and Alessandro Righi. 1999. Are prions a relic of an early stage of peptide evolution? *Peptides* 20: 1513–1516.
Rodríguez-Ezpeleta, Naiara, and Hervé Phillipe. 2006. Plastid origin: Replaying the tape. *Current Biology* 16: R53–R56.
Rottem, Shlomo, and Yehudith Naot. 1998. Subversion and exploitation of host cells by mycoplasmas. *Trends in Microbiology* 6: 436–440.
Roughgarden, Joan. 2009. *The genial gene: Deconstructing Darwinian selfishness*. Berkeley: University of California Press.
Ruiz-Mirazo, Kepa, Juli Peretó, and Alvaro Moreno. 2004. A universal definition of life: Autonomy and open-ended evolution. *Origins of Life and Evolution of the Biosphere* 34: 323–346.
Sapp, Jan. 1994. *Evolution by association: A history of symbiosis*. New York: Oxford University Press.
Schuch, Raymond, and Vincent Fischetti. 2009. The secret life of the anthrax agent Bacillus anthracis: Bacteriophage-mediated ecological adaptations. *PLoS One* 4(8): e6532. doi:10.1371/journal.pone.0006532.

Schulman, Rebecca, and Erik Winfree. 2007. How crystals that sense and respond to their environments could evolve. *Natural Computing*. doi:10.1007/s11047-007-9046-8.
Scola, La, Desnues Bernard, Pagnier Chriselle, Robert Isabelle, Barrassi Catherine, Fournous Lisa, Merchat Ghislain, Suzan-Monti Michèle, Forterre Marie, Koonin Patrick, V. Eugene, and Didier Raoult. 2008. The virophage as a unique parasite of the giant mimivirus. *Nature* 455: 100–104.
Segré, Daniel, and Doron Lancet. 2000. Composing life. *EMBO Reports* 1: 217–222.
Seufferheld, Manfredo, Mauricio Viera, Felix A. Fuiz, Claudia O. Rodrigues, Silvia N.J. Moreno, and Roberto Docampo. 2003. Identification of organelles in bacteria similar to acidocalcisomes of unicellular eukaryotes. *Journal of Biological Chemistry* 278: 29971–29978.
Shapiro, James A. 1998. Thinking about bacterial populations as multicellular organisms. *Annual Review of Microbiology* 52: 81–104.
Shorter, James, and Susan Lindquist. 2005. Prions as adaptive conduits of memory and inheritance. *Nature Reviews Genetics* 6: 435–450.
Smith, Hamilton O., Clyde A. Hutchison III, Cynthia Pfannkoch, and J.Craig Venter. 2003. Generating a synthetic genome by whole genome assembly: X174 bacteriophage from synthetic nucleotides. *Proceedings of the National Academy of Sciences USA* 100: 15440–15445.
Sober, Elliot, and David Sloan Wilson. 1998. *Unto others: The evolution and psychology of unselfish behavior*. Cambridge, MA: Harvard University Press.
Sørensen, Søren J., Mark Bailey, Lars H. Hansen, Niels Kroer, and Stefan Wuertz. 2005. Studying plasmid horizontal transfer in situ: A critical review. *Nature Reviews Microbiology* 3: 700–710.
Soto, Claudio, and Gabriela P. Saborío. 2001. Prions: Disease propagation and disease therapy by conformational transmission. *Trends in Molecular Medicine* 7: 109–114.
Stanley, Wendell Meredith. 1941. Some chemical, medical and philosophical aspects of viruses. *Science* 93: 145–151.
Stanley, Wendell Meredith. 1957. On the nature of viruses, cancer, genes, and life – A declaration of dependence. *Proceedings of the American Philosophical Society* 101: 317–324.
Stoodley, Paul, Karin Sauer, David J.G. Davies, and J.William Costerton. 2002. Biofilms as complex differentiated communities. *Annual Review of Microbiology* 56: 187–209.
Stouthamer, Richard, Johannes A.J. Breeuwer, and Gregory D.D. Hurst. 1999. *Wolbachia pipientis*: Microbial manipulator of arthropod reproduction. *Annual Review of Microbiology* 53: 71–102.
Suttle, Curtis A. 2005. Viruses in the sea. *Nature* 437: 356–361.
Szathmáry, Eörs. 2006. The origin of replicators and reproducers. *Philosophical Transactions of the Royal Society B* 361: 1761–1776.
Tamas, Ivica, Lisa M. Klasson, Jonas P. Sandström, and Siv G.E. Andersson. 2001. Mutualists and parasites: How to paint yourself into a (metabolic) corner. *FEBS Letters* 498: 135–139.
Theissen, Ursula, and William Martin. 2006. The difference between organelles and endosymbionts. *Current Biology* 16: R1016–R1018.
Thomas, Christopher M. 2000. Paradigms of plasmid organization. *Molecular Microbiology* 37: 485–491.
Thomas, Christopher M. 2006. Transcription regulatory circuits in bacterial plasmids. *Biochemical Society Transactions* 34: 1072–1074.
Timmis, Jeremy N., Michael A. Ayliffe, Chun Y. Huang, and William Martin. 2004. Endosymbiotic gene transfer: Organelle genomes forge eukaryotic chromosomes. *Nature Reviews Genetics* 5: 123–135.
True, Heather L., and Susan L. Lindquist. 2000. A yeast prion provides a mechanism for genetic variation and phenotypic diversity. *Nature* 407: 477–483.
Tumpey, Terrence M., Chriostopher F. Basler, Patricia V. Aguilar, Hui Zeng, Alicia Solórzano, David E. Swayne, Nancy J. Cox, Jacqueline M. Katz, Jeffrey K. Taubenberger, Peter Palese, and Adolfo Garcia-Sastre. 2005. Characterization of the reconstructed 1918 Spanish influenza pandemic virus. *Science* 310: 77–80.
Uptain, Susan M., and Susan Lindquist. 2002. Prions as protein-based genetic elements. *Annual Review of Microbiology* 56: 703–741.
Valdivia, Raphael H., and Joseph Heitman. 2007. Endosymbiosis: The evil within. *Current Biology* 17: R408–R410.

van Helvoort, Ton. 1992. Bacteriological and physiological research styles in the early controversy on the nature of the bacteriophage phenomenon. *Medical History* 36: 243–270.

van Niftrik, Laura A., John A. Fuerst, Jaap Sinninghe S. Damsté, J.Gijs Kuenen, Mike S.M. Jetten, and Marc Strous. 2004. The ammoxosome: An introcytoplasmic compartment in anammox bacteria. *FEMS Microbiology Letters* 233: 7–13.

Villarreal, Luis P. 2004. Can viruses make us human? *Proceedings of the American Philosophical Society* 148: 296–323.

von Braun, Schmidt Serena, and Enrico Schleiff. 2007. Movement of endosymbiotic organelles. *Current Protein & Peptide Science* 8: 426–438.

Walsh, James Bruce. 1992. Intracellular selection, conversion bias, and the expected substitution rates of organelle genes. *Genetics* 130: 939–946.

Warren, Graham, and William Wickner. 1996. Organelle inheritance. *Cell* 84: 395–400.

Weber, Bruce H. 2007. Emergence of life. *Zygon* 42(4): 837–856.

Weeks, Andrew R., K.Tracy Reynolds, and Ary A. Hoffman. 2002. *Wolbachia* dynamics and host effects: What has (and has not) been demonstrated? *Trends in Ecology & Evolution* 17: 257–262.

Wegrzyn, Grzegorz. 2005. What does "plasmid biology" currently mean? Summary of the plasmid biology 2004 meeting. *Plasmid* 53: 14–22.

Weinbauer, Markus G. 2004. Ecology of prokaryotic viruses. *FEMS Microbiology Reviews* 28: 127–181.

Weinbauer, Markus G., and Fereidoun Rassoulzadegan. 2004. Are viruses driving microbial diversification and diversity? *Environmental Microbiology* 6: 1–11.

Weissmann, Charles. 2004. The state of the prion. *Nature Reviews Microbiology* 2: 861–871.

Weissmann, Charles, Masato Enari, Peter-Christian Klöhn, Daniela Rossi, and Eckhard Flechsig. 2002. Transmission of prions. *Journal of Infectious Diseases* 186(Suppl. 2): S157–S165.

Wernegreen, Jennifer J. 2002. Genome evolution in bacterial endosymbionts of insects. *Nature Reviews Genetics* 3: 850–861.

Wernegreen, Jennifer J. 2004. Endosymbiosis: Lessons in conflict resolution. *PLoS Biology* 2(3): 0307.

Werren, John H. 1997. Biology of *Wolbachia*. *Annual Review of Entomology* 42: 587–609.

Wickner, Reed B., Herman K. Edskes, Eric D. Ross, Michael M. Pierce, Ulrich Baxa, Andreas Brachmann, and Frank Shewmaker. 2004. Prion genetics: New rules for a new kind of gene. *Annual Review of Genetics* 38: 681–707.

Wickner, Reed B., Herman K. Edskes, Frank Shewmaker, and Toru Nakayashiki. 2007. Prions of fungi: Inherited structures and biological roles. *Nature Reviews Microbiology* 5: 611–618.

Wilhelm, Steven W., and Curtis A. Suttle. 1999. Viruses and nutrient cycles in the sea. *BioScience* 49: 781–788.

Wilson, Jack A. 2000. Ontological butchery: Organism concepts and biological generalizations. *Philosophy of Science* 67(Proceedings): S301–S311.

Wimmer, Eckard. 2006. The test-tube synthesis of a chemical called poliovirus. *EMBO Reports* 7(Special Issue): S3–S9.

Woese, Carl R. 2005. Evolving biological organization. In *Microbial phylogeny and evolution: Concepts and controversies*, ed. Jan Sapp, 99–117. Oxford: Oxford University Press.

Woese, Carl R., Gary J. Olsen, Michael Ibba, and Dieter Söll. 2000. Aminoacyl-tRNA synthetases, the genetic code, and the evolutionary process. *Microbiology and Molecular Biology Reviews* 64: 202–236.

Wu, Dongying, Sean C. Daugherty, Van Aken, E. Susan, Grace H. Pai, Kisha L. Watkins, Hoda Khouri, Luke J. Tallon, Jennifer M. Zaborsky, Helen E. Dunbar, Phat L. Tran, Nancy A. Moran, and Jonathan A. Eisen. 2006. Metabolic complementarity and genomics of the dual bacterial symbiosis of sharpshooters. *PLoS Biology* 4(6): e188.

Zhuravlev, Yuri N., and Vladik A. Avetisov. 2006. The definition of life in the context of its origins. *Biogeosciences* 3: 281–291.

Chapter 14
Addressing the Vitalist's Challenge to Mechanistic Science: Dynamic Mechanistic Explanation

William Bechtel

Abstract Vitalists, especially in the nineteenth century, correctly objected that mechanists' explanations in biology lacked the resources to explain important features of biological phenomena. As some mechanists, especially Claude Bernard, recognized, the key to addressing these objections was to incorporate in mechanistic explanations the contribution of organization found in living systems. In particular, it is necessary to understand how non-sequential organization (combined with nonlinear operations) enables mechanisms to exhibit the sort of complex behavior, including endogenously generated behavior, exhibited by living organisms. Non-sequential organization poses a serious problem for human understanding, which characterizes the functioning of mechanisms qualitatively in a step-by-step manner. To understand the effects of non-sequential organization between nonlinear operations requires developing mathematical equations to represent the operations and computational simulations using these equations to determine how various components of the mechanism change depending on their own state and those of other components of the mechanism. Further, analyzing the results of these simulations requires appropriate representations such as the phase-space representations employed in dynamical systems theory. Fortunately, mechanistic science can be coupled with dynamical modeling to yield dynamic mechanistic explanations such as those being proposed in systems biology. These hold the promise of explaining the features of biological phenomena on which the vitalists appropriately focused attention.

Keywords Dynamics • Mechanistic explanation • Network motifs • Non-sequential organization • Vitalism

W. Bechtel (✉)
Department of Philosophy, Center for Chronobiology and Interdisciplinary Programs in Science Studies and Cognitive Science, University of California, San Diego, 9500 Gilman Drive, La Jolla, CA 92093-0119, USA
e-mail: bill@mechanism.ucsd.edu

S. Normandin and C.T. Wolfe (eds.), *Vitalism and the Scientific Image in Post-Enlightenment Life Science, 1800–2010*, History, Philosophy and Theory of the Life Sciences 2, DOI 10.1007/978-94-007-2445-7_14,

1 Introduction

The vitalist-mechanist controversy of the eighteenth and nineteenth centuries is often portrayed in terms of the progressive mechanists being opposed by the reactionary vitalists. As far as the mechanists were concerned, the picture is basically correct – the mechanists were charting a new path, one that would prove immensely productive in generating biological knowledge. They were, in carrying out the program that Descartes had envisaged but only pursued speculatively, showing how many of the phenomena exhibited in living organisms, could be explained in terms of the component parts of those organisms carrying out component operations in much the same manner as is the case in human-engineered machines. Many of the vitalists, however, were astute critics, recognizing the limitations of the mechanistic accounts of their day. The limitations were not just incidental but went to the core of the mechanistic project as it was pursued – they recognized that the mechanist accounts lacked the resources to account for some of the most fundamental features of living organisms. Unlike the human-engineered machines that provide the model for mechanistic accounts, organisms build, sustain, and repair themselves. In doing this they are not just reactive – they are endogenously active.

While correctly identifying the complexity of the phenomena associated with life, the vitalists were in worse shape than the mechanists they criticized when it came to explanation, for they had no research program to pursue to explain the distinctive features of living systems. Those vitalists who availed themselves of non-mechanistic forces or powers did not provide explanations but merely labeled the difference. Ultimately, I will argue, mechanistic biology can and is being extended to address the vitalists' objections. This requires taking seriously the importance of the way organisms are organized and, as is currently being done in parts of systems biology, drawing upon mathematical tools such as those of dynamical systems theory to understand how appropriate modes of organization can give rise to endogenously active mechanisms.

The challenges to mechanistic biology take on new currency in the wake of the emergence of the new mechanistic philosophy of science. As I will argue, the accounts of the new mechanists provide an accurate philosophical analysis of mechanistic biology as it has been practiced. But while noting the importance of organization, they tend to focus on rather simple modes of organization, emphasizing sequential execution of operations from "start or set-up to finish or termination conditions." I will contrast this *basic* conception of mechanism with one that recognizes that the organization of biological mechanisms is often nonsequential and that this gives rise to ways of behaving (especially when the component operations are nonlinear) that cannot readily be captured in the qualitative and intuitive analyses offered by the basic accounts. Rather, they require developing mathematical representations and invoking the resources of dynamical systems theory to appreciate how such mechanisms will behave. Abrahamsen and I refer to accounts integrating mechanistic decomposition of systems into parts and operations with the quantitative tools provided by dynamical systems theory as *dynamic mechanistic explanations*.

I begin in Sect. 2 with a characterization of mechanistic biology as it developed in the nineteenth and twentieth centuries and in Sect. 3 develop the account of basic mechanistic explanation, showing how it not only correctly characterizes mechanistic science but also offers significant advantages over the previously dominant tradition in philosophy of science that appealed primarily to laws in explanation. In Sect. 4 I introduce what I take to be the most serious vitalist challenge to mechanistic biology – the fact that living systems seem to resist external forces imposed on them – and describe a first step in the mechanist's response – the introduction of a mode of non-sequential organization, negative feedback, to explain homeostasis. In Sect. 5 I explore a further response, one that recognizes that living systems are endogenously active and that such activity serves to maintain the autonomous existence of living organisms. Finally, I show in Sect. 6 such endogenously active systems require non-sequential modes of organization, and that understanding the effects of such organization in mechanisms requires mathematical representation and modeling that results in dynamic mechanistic explanations. While the project of understanding the behavior of non-sequential organization through mathematical modeling is still in its infancy, I describe some of the tools being developed that promise to enable the mechanists to finally offer explanations that address the vitalist's challenge.

2 Mechanism's Forte: Identifying Parts and Operations

In articulating the original mechanistic philosophy Descartes attempted to explain all features of material objects in terms of their being composed of corpuscles, each with a particular size, shape, and motion. His speculative account of magnetism in terms of screw-shaped corpuscles whose motion pulled the attracted material to the magnet both exemplify his strategy and its limitations. Corpuscles were presumed to be too small to be seen and as a result researchers could not determine their properties through empirical inquiry. Moreover, the catalog of acceptable properties was too limiting to provide plausible accounts of the phenomena exhibited of living systems. But the machine metaphor was potent. Just as machines, from the simple classical machines such as the lever and pulley to the machines being developed by Descartes' contemporaries (weapons often exhibit the most inventiveness, but other devices such as flush toilets were being developed; Da Vinci's many inventions are illustrative), provided a model of how phenomena might be generated by the coordinated action of parts that carried out different tasks. By reversing the process through which the machine was put together – decomposing it into its parts and examining the operations they could perform – someone other than the inventor could explain how a machine worked. Moreover, the parts from which such a machine might be made, or into which it could be decomposed, need not be primitive parts, but simply anything, including other machines, that were already available.

One of the early successes of developing mechanistic explanations of biological phenomena actually preceded Descartes. Harvey (1628) offered a mechanistic account of the circulation of blood, with the heart serving as a pump and the valves serving to restrict the flow of blood in one direction.[1] Ironically, Descartes dissented from Harvey and advanced his own mechanistic explanation in which the lungs were construed as a furnace that heated and thereby expanded the blood and so forced it out to the tissues. Descartes offered other speculative proposals for physiological mechanisms. Intrigued by the statues in the Royal Gardens that moved by hydraulic processes, for example, he proposed that the nervous system is a hydraulic mechanism in which animal spirits are forced to flow through one set of nerves from the senses to the ventricles of the brain and then, in another set of nerves, back to the muscles, causing them to move.

Many of the mechanistic proposals advanced in physiology in the 150 years after Descartes were also highly speculative. But by the end of the eighteenth century researchers were developing experimental techniques that provided empirical evidence about the component parts of organisms and how they behaved on which they could base their proposals. I will briefly review two of the great success cases of mechanistic biology over the two ensuing centuries.

The first involved identifying the chemical constituents of organisms and appealing to reactions involving them to explain physiological phenomena. This project had actually begun in the era of phlogiston chemistry, but took new form with Lavoisier's demonstration that oxygen was a basic element. He generalized his account to other elements, several of which – oxygen, carbon, and hydrogen – he also showed to be primary constituents of living organisms. Berthollet (1780) soon added nitrogen. Based on their distinctive ratios of these basic elements, Prout (1827) proposed and Liebig (1831) refined the catalog of proteins, carbohydrates, and fats as the basic types of animal nutrients and initiated inquiry into how they were used in animals. The result was that during the nineteenth century proposed chemical theories could appeal either to basic elements or to proteins, carbohydrates, and fats in their theorizing about the chemical events in organisms. Liebig himself worked at both levels. Reasoning that plants already synthesized proteins, carbohydrates, and fats,[2] he hypothesized that animals only carried out catabolic reactions. More specifically, he hypothesized that proteins were incorporated into the body and used for muscle contraction whereas fats and carbohydrates were burned to produce animal heat (Liebig 1842). This proposal was effectively refuted by physiological experiments (Fick and Wislicenus 1866; Frankland 1866), but this led to ever more intense inquiry into the actual processes in which foodstuffs were utilized in animals.

In his accounts of both the synthesis of organic compounds in plants and their breakdown in animals Liebig had stressed the importance of balanced chemical equations, but in constructing these he and subsequent chemists for most of the nineteenth century could only work at the level of chemical elements and speculate

---

[1]As Hall (1969) argues (Chapter 17), Harvey was not a mechanist at a micromechanical level; for example, he construed blood as the vital fluid that directs the organization of descendent organisms in reproduction.

[2]He offered chemical accounts of how they did so in Liebig (1840).

about how they were added or removed from compounds in hypothesized reactions. A new level of explanation emerged towards the end of the nineteenth century with the identification of chemical groups (hydroxyl, phosphate, etc.); this enabled physiological chemists to identify a set of fundamental reactions (oxidations, reductions, phosphorylations) that could figure in physiological processes. Once Buchner (1897) showed that fermentation could be achieved in a cell-free preparation, researchers began trying to identify a pathway of reactions from glucose to alcohol or lactic acid. The initial efforts (Neuberg and Kerb 1914) encountered a number of empirical obstacles but with the determination that the intermediates were phosphorylated compounds (Embden et al. 1933), researchers soon succeed in constructing an empirically well-supported account (Meyerhof and Lohmann 1934; Meyerhof and Kieslling 1936; cf. Bechtel 2006, for a discussion of this history).

A second major advance in mechanistic science was the discovery of cells as basic living units and mechanistic accounts of their function. Although reports of cells and theorizing about them goes back to Hooke's (1665) identification of cells in cork and Leeuwenhoek's (1719) identification of single-celled organisms, the microscopes available up through the early years of the nineteenth century introduced sufficient spherical and chromatic aberrations that one must assume that most of the cells researchers claimed to see during this period were artifacts. With improved microscopes that reduced these distortions, Brown (1833) observed the nucleus in orchids and this provided one of the foundations on which Schleiden (1838) argued that plant tissue generally consisted of cells. Schwann (1839) both extended the claim to animal tissue and developed a theoretical argument that cells were the fundamental unit in which the processes of life transpired. Although his insistence on cells as fundamental units suggested to some that he was a closet vitalist (Schwann 1836, had argued that cells were necessary for fermentation, drawing the wrath of Wöhler and Leibig), a critical feature of Schwann's account of cells was that they were formed by a mechanical process of chemical precipitation analogous to crystal formation. He attributed the fact that some reactions only seemed to occur within cells to the distinctive chemical constitution they acquired through this precipitation process. When von Mohl (1835) and ultimately Virchow (1855) made a compelling case that cells arise by division of existing cells, it was they who seemed to abandon the attempts to ground cell formation in mechanical processes. A mechanistic account of these processes had to await further improvements in the optics of microscopes and the introduction of stains that revealed the complex behavior of chromosomes within the nucleus (Fol 1873; Flemming 1878). While there were controversial claims as to other organelles within cells, especially mitochondria (Altmann 1890) and the Golgi apparatus (Golgi 1898), the question of whether these were artifacts was unresolved until the application of the electron microscope.[3] Determining what function mitochondria, the endoplasmic reticulum, the Golgi apparatus, etc., played required relating the observations made using

[3]One of the pioneers in electron microscopy continued to challenge the reality of the Golgi apparatus (Palade and Claude 1949a, b) until research in his own laboratory revealed its role in packaging newly synthesized proteins for excretion (Jamieson and Palade 1966).

electron micrographs with techniques such as cell fractionation that could localize enzymes in particular structures. With these new tools, though, cell biology developed, offering mechanistic explanations of a host of cellular phenomena by identifying basic operations with organelles and determining the enzymes within them that catalyzed the critical reactions (cf. Bechtel 2006, for discussion of this history). The understanding of oxidative phosphorylation as resulting from the interaction of cytochromes with the inner mitochondrial membrane that allowed for the generation of a proton gradient that drove ATPase in reverse to synthesize ATP represents a crowning achievement of this endeavor.

These are just two of the research fields in which mechanistically oriented biologists made great progress in the nineteenth and twentieth centuries. To those engaged in such progressive research, the vitalist's concerns that living processes might not be susceptible to mechanistic explanation seemed of little point. Some problems proved more challenging than others, but eventually each seemed to give way to a mechanistic account.

3 The New Mechanistic Philosophy of Science

During the period when mechanistic biology was achieving great successes in the twentieth century, the dominant philosophical accounts of science provided little illumination. Highly influenced by Newton's success in articulating laws of motion that could subsume many phenomena from the terrestrial to the astronomical (consider, for example, the application of these laws to the motion of the pendulum), and comparable discovery of laws in other domains (the laws of thermodynamics, Olm's law and the Nernst equation for electrodynamics), the natural focus of philosophers was on laws of nature. Explanation on the accounts they offered involved showing how a given phenomenon represented an application of a law of nature. This became enshrined in the deductive-nomological (D-N) account advanced and defended by Hempel (1965, 1966) according to which one explained a phenomena by deriving a description of it from statements specifying one or more laws and initial conditions:

Law(s) (Newton's law of attraction)
<u>Initial conditions (distance of object from the center of the earth)</u>
∴Phenomenon to be explained (object accelerates at 32 ft/s^2)

While philosophers offered many illustrative examples from the physical sciences that fit this model, the attempts to provide examples from biology were generally uncompelling. A major reason for this is that there are few distinctively biological laws. When biologists speak of explaining a phenomenon, they only occasionally avert to laws (typically those of physics and chemistry[4]) but commonly refer to

[4]Laws of physics and chemistry are, of course, widely employed in explaining biological phenomena. While they are certainly important, a crucial question is whether it is the laws that are performing the crucial explanatory work. For a sustained argument that they are the vehicle of explanation in biology, cf. Weber (2005).

discovering or identifying the mechanism responsible for the phenomenon. Wimsatt (1976) was one of the first philosophers to draw attention to this[5] and beginning in the 1990s a number of philosophers of biology and neuroscience began to focus on the role of mechanisms in explanations in these fields. Building on Wimsatt's insight as well as those of Simon (1962) and Kauffman (1971), Bechtel and Richardson (1993/2010) emphasized the processes by which scientists decompose mechanisms structurally into their parts and functionally into their operations. Emphasizing the same distinction between parts and operations (which they call *entities* and *activities*), Machamer et al. (2000, 3) proposed their widely cited characterization of mechanisms: "Mechanisms are entities and activities organized such that they are productive of regular changes from start or set-up to finish or termination conditions."[6] Skipper and Millstein (2005) coined the label *new mechanistic philosophy of science* for this perspective.

All philosophical accounts of mechanisms emphasize that the parts and operations are organized and part of what this organization requires is that the different parts and operations are connected to each other, achieving what Machamer, Darden, and Craver speak of as "productive continuity." But the last phrase of their characterization of a mechanism – "from start or set-up to finish or termination conditions" – goes beyond insisting on productive continuity to impose a sequential ordering on the operations in the mechanism. This requirement, as we will see, is extremely problematic, and is a large part of what makes explanations that fit their characterization insufficient to explain many biological phenomena. Although the phrase is not employed in other characterizations of mechanism,[7] it does reflect the practices of many mechanistic biologists, who attempt to envisage sequentially the qualitative changes occurring in the mechanisms they investigate. More fundamentally, it also reflects the sequential nature of human mental processes that are employed in the attempt to understand the functioning of mechanisms. In particular, scientists rely on their processes of imagination, and research in cognitive neuroscience (cf. e.g. Kosslyn 1994) has revealed that in imagination we redeploy perceptual processes and so imagine changes sequentially.

Although in what follows I will focus on the shortcomings of this approach (which I call the *basic mechanistic approach*), it not only characterizes much of mechanistic biology but also provides a foundation on which one can fruitfully recast many of the fundamental issues in philosophy of science. I note three here. First, while in philosophical accounts laws are typically represented in propositions[8]

[5] He contended: "At least in biology, most scientists see their work as explaining types of phenomena by discovering mechanisms, rather than explaining theories by deriving them from or reducing them to other theories, and *this* is seen by them as reduction, or as integrally tied to it" (Wimsatt 1976, 671).

[6] For related characterizations of mechanism, cf. Glennan (1996, 2002) and Thagard (2003, 2006).

[7] In particular, it was not part of Bechtel and Richardson's account, as part of their emphasis was on how research that started trying to fit parts and operations into a sequential order often lead researchers to abandon the quest and embrace more complex modes of organization. Bechtel and Abrahamsen speak of the "orchestrated functioning of the mechanism" (Bechtel and Abrahamsen 2005, 423).

[8] In most scientific contexts they are represented as equations.

and reasoning from laws is viewed as logical inference, mechanisms are often best represented in diagrams, and reasoning about them conducted in mental or computational simulations. In simple cases the diagram may just project a sequence of operations and the scientist can simulate the mechanism by rehearsing these in her head. In more complex cases, the diagram can represent multiple simultaneous reactions and when simulating these exceeds human capacity, the quantities depicted in them can be represented in equations and these used to conduct computational simulations. Second, while in philosophical accounts, reduction consists in the derivation of laws of one science from those of another (Nagel 1961), mechanistic inquiry is reductionistic in another sense – it focuses on the constituent parts and operations of a mechanism. Such reductionistic mechanistic research does not promise to explain everything at the lowest level – rather, it recognizes that to explain the behavior of a mechanism the scientists must also determine how the mechanism as an organized whole engages other entities in its environment. If there is systemic organization between these entities and the mechanism, research must analyze it as well to understand how the mechanism will behave over time. As a result, mechanistic explanations are always multi-level accounts, integrating information about parts, operations, and organization within the mechanism with how the mechanism is situated in an environment (Bechtel and Abrahamsen 2009; Craver 2002, 2007; Craver and Bechtel 2007). Finally, whereas the nomological tradition largely eschewed philosophical analysis of scientific discovery (for exceptions, cf. Langley et al. 1987; Thagard 1988), the mechanistic approach has made discovery a focal concern and identified a range of strategies scientists employ in discovering mechanisms (Bechtel and Richardson 1993/2010; Craver and Darden 2001; Darden 2006). (For further discussion of these and other differences between mechanistic and D-N explanations, cf. Bechtel and Abrahamsen 2005.)

Below I will argue that the basic account of mechanistic explanation must be enriched in important respects to handle the vitalist's challenge. But for now it is important to note that it fits well the main threads in mechanistic science that I highlighted in Sect. 1. Two major accomplishments of these mechanistic sciences were (1) to identify the parts of organisms, such as the parts of cells and the various types of molecules out of which they are composed, and (2) to determine the sorts of operations each performed. In the case of cell division, once the chromosomes were discovered, researchers identified the sequence of operations occurring in cell division. In the case of fermentation, after Buchner demonstrated that it could be accomplished in a cell-free extract, researchers sought the three carbon compounds into which glucose could be broken and these transformed in steps in the generation of alcohol or lactic acid. The same was true of other biochemical reactions, and when electron microscopy and cell fractionation were developed, researchers took advantage of them to determine where in the cell various reactions occurred and how internal membranes contributed to phenomena such as oxidative phosphorylation. The triumph of mechanistic science has been the discovery of legions of parts and determination of the operations they perform in generating important phenomena exhibited in living organisms.

4 The Vitalist's Challenge and Initial Responses: Homeostasis and Negative Feedback

As mechanistic biologists were seeing the promise of developing mechanistic explanations of phenomena exhibited by living organisms and eagerly trying to exploit it, vitalists were pointing out features of life that seemed beyond the scope of mechanistic explanation. Xavier Bichat provides an illuminating example. In part he seemed to embrace the mechanist project as he went so far as decomposing human organs into 21 different types of tissue and appealed to the different properties of these tissues to explain the properties of the various organs. But for him the efforts at decomposition could go no further – in particular, one could not explain the properties of the tissues by decomposing them into their chemical parts and operations. He offered two reasons for rejecting such further decomposition. He claimed, first, that the behavior of living organisms is too variable to be explained mechanically: "The instability of vital forces marks all vital phenomena with an irregularity which distinguishes them from physical phenomena [which are] remarkable for their uniformity" (Bichat 1805, 81). Second, he claimed that this behavior involves distinctive forces that actively resist those forces operative in the inorganic world that would destroy the living tissue if left unopposed: "life is the sum of all those forces which resist death."

For the most part, vitalists and mechanists talked past each other. Mechanists focused on phenomena for which it seemed possible to develop mechanistic explanations and saw in their success reason to carry on. As I noted above, some problems proved temporarily challenging, but these could be safely put aside until new strategies or techniques made it possible to pursue them. Thus, for the most part, mechanists did not address the vitalists' challenges – they ignored them. On this score Bernard (1865) was an exception – he took the vitalists' challenges, especially those of Bichat, seriously enough to address them. One of his objectives was to defend a deterministic physiology. For him a phenomenon was explained only when it could be demonstrated to occur in all situations in which the conditions specified in the explanation were satisfied and accordingly he directly responded to Bichat's claims about the irregularity of the behavior of living systems. In part his strategy was to show how to set up experiments that always produced the same result. But another part of his response was to provide a framework that explained the apparent indeterminacy in the behavior of organisms. He distinguished two environments – the internal environment in which the organs of an organism functioned and the external environment in which the organism as a whole functioned. The variability in the response of the organism to an external stimulus was due, he claimed, to the failure to properly consider the conditions in the internal environment. When one insured that the same conditions were maintained in the internal environment, responses to conditions in the external environment became deterministic.

The distinction between the internal and external environment also enabled Bernard to begin to develop a response to Bichat's claims about living organisms as resisting death. It is important for the functioning of components of the organism that the conditions in the internal environment are kept relatively constant. A prominent example from his research was the process by which the liver converts glycogen to glucose whenever concentrations of glucose in the blood decline. This maintains a constant condition in which there is a supply of glucose to provide energy for energy-demanding operations. From examples such as these Bernard concludes: "all the vital mechanisms, however varied they may be, have only one object, that of preserving constant the conditions of life in the internal environment" (Bernard 1878, 121, translated in Cannon 1929, 400). As a result of these processes directed at maintaining the constancy of the internal environment, organisms appear to resist the natural processes that would result in death (e.g. those that would deplete them of available glucose). Walter Cannon (1929) named this capacity *homeostasis* (from the Greek words for *same* and *state*).

Cannon identified a number of processes by which organisms maintain homeostasis. Some were as simple as accumulating surplus supplies in selected tissues (e.g. water in muscle or skin), or by converting them into a different form (e.g. glucose into glycogen) from which reconversion in time of need is possible. Others involved altering the rate of continuous processes (e.g. changing the rate of blood flow by modifying the size of capillaries to maintain uniform temperature). Cannon noted such control mechanisms are regulated by the autonomic nervous system.

What underlies the processes Cannon identified is a mode of organization known as negative feedback, a mode of organization whose importance was only coming to be generally recognized at the time Cannon was working. Although most people are today familiar with the notion of negative feedback, it was an extremely challenging concept for humans to grasp and many still fail to grasp its full consequences. As far as we can tell, it was first introduced by Ksebios in about 270 BCE as part of his design for a water clock. To keep time accurately, he needed to ensure that water entered the recording vesicle at a constant rate. He accomplished this by inserting a second vesicle between the water intake and the recording vesicle, and added a floating plug that would rise up and close the intake when water reached the target level. The plug would drop to open the intake as soon as the water fell below the target. The solution in which behavior (closing the water supply) was dependent on a downstream effect (the rise of water above its target level) is elegant, but it was not readily generalized. When a way to regulate other technologies such as windmills and furnaces was required, the procedure of inserting something that would alter a process in light of its effect had to be invented *de novo* (Mayr 1970). So when in 1788 Watt needed a governor to regulate the flow of steam in a steam engine to ensure that the appliances ran at a constant rate, he once again invented the process of negative feedback. In this instance the solution (discussed further below) attracted the attention of Maxwell, who provided a mathematical analysis of governors (Maxwell 1868). Maxwell's analysis, the increasingly urgent need for control in engineered mechanisms, and the discovery that biological systems often relied on the same procedure to regulate themselves, combined to make negative feedback a general principle of organization that was championed by the cyberneticists as a general control architecture for biological as well as social and engineered systems (Wiener 1948).[9]

What the legacy of theorizing from Bernard through the cyberneticists served to make clear is that the key to addressing many of the limitations of basic accounts of mechanistic explanation is attention to modes of organization beyond sequential organization. As momentous as is the step from sequential organization to negative feedback in allowing for the maintenance of a desired state within a mechanism, it represents only the first step beyond sequential organization, and many others are required to address the concerns of the vitalists and to demonstrate that mechanism is capable of explaining the distinctive features of living systems.

5 Beyond Cybernetics: Endogenous Activity and Autonomous Systems

In some ways a negative feedback system is quite easy to understand. Consider the governor Watt designed (Fig. 14.1). Watt needed to insert within the steam engine some means of detecting whether the engine was running too fast or too slow that could be used to change the setting of the steam valve. He inserted a spindle that would revolve proportionally to the speed of the engine and attached to it arms with weights at the end that were free to move out or in by centrifugal force. When the engine sped up, the spindle rotated faster, and the arms moved out; when it slowed down, the spindle rotated more slowly, and the arms dropped. Now all he needed was a linkage mechanism that worked to close a valve on the steam supply when the arms moved out and open it when the arms moved in. From this verbal description and the diagram, most people are able to understand how the governor works. But there is an additional feature of negative feedback systems that such a verbal account fails to capture but with which engineers are only too familiar. Such systems oscillate. Sometimes when the system is allowed to run long enough it will reach a stable configuration; for instance, the valve will reach the setting at which the arms of the governor will cease to move. This constitutes a dampened oscillator, but even it generates

[9]Wiener himself had been engaged with Bigelow in the 1930s in the design of negative feedback systems for controlling anti-aircraft fire. In this endeavor they encountered a serious challenge: if the feedback signal was at all noisy and the system responded too quickly, feedback caused it to go into uncontrollable oscillations. Through consulting Mexican physiologist Rosenblueth, they learned of similar behavior in human patients with damage to the cerebellum and came to recognize the importance of dampening the feedback signal to achieve reliable control. Inspired by the ability of negative feedback to direct a mechanism toward a target, Rosenbleuth et al. (1943) argued that it provided a framework for employing such notions as teleology and purpose without invoking vitalism. Wiener secured support from the Macy Foundation for a series of twice-yearly conferences known as the Conference for Circular Causal and Feedback Mechanisms in Biological and Social Systems that attracted some of the leading theorists in biology and the social sciences; after he coined the term *cybernetics* from the Greek term for steersman, the movement adopted it as its name.

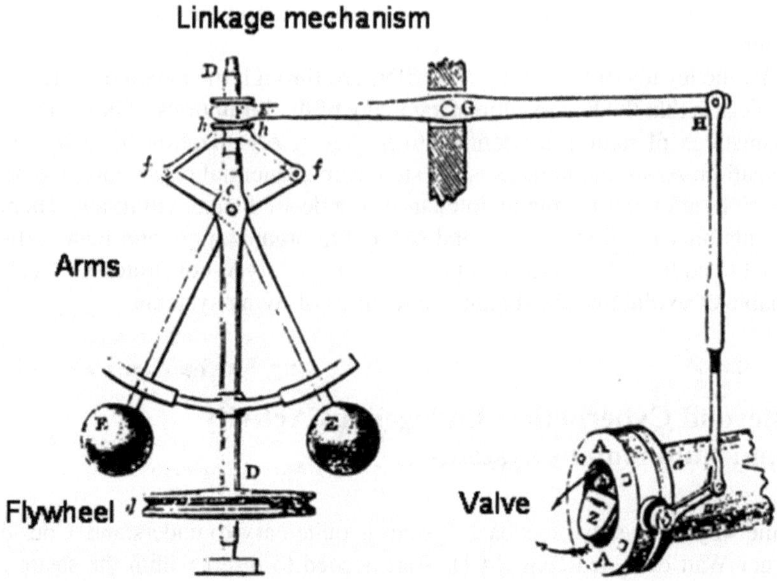

Fig. 14.1 A schematic representation of the governor James Watt designed for his steam engine. The speed of the flywheel determines how far out the angle arms move by centripetal force. They are in turn linked to the valve in such a way that when the flywheel is turning too quickly, the steam supply is reduced, and when it is turning too slowly, the steam supply is increased (Drawing reproduced from Farey 1827)

interesting dynamics as it approaches the stable state, described in dynamical systems theory as a *fixed attractor*.

Many negative feedback systems, however, fail to dampen but continue to oscillate. This is especially the case if there are delays within the system as in the case of a furnace controlled by a thermostat. The temperature will drop below the target temperature before the furnace begins to produce heat, and will rise above it before the furnace is turned off. In many human engineered systems such oscillations are considered undesirable, and engineers strive to minimize them. But biological systems often make use of such oscillations. I briefly note three such systems:

1. *Circadian rhythms.* Organisms from cyanobacteria to us depend on internal clocks that represent time of day and regulate a great variety of physiological and behavioral processes (Bechtel 2011). The operations required for these activities often require extensive time to prepare and organisms rely on being able to act in advance on the basis of tracking time endogenously. They do so through oscillatory processes that utilize multiple feedback loops (discussed further below).
2. *Central pattern generators.* Although the standard view of motor activities is that they are responses to sensory stimuli, pioneering research on decorticate preparations by Graham Brown (1911, 1914) showed that stepping behavior occurs without input from cortex. Wilson (1961) showed that the isolated locust

nervous system generated rhythmic output comparable to that produced in flight. Although in many cases the details of the mechanism are still not known, the paradigm cases involve what is referred to as a *half-center oscillator* involving two neurons that reciprocally inhibit each other in such a manner that after one neuron is inhibited by the other, it gradually depolarizes, reaches threshold, and begins to fire, now inhibiting the other (Hooper 2001).

3. *Brain rhythms.* Research beginning with Berger (1929) revealed electrical activity that can be detected by electrodes placed on the scalp. The dominant frequency changes depending on the state of activity or sleep (alpha rhythms between 8 and 12 Hz dominate while sitting quietly, higher frequency rhythms are more evidence while performing cognitive activities, and slower rhythms while in various stages of sleep) but there is always oscillatory activity. More recently synchronized oscillatory activity at even lower frequencies (<.1 Hz) across networks of areas that are elicited together in task conditions have been detected in fMRI while subjects are at rest (Fox and Raichle 2007). These rhythms are increasingly being recognized as playing important roles in various cognitive activities (cf. Buzsáki 2006).

In Abrahamsen and Bechtel (2012) we speak of systems that continue to generate activity, such as oscillations, independently of input stimuli as *endogenously active*. Biological organisms are primary examples of endogenously active systems. The examples of circadian, motor, and brain rhythms are all instances of endogenous activity. But endogenous activity is far more ubiquitous. It is a notorious feature of humans that they have a very hard time sitting still. And it takes extensive training to stop one's mind from spontaneously generating thoughts. The active state is seemingly the default state. But endogenous activity is not limited to humans. Look at a bacterial cell in a microscope – it is constantly in motion, and it is a challenge to keep it within the range of view. There are times when organisms seem less active than at others, and there are organisms whose activity seems minimal. A sleeping or hibernating animal exhibits fewer spontaneous movements, and is harder to arouse with stimuli. But even during sleep or hibernation, fundamental metabolic activities are continuing. Moreover, there is ongoing slow-wave oscillatory activity in the brain that can be detectable by EEG (and is regarded as the signature of sleep in mammals). Plants seem not to move except to grow, but time-lapse photography can reveal their movement to orient with the sun, and this continues even in the absence of exogenous cues such as sunlight. And again, basic metabolic activities – including the transport of water, capturing energy, and building new tissue – are ongoing.

Endogenous activity is not a merely incidental feature of living organisms. It is crucial for living organisms to maintain themselves as living systems. Unlike many physical objects, which exist in relatively stable configurations due to the strength of the physical and chemical bonds that hold their parts together, living organisms are not stable structures. Nonetheless, they maintain themselves by continually executing the operations needed to build or rebuild themselves. This was the point of Bichat's characterization of life as resisting death, a point that was developed in

Varela's conception of the identity of living system as constituted by the organization which they maintain "through the active compensation of deformations" (Varela 1979, 3). In articulating this idea Varela expands upon Cannon's notion of homeostasis to develop his own concept of *autopoiesis*. The extension involves two steps: (i) "making every reference for homeostasis internal to the system itself through mutual interconnections of processes" and (ii) "positing this interdependence as the very source of the system's identity as a concrete unity which we can distinguish" (12–13). In other words, all homeostatic operations in organisms are caused from within the system and it is the continued existence of the set of causally dependent processes that constitutes the continued existence of the system. Varela then provides his canonical characterization of autopoiesis:

> An autopoietic system is organized (defined as a unity) as a network of processes of production (transformation and destruction) of components that produces the components that: (1) through their interactions and transformations continuously regenerate and realize the network of processes (relations) that produce them; and (2) constitute it (the machine) as a concrete unity in the space in which they exist by specifying the topological domain of its realization as such a network. (Varela 1979, 13; cf. also Maturana and Varela 1980).

For current purposes, the critical feature in Varela's notion of autopoiesis is that the causal processes that build and rebuild the organism originate within it; they are not just the effects of external forces impinging on it.

Varela goes on to characterize autopoietic systems as autonomous systems: "Autopoietic machines are autonomous: that is, they subordinate all changes to the maintenance of their own organization, independently of how profoundly they may be otherwise transformed in the process" (ibid., 15). He later elaborates

> Autonomous systems are mechanistic (dynamical) systems defined as a unity by their organization. We shall say that autonomous systems are organizationally closed. That is, their organization is characterized by processes such that (1) the processes are related as a network, so that they recursively depend on each other in the generation and realization of the processes themselves, and (2) they constitute the system as a unity recognizable in the space (domain) in which the processes exist (Varela 1979, 55).

The concept of autonomy requires features of endogenous activity I noted earlier – operations that build and maintain the living organism are controlled from within it and are not merely responses to conditions presented to it. The concept thus provides a helpful way to characterize what is distinctive about living systems and why they are able to resist death. In calling living systems autonomous, one is not denying that they interact with their environments in many ways and are dependent on these interactions. What autonomy does draw attention to is that these interactions with the environment are in part regulated from within the organism so that the organism directs the interactions in ways that maintain its own identity as an organism.

In characterizing autopoietic systems as autonomous, Varela does not draw attention to one of the most noteworthy features of such systems – a feature which in fact necessitates the internal regulation of activity. This is the fact that biological organisms are, as highly organized systems, far from thermodynamic equilibrium with their environments. As such they dissipate Gibbs free energy. Moreover, since they are chemical systems, not solid systems, such dissipation will be relatively rapid. As a result, organisms must draw free energy as well as matter from sources in their environments and after consuming the energy and the useful matter, release waste products, now in a lower energy state, back into their environment. The free energy and material that is secured from the environment must be utilized to build and repair the organism (restoring what was lost through dissipation), and this requires channeling it appropriately to the reactions that carry out the synthesis and repair. In animals captured free energy is typically stored in the phosphate bonds of ATP that can then be broken down in energy demanding operations. Managing the creation and distribution of these molecules is thus a critical task.

Kepa Ruiz-Mirazo and Alvaro Moreno (2004) have made these energetic considerations central to their account of autonomy. They begin with the recognition that as organized systems, living organisms are far from thermodynamic equilibrium and that preserving that organization entails maintaining themselves far from equilibrium (cf. Schrödinger 1944). Many of the chemical reactions required to maintain such a system are endergonic (require free energy) and so must be coupled with those that liberate energy from another source (exergonic reactions). In order to maintain themselves far from equilibrium, Ruiz-Miorazo and Moreno focuses on how the system *manages* the flow of energy so as to provide for its own construction and reconstruction. The membrane presents one point of management, determining what gets in and out of the system. The metabolic pathways that extract energy and raw materials and then synthesize constituents of the organisms own structure are another. Focusing on these management processes, they characterize *basic autonomy* as:

> the capacity of a system to *manage* the flow of matter and energy through it so that it can, at the same time, regulate, modify, and control: (i) internal self-constructive processes and (ii) processes of exchange with the environment. Thus, the system must be able to generate and regenerate all the constraints – including part of its boundary conditions – that define it as such, together with its own particular way of interacting with the environment (Ruiz-Mirazo and Moreno 2004, 240; cf. also Ruiz-Mirazo et al. 2004, 330).

I contend that with the concept of autonomy theorists such as Ruiz-Mirazo and Moreno have captured the phenomenon that vitalists thought defied mechanistic explanation. But they have not just relabeled that phenomena – they have shown what is required of an organism, consisting of mechanisms, to realize it. The key is coordinated organization so that the various operations performed within mechanisms and by overall mechanisms are directed at building and maintaining the overall identity of the organism as an entity that, as a result of that very organization, is far from equilibrium with its environment. The challenge for the mechanist is to develop appropriate ways to understand how such organization generates the requisite effects.

6 From Basic to Dynamic Mechanistic Explanations

I turn now to the implications of this perspective on biological organisms as endogenously active autonomous systems for the sorts of mechanisms found in biology and the challenges for scientists in understanding them. A mechanism that operates

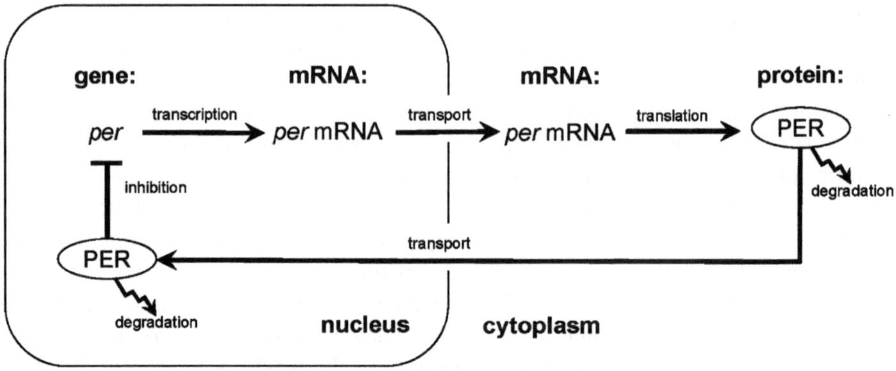

Fig. 14.2 The mechanism for generating circadian rhythms through a transcription, translation feedback loop as proposed by Hardin et al. (1990)

sequentially to carry out operations only when their start conditions are realized and stops when their termination conditions are reached cannot generate the sustained, endogenous behavior that is required to keep the organism far from equilibrium. Only a mechanism with cyclic organization, such as a feedback system, has the capacity to keep itself going by supplying the various operations with the conditions needed for their operation. As a result, it is not surprising that the organization found in biological organisms (either in an organism as a whole or within its individual mechanisms) is nonsequential (Bechtel and Abrahamsen 2011).

Understanding how a nonsequentially organized mechanism behaves is challenging. In mentally rehearsing individual operations, researchers lose track of how the functioning of other components affects these operations. As I noted, even with simple negative mechanisms researchers cannot determine whether the mechanism will generate ongoing oscillations or settle into a stable state. The project of identifying the parts and characterizing the operations of a mechanism explanation must be complemented by that of mathematical and computational modeling to show how such a mechanism will operate. The result is dynamic mechanistic explanation (Bechtel and Abrahamsen 2010).

I illustrate the project of dynamic mechanistic explanation first in the case of simple negative feedback. Figure 14.2 shows the mechanism Hardin et al. (1990) proposed for the circadian clock in animals. In it a gene, *period* or *per* is transcribed into its mRNA, which then is transported to the cytoplasm where it is translated into the corresponding protein PER. PER then is transported back into the nucleus and in some manner impairs its own transcription. (Although many additional constituents have been identified in the two subsequent decades, this is still thought to constitute the core intracellular mechanism underlying circadian rhythms.)

A critical question is whether such a mechanism could generate sustained oscillations of approximately 24 h, or will it dampen to a steady state. To address this question, Goldbeter (1995) first represented the mechanism in terms of variables and parameters and then wrote a set of five differential equations to characterize the

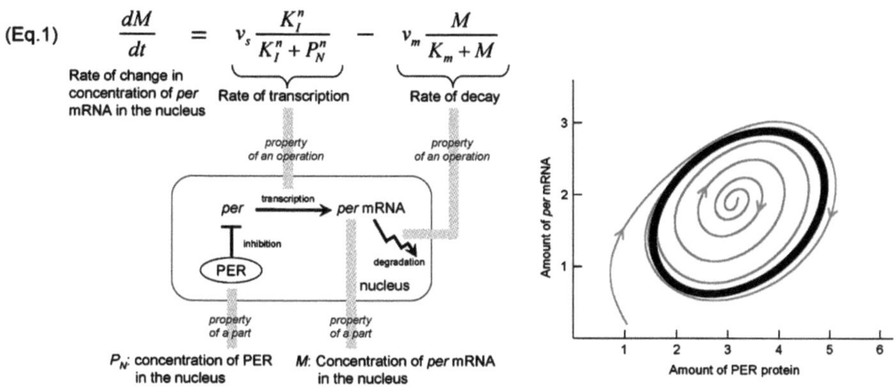

Fig. 14.3 On the *left* the first equation in Goldbeter's (1995) model is related to the relevant components of the mechanism, as illustrated on the *right* of Fig. 14.2. On the *right* is the limit cycle generated from the model with what Goldbeter claimed were plausible parameter values

functioning of the mechanism. The left side of Fig. 14.3 shows the relation between one of these equations and an operation shown in Fig. 14.2. Goldbeter was then able to numerically simulate the functioning of the mechanism and show that with plausible parameter values it would produce what in dynamical systems terms is known as a *limit cycle*: as shown on the right of Fig. 14.3, the values of two parameters (*per* mRNA and PER) change over time but approach a cycle in which the rise in the quantity of *per* mRNA is followed by the rise in the quantity of PER. Only then does the quantity of *per* mRNA drop, followed in turn by the drop in the quantity of PER.

With nonsequential organization, especially when the operations are nonlinear, any change could potentially affect the behavior of the mechanism. Thus, the discovery of additional components in animal circadian clocks and the determination that there are probably three feedback loops that interact with each other, demanded further modeling endeavors with much more complex models to assess how such a mechanism would behave (cf. e.g. Leloup and Goldbeter 2008).

A cursory examination of the current understanding of biological mechanisms reveals that the circadian case is far from exceptional and that sequential execution of operations is the exception rather than the rule. One of the products to emerge from mechanistic science in the past two decades has been the development of massive databases identifying the components (genes, proteins, neurons, etc.) of even relatively simply biological organisms such as bacteria. Although our knowledge of the operations in which these participate is far from complete, when they are presented in networks in which nodes represent the components and edges the interactions, one gains an appreciation of the complexity of the system. The overall organization is far from sequential. A returning vitalist might become convinced that they were right all along (cf. Kirschner et al. 2000). The mechanistic project of decomposing the mechanism into its parts and operations simply cannot show how the parts will behave as a result of receiving various inputs from across the network and how such a network as a whole will behave. Even if science could identify all the parts, determine all the operations in which they participated, and from equations simulate the operation of the whole organism, the modern day vitalist might complain that one would have no explanation of why the organism behaved as it did. In identifying such complexity, mechanism might have finally generated its own undoing.

While the task is daunting, the prospects for dynamic mechanistic explanation are a good deal more promising than this suggests. There appears to be order in the complexity, and this order is precisely what is required for dynamic mechanistic explanation to make progress. Let us consider first the overall structure of the sorts of large networks researchers encounter. Initially mathematicians in the field of graph theory focused on two sorts of networks, regular lattices and randomly (or completely) connected graphs. Two important measures were developed for characterizing these two forms of organization: a clustering coefficient which indicates to how many of its neighbors an individual node is typically connected and a characteristic path length which indicates the average shortest path between two randomly selected nodes. Regular lattices have, not surprisingly, both a high clustering coefficient and characteristic path length while random networks have both a low clustering coefficient and characteristic path length.

While regular lattices and random networks are the easiest to analyze mathematically, Watts and Strogratz (1998) found that many real-world networks fall into a class that lies between them, which they designated *small worlds*. In these networks most connections are between neighbors, thus generating a high clustering coefficient close to that of a regular lattice. A few connections are long-distance, though; adding just a few of these radically reduces the characteristic path length so that it approximates that of a random network. Thus, small worlds exhibit both local clusters of nodes and high levels of integration across the whole network. These conditions, as Watts and Strogatz indicated in their initial paper, are precisely what is desired for processing information.[10]

The occurrence of local clusters or modules brings us back within the context of mechanistic explanation, which emphasizes decomposition of the mechanism. Local clusters can perform specific operations even as this performance is modulated by the long-distance connections that integrate them into a larger small-world network. But what is the organization of these local clusters? A common finding is that within large networks, such as a cell signaling system or the mammalian brain, the local clusters themselves have small-world organization, allowing, as in the mechanist project, for further decomposition. But how are components, at whatever level of organization on which one focuses (individual molecules or neurons, or small-world networks of these) organized into the next larger structure? Here a

[10]Subsequent research by Barabási and his colleagues (Barabási and Albert 1999; Barabási and Bonabeau 2003) has shown that in many real world networks the number of edges per node is not randomly distributed but follows a power-law such that most nodes have few connections but a very few are highly connected. These constitute hubs.

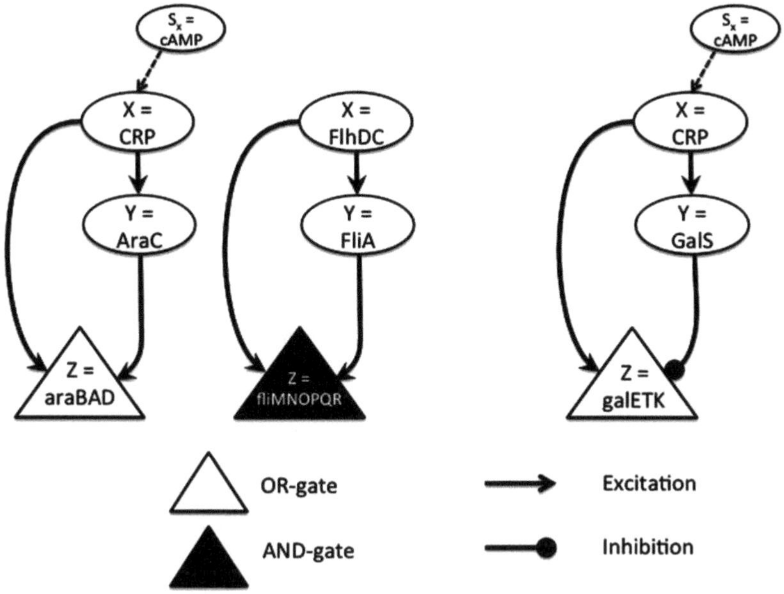

Fig. 14.4 *Left*: Two consistent feedforward loops of type 1, one employing an AND-gate and one an OR-gate in the regulation of Z. *Right*: Inconsistent feedforward loop of type 1

complementary tool is being developed – the analysis of motifs – which offers great insight into how mechanisms are organized.

Motifs are characterized as patterns of connections between components that occur far more often than would be expected in a randomly generated network with the same number of nodes and edges (Alon 2007). Alon and his collaborators developed an algorithm for detecting motifs and applied it first to the transcription factors and the operons they regulate in *Escherichia coli* (Shen-Orr et al. 2002). What is particularly interesting is that analysis of these motifs can reveal how small networks with these designs will behave. The feedforward loops they identified are especially interesting and will serve to illustrate their approach.[11] In feedforward loops an initial transcription factor X regulates a second transcription factor Y, and X and Y jointly regulate operon Z. Regulation at any step can be positive or negative. Although there are eight possible combinations of positive and negative connections, only the consistent loop illustrated with two examples on the left and the inconsistent loop illustrated with an example on the right in Fig. 14.4 are realized in the *E. coli* network much more often than they occur in randomly-generated networks. (A loop is consistent if the two routes to the output both operate in the same manner – accelerate or

[11] Although these motifs do not involve feedback and so cannot generate on their own oscillations or other complex behavior which is crucial for maintaining endogenously active mechanisms, they do generate interesting behaviors which require mathematical analysis.

inhibit, and inconsistent if they operate in the opposite manner – one accelerates and one inhibits.) To analyze these networks, Alon and his colleagues treat the regulation of Z as involving either an OR and an AND operation, which are sufficiently close approximations of the continuous functions that occur in actual operons not to affect the accuracy of the analysis.

The consistent feedforward loop functions as what Alon characterizes as a sign-sensitive delay. In the example feedforward loop with an AND-gate (Mangan et al. 2003) (employed in the system for converting to use of arabinose as a fuel source), when the input cAMP is supplied to X it begins to generate its product CRP (cyclic response protein). After a sufficient quantity is generated to exceed a threshold, Y = AraC begins to be synthesized, but only when it reaches a threshold does Z = araBAD begin to be synthesized. The production of the output araBAD is delayed by the time it takes for sufficient Y to accumulate (approximately 20 min). Alon proposes that the utility of this is to prevent responses to spurious pulses of cAMP, which can last about the same period as the delay. This can be important since vital resources may be utilized in generating araBAD when it is not needed. With an AND-gate there is no delay in terminating the response but when an OR-gate is employed (as in the mechanism controlling the generation of the flagellum motor), the response will continue until the supply of Y (FliA) is also depleted (Kalir et al. 2005). This allows fliLMNOPQR to continue to be generated even during a temporary interruption of the input signal (active FlhDC), leading Alon to characterize it as a persistence detector. It, however, operates with no delay after the initially appearance of the input signal. The inconsistent feedforward loop using an OR gate shown on the right (used in the mechanism regulating galactose utilization) results in pulses of the output product galETK due to the fact that the production of CRP initially begins generation of galETK, but after GalS reaches threshold, it inhibits galETK production (Mangan et al. 2006). A further consequence of this arrangement is that the output can initially be produced at a very high rate without significantly overshooting the target because of the subsequent suppression by the intermediate.

I have characterized the feedforward loop motif in qualitative terms since as they are sufficiently simple that one can understand these behaviors in such a manner. However, the importance of computational modeling is not thereby negated as the details of how any particular mechanism employing the motif will behave depend upon the particular parameter values found in an actual mechanism. In the above descriptions I assumed that the parameter values were such that the direct path to Z produced its effect before the indirect path. If that were violated, the mechanism would behave differently. Moreover, under particular parameter values these motifs can generate more complex behavior, and Alon and his colleagues are still engaged in such a process of discovery, in part by experimenting with additional biological mechanisms exhibiting the motif and in part by sampling parameter values in computational models. Further, while the feedforward motif does not employ feedback, others of the motifs they have identified do, including motifs involving single nodes that feedback positively or negatively on themselves. Multiple motifs can be combined. Although apparently not in bacterial transcription networks, feedforward loops can be included within larger networks employing feedback.

The strategy of identifying motifs and analyzing how they affect the operation of a mechanism provides a means to generate understanding of how organization affects the behavior of biological mechanisms (see Levy and Bechtel forthcoming). Motif analysis, together with the analysis of the overall connectivity pattern of larger networks, and the more complex dynamics that arise with feedback loops, are instances of dynamic mechanistic explanation – explanation that focuses on modes of organization and uses computational modeling and dynamical systems tools to analyze the consequences of such organization for the behavior of a mechanism.

Having introduced the framework of dynamic mechanistic explanation and sketched some of the ways it can be pursued, I finish with a brief discussion of some of the issues to which it gives rise. In one respect dynamic mechanistic explanations are closer to D-N explanations than are basic mechanistic explanations. In characterizing the effects of different modes of organization they are generalizing beyond a specific mechanism to the generic features of such organization. Thus, the focus is on networks in which one considers only the connectivity represented by the edges, not the specific features of the components that occupy the nodes, and tries to ascertain how such a network will behave. For the purposes of the analysis, it does not matter whether the occupants are genes, proteins, neurons, neural columns, brain regions, people, or social groups. There are parameter values in the model that must be satisfied before it applies to any given mechanism, but often the focus is on developing models that are robust over a broad range of parameter values. Thus, in utilizing the mathematical equations developed in the dynamical models, one is applying them to specific mechanisms, much as the D-N model applies laws to different sets of initial conditions. The mathematical equations one uses in dynamic models, however, are not presented as general laws. While they are not specific to individual mechanisms, they are specific to particular modes of organization (change a node or an edge in a motif, and a different set of equations will be required). Moreover, unlike D-N accounts, dynamic mechanistic explanation is still anchored in the mechanism in that the mathematical representation is characterizing how a mechanism with a specific form of organization will behave.[12]

7 Conclusions

The challenges of vitalists such as Bichat to mechanism, despite the evident successes of mechanistic explanations for many biological phenomena, have been my focus. My contention has been that the phenomena to which vitalists appealed in objecting to

---

[12]The approach is thus different from that that advocated by some proponents of dynamical systems theory in psychology who propose dynamical explanation as an alternative to mechanistic explanation (Chemero and Silberstein 2008; Stepp et al. 2011). For responses to the proposal to cleave dynamics from mechanism, see Kaplan and Bechtel (2011) and Kaplan and Craver (2011). In Bechtel and Abrahamsen (2010) we argue that many computational models in cognitive science, especially neural network models, are mechanistic in spirit, but since little is known about the components and their actual operations, the models remain hypothetical and not accounts of actual mechanisms.

mechanism are ones entwined with endogenous activity and autonomy and that biological organisms generate activity endogenously and acquire autonomy as a result of the modes of organization exhibited in them. A mechanism or an organized system of mechanisms that can maintain itself in a far from equilibrium condition is one that is organized so as to insure the appropriate relations between operations so that it can capture energy and deploy it in the continuing endeavor to construct and repair itself. Understanding the effects of such organization requires going beyond the approach of basic mechanistic explanation in which scientists attempt to simulate the sequential operation of parts in their heads. Rather, they must invoke a different set of tools, characterizing the mechanism mathematically and performing and analyzing computational simulations. Such an approach integrates the tools of traditional mechanistic science that facilitate decomposing the mechanism and generating basic mechanistic explanations with tools of dynamical analysis that facilitate recomposing the complexly organized mechanism in simulation. This integration of approaches is what Abrahamsen and I refer to as *dynamic mechanistic explanation*. Only such an integrated approach has the resources to deal with the type of mechanisms that can account for the distinctive features of living systems to which the vitalists drew attention.

Acknowledgment The paper was partly written while I was a fellow at the Institute for Advanced Studies of the Hebrew University of Jerusalem. I am most grateful for their support and the valuable exchanges with colleagues there.

References

Abrahamsen, Adele, and Bechtel William. 2012. From reactive to endogenously active dynamical conceptions of the brain. In *Philosophy of behavioral biology*, eds. T. Reydon and K. Plaisance, 329–366. New York: Springer.

Alon, Uri. 2007. Network motifs: Theory and experimental approaches. *Nature Reviews Genetics* 8: 450–461. doi:[10.1038/nrg2102].

Altmann, Richard. 1890. *Die Elementaroganismen und ihre Beziehungen zu den Zellen*. Leipzig: von Veit.

Barabási, Albert-László, and Réka Albert. 1999. Emergence of scaling in random networks. *Science* 286: 509–512.

Barabási, Albert-László, and Eric Bonabeau. 2003. Scale-free networks. *Scientific American* 50–59.

Bechtel, William. 2006. *Discovering cell mechanisms: The creation of modern cell biology*. Cambridge: Cambridge University Press.

Bechtel, William. 2011. Representing time of day in circadian clocks. In *Knowledge and representation*, ed. A. Newen, A. Bartels, and E.-M. Jung, 129–162. Palo Alto: CSLI Publications.

Bechtel, William, and Adele Abrahamsen. 2005. Explanation: A mechanist alternative. *Studies in History and Philosophy of Biological and Biomedical Sciences* 36: 421–441.

Bechtel, William, and Adele Abrahamsen. 2009. Decomposing, recomposing, and situating circadian mechanisms: Three tasks in developing mechanistic explanations. In *Reduction and elimination in philosophy of mind and philosophy of neuroscience*, ed. H. Leitgeb and A. Hieke, 173–186. Frankfurt: Ontos Verlag.

Bechtel, William, and Adele Abrahamsen. 2010. Dynamic mechanistic explanation: Computational modeling of circadian rhythms as an exemplar for cognitive science. *Studies in History and Philosophy of Science Part A* 41: 321–333.

Bechtel, William, and Adele Abrahamsen. 2011. Complex biological mechanisms: Cyclic, oscillatory, and autonomous. In *Philosophy of complex systems. Handbook of the philosophy of science*, 10th ed, ed. C.A. Hooker, 257–285. New York: Elsevier.
Bechtel, William, and Robert C. Richardson. 1993/2010. *Discovering complexity: Decomposition and localization as strategies in scientific research*. Cambridge, MA: MIT Press. 1993 edition published by Princeton University Press.
Berger, Hans. 1929. Über daas Elektroenkephalogramm des Menschen. *Archiv für Psychiatrie und Nervenkrankheiten* 87: 527–570.
Bernard, Claude. 1865. *An introduction to the study of experimental medicine*. New York: Dover.
Bernard, Claude. 1878. *Leçons sur les phénomènes de la vie communs aux animaux et aux végétaux*. Paris: Baillière.
Berthollet, Claude-Louis. 1780. Recherches sur la nature des substances animales et sur leurs rapports avec les substances végétales. *Mémoires de l'Académie royale des sciences* 1: 120–125.
Bichat, Xavier. 1805. *Recherches Physiologiques sur la Vie et la Mort*, 3rd ed. Paris: Machant.
Brown, R. 1833. On the organs and mode of fecundation in Orchideae and Asclediadeae. *Transactions of the Linnean Society* 16: 685–745.
Büchner, E. 1897. Alkoholische Gärung ohne Hefezellen (Vorläufige Mittheilung). *Berichte der deutschen chemischen Gesellschaft* 30: 117–124.
Buzsáki, György. 2006. *Rhythms of the brain*. Oxford: Oxford University Press.
Cannon, Walter B. 1929. Organization of physiological homeostasis. *Physiological Reviews* 9: 399–431.
Chemero, Anthony, and Michael Silberstein. 2008. After the philosophy of mind: Replacing scholasticism with science. *Philosophy of Science* 75: 1–27.
Craver, Carl F. 2002. Interlevel experiments and multilevel mechanisms in the neuroscience of memory. *Philosophy of Science* 69: S83–S97.
Craver, Carl F. 2007. *Explaining the brain: What a science of the mind-brain could be*. New York: Oxford University Press.
Craver, Carl F., and William Bechtel. 2007. Top-down causation without top-down causes. *Biology and Philosophy* 22: 547–563.
Craver, Carl F., and Lindley Darden. 2001. Discovering mechanisms in neurobiology: The case of spatial memory. In *Theory and method in neuroscience*, ed. P.K. Machamer, R. Grush, and P. McLaughlin, 112–137. Pittsburgh: University of Pittsburgh Press.
Darden, Lindley. 2006. *Reasoning in biological discoveries: Essays on mechanisms, interfield relations, and anomaly resolution*. Cambridge: Cambridge University Press.
Embden, Gustav, H.J. Deuticke, and G. Kraft. 1933. Über die intermediaren Vorgänge bei der Glykolyse in der Muskulatur. *Klinische Wochenschrift* 12: 213–215.
Farey, John. 1827. *A treatise on the steam engine: Historical, practical, and descriptive*. London: Longman, Rees, Orme, Brown, and Green.
Fick, Adolf, and Johannes Wislicenus. 1866. On the origin of muscular power. *Philosophical Magazine & Journal of Science London, 4th ser.* 31: 485–503.
Flemming, Walther. 1878. Zur Kenntnis der Zelle und ihrer Theilungserscheinungen. *Schriften des naturwissenschaftlicher Verein für Schleswig-Holstein* 3: 23–27.
Fol, Hermann. 1873. Le premier développement de l'oeuf chez les Géronidés. *Archives des sciences physiques et naturelles, 2nd series* 48: 335–340.
Fox, Michael D., and Marcus E. Raichle. 2007. Spontaneous fluctuations in brain activity observed with functional magnetic resonance imaging. *Nature Reviews Neuroscience* 8: 700–711.
Frankland, Edward. 1866. On the source of muscular power. *Royal Institution of Great Britain. Notices of the proceedings at the meetings of the members* 4: 661–685.
Glennan, Stuart. 1996. Mechanisms and the nature of causation. *Erkenntnis* 44: 50–71.
Glennan, Stuart. 2002. Rethinking mechanistic explanation. *Philosophy of Science* 69: S342–S353.
Goldbeter, Albert. 1995. A model for circadian oscillations in the *Drosophila* Period protein (PER). *Proceedings of the Royal Society of London B: Biological Sciences* 261: 319–324.
Golgi, Camillo. 1898. Intorno alla struttura delle cellule nervose. *Bollettino della Società Medico-Chirurgica di Pavia* 13: 3–16.

Graham Brown, T. 1911. The intrinsic factors in the act of progression in the mammal. *Proceedings of the Royal Society of London Series B, Containing Papers of a Biological Character* 84: 308–319.
Graham Brown, T. 1914. On the nature of the fundamental activity of the nervous centres; together with an analysis of the conditioning of rhythmic activity in progression, and a theory of the evolution of function in the nervous system. *The Journal of Physiology* 48: 18–46.
Hall, T.S. 1969. *Ideas of life and matter; studies in the history of general physiology, 600 B.C.-1900 A.D.* Chicago: University of Chicago Press.
Hardin, P.E., J.C. Hall, and M. Rosbash. 1990. Feedback of the *Drosophila period* gene product on circadian cycling of its messenger RNA levels. *Nature* 343: 536–540.
Harvey, William. 1628. *Exercitatio anatomica de motu cordis et sanguinis in animalibus.* Francofurti: Sumptibus Guilielmi Fitzeri.
Hempel, Carl G. 1965. Aspects of scientific explanation. In *Aspects of scientific explanation and other essays in the philosophy of science*, ed. C.G. Hempel, 331–496. New York: Macmillan.
Hempel, Carl G. 1966. *Philosophy of natural science.* Englewood Cliffs: Prentice-Hall.
Hooke, Robert. 1665. *Micrographia: Or some physiological descriptions of minute bodies made by magnifying glasses with observations and inquiries thereupon.* London: John Martin and James Allestry.
Hooper, S.L. 2001. Central pattern generators. In *eLS*. Chichester: Wiley.
Jamieson, J.D., and G.E. Palade. 1966. Role of the Golgi complex in the intracellular transport of secretory proteins. *Proceedings of the National Academy of Sciences (USA)* 55: 424–431.
Kalir, Shiraz, Shmoolik Mangan, and U. Alon. 2005. A coherent feed-forward loop with a SUM input function prolongs flagella expression in Escherichia coli. *Molecular Systems Biology* 1. doi:[10.1038/msb4100010].
Kaplan, D.M., and William Bechtel. 2011. Dynamical models: An alternative or complement to mechanistic explanations? *Topics in Cognitive Science* 3: 438–444.
Kaplan, D.M., and Carl Craver. 2011. The explanatory force of dynamical and mathematical models in neuroscience: A mechanistic perspective. *Philosophy of Science* 78: 601–627.
Kauffman, Stuart. 1971. Articulation of parts explanations in biology and the rational search for them. In *PSA 1970*, ed. R.C. Bluck and R.S. Cohen, 257–272. Dordrecht: Reidel.
Kirschner, Marc, John Gerhart, and Tim Mitchison. 2000. Molecular "Vitalism". *Cell* 100: 79–88.
Kosslyn, Stephen M. 1994. *Image and brain: The resolution of the imagery debate.* Cambridge, MA: MIT Press.
Langley, P., H.A. Simon, G.L. Bradshaw, and J.M. Zytkow. 1987. *Scientific discovery: Computational explorations of the creative process.* Cambridge: MIT Press.
Leeuwenhoek, Antoni V. 1719. *Epistolae physiologicae super compluribus naturae arcanis.* Delft: Beman.
Leloup, Jean-Christophe, and Albert Goldbeter. 2008. Modeling the circadian clock: From molecular mechanism to physiological disorders. *BioEssays* 30: 590–600.
Levy, Arnon, and William Bechtel. forthcoming. Abstraction and the organization of mechanisms. 80.
Liebig, Justus. 1831. Ueber einen neuen Apparat zur Analyse organischer Körper, und über die Zusammensetzung einiger organischer Substanzen. *Annalen der Physik und Chemie* 21: 1–43.
Liebig, Justus. 1840. *Organic chemistry in its applications to agriculture and physiology.* London: Taylor and Walton.
Liebig, Justus. 1842. *Animal chemistry: Or organic chemistry in its application to physiology and pathology.* Cambridge: John Owen.
Machamer, Peter, Lindley Darden, and Carl F. Craver. 2000. Thinking about mechanisms. *Philosophy of Science* 67: 1–25.
Mangan, Shmoolik, Alon Zaslaver, and Uri Alon. 2003. The coherent feedforward loop serves as a sign-sensitive delay element in transcription networks. *Journal of Molecular Biology* 334: 197–204.
Mangan, Shmoolik, Shalev Itzkovitz, Alon Zaslaver, and Uri Alon. 2006. The incoherent feed-forward loop accelerates the response-time of the gal system of *Escherichia coli. Journal of Molecular Biology* 356: 1073–1081.

Maturana, Humberto R., and Francisco J. Varela. 1980. Autopoiesis: The organization of the living. In *Autopoiesis and cognition: The realization of the living*, ed. H.R. Maturana and F.J. Varela, 59–138. Dordrecht: Reidel.
Maxwell, James C. 1868. On governors. *Proceedings of the Royal Society of London* 16: 270–283.
Mayr, Otto. 1970. *The origins of feedback control*. Cambridge, MA: MIT Press.
Meyerhof, Otto, and Wilhelm Kiesslling. 1936. Über den Hauptweg der Milchsäurebildung in der Muskulatur. *Biochemische Zeitschrift* 263: 83–113.
Meyerhof, Otto, and Karl Lohmann. 1934. Über die enzymatische Gleichgewichtsreaktion zwischen Hexosediphosphorsäure und Dioxyacetonphosphorsäure. *Die Naturwissenschaften* 22: 220.
Nagel, Ernest. 1961. *The structure of science*. New York: Harcourt, Brace.
Neuberg, Carl, and Johannes Wolfgang Kerb. 1914. Über zukerfreie Hefegärungen. *Biochemische Zeitschrift* 58: 158–170.
Palade, Georges E., and Albert Claude. 1949a. The nature of the Golgi apparatus. I. Parallelism between intercellular myelin figures and Golgi apparatus in somatic cells. *Journal of Morphology* 85: 35–69.
Palade, Georges E., and Albert Claude. 1949b. The nature of the Golgi apparatus. II. Identification of the Golgi apparatus with a complex of myelin figures. *Journal of Morphology* 85: 71–111.
Prout, William. 1827. On the ultimate composition of simple alimentary substances; with some preliminary remarks on the analysis of organised bodies in general. *Philosophical Transactions of the Royal Society of London* 117: 355–388.
Rosenblueth, Arturo, Norbert Wiener, and Julian Bigelow. 1943. Behavior, purpose, and teleology. *Philosophy of Science* 10: 18–24.
Ruiz-Mirazo, Kepa, and Alvaro Moreno. 2004. Basic autonomy as a fundamental step in the synthesis of life. *Artificial Life* 10: 235–259.
Ruiz-Mirazo, Kepa, Juli Peretó, and Alvaro Moreno. 2004. A universal definition of life: Autonomy and open-ended evolution. *Origins of Life and Evolution of the Biosphere* 34: 323–346.
Schleiden, Matthias Jacob. 1838. Beiträge zur phytogenesis. *Archiv für Anatomie, Physiologie und wissenschaftliche Medecin* 137–176.
Schrödinger, Erwin. 1944. *What is life? The physical aspect of the living cell*. Cambridge: Cambridge University Press.
Schwann, Theodor. 1836. Über das Wesen des Verdauungsprocesses. *Archiv für Anatomie, Physiologie und wissenschaftliche Medecin* 90–138.
Schwann, Theodor. 1839. *Mikroskopische Untersuchungen über die Übereinstimmung in der Struktur und dem Wachstrum der Theire und Planzen*. Berlin: Sander.
Shen-Orr, Shai S., Ron Milo, Shmoolik Mangan, and Uri Alon. 2002. Network motifs in the transcriptional regulation network of Escherichia coli. *Nature Genetics* 31: 64–68. doi:[10.1038/ng881].
Simon, Herbert A. 1962. The architecture of complexity: Hierarchic systems. *Proceedings of the American Philosophical Society* 106: 467–482.
Skipper, Robert A., and Roberta L. Millstein. 2005. Thinking about evolutionary mechanisms: Natural selection. *Studies in History and Philosophy of Biological and Biomedical Sciences* 36: 327–347.
Stepp, Nigel, Anthony Chemero, and Michael T. Turvey. 2011. Philosophy for the rest of cognitive science. *Topics in Cognitive Science* 3: 425–437.
Thagard, Paul. 1988. *Computational philosophy of science*. Cambridge, MA: MIT Press/Bradford Books.
Thagard, Paul. 2003. Pathways to biomedical discovery. *Philosophy of Science* 70: 235–254.
Thagard, Paul. 2006. *Hot thought: Mechanisms and applications of emotional cognition*. Cambridge, MA: MIT Press.
Varela, Francisco J. 1979. *Principles of biological autonomy*. New York: Elsevier.
Virchow, Rudolf. 1855. Cellular pathology. In *Disease, life, and man: Selected essays of Rudolf Virchow*, ed. L. Rather, 71–101. Stanford: Stanford University Press.
von Mohl, Hugo. 1835. Über die Vermehrung der Pflanzenzellen durch Teilung. Dissert., Tübingen.

Watts, Duncan J., and Steven H. Strogatz. 1998. Collective dynamics of small worlds. *Nature* 393: 440–442.

Weber, Marcel. 2005. *Philosophy of experimental biology*. Cambridge: Cambridge University Press.

Wiener, Norbert. 1948. *Cybernetics: Or, control and communication in the animal and the machine*. New York: Wiley.

Wilson, Donald M. 1961. The central nervous control of flight in a locust. *Journal of Experimental Biology* 38: 471–490.

Wimsatt, William C. 1976. Reductionism, levels of organization, and the mind-body problem. In *Consciousness and the brain: A scientific and philosophical inquiry*, ed. G. Globus, G. Maxwell, and I. Savodnik, 202–267. New York: Plenum Press.

Index[1]

[1](Note: The page members are in **BOLD** face denote that the index term which have more than 200 instances and listed only first instance of each chapter.)

S. Normandin and C.T. Wolfe (eds.), *Vitalism and the Scientific Image in Post-Enlightenment Life Science, 1800–2010*, History, Philosophy and Theory of the Life Sciences 2, DOI 10.1007/978-94-007-2445-7,

N

O

MIX
Papier aus verantwortungsvollen Quellen
Paper from responsible sources
FSC® C105338

If you have any concerns about our products,
you can contact us on
ProductSafety@springernature.com

In case Publisher is established outside the EU,
the EU authorized representative is:
Springer Nature Customer Service Center GmbH
Europaplatz 3, 69115 Heidelberg, Germany

Printed by Libri Plureos GmbH
in Hamburg, Germany